Advanced Engineering Mat
Applications Guide

Raymond N. Laoulache

Department of Mechanical Engineering
University of Massachusetts Dartmouth

John M. Rice

Department of Mechanical Engineering
University of Massachusetts Dartmouth

WILEY

VP AND EXECUTIVE PUBLISHER	Laurie Rosatone
SENIOR ACQUISITIONS EDITOR	David Dietz
ACQUISITIONS EDITOR	Shannon Corliss
SPONSOSING EDITOR	Mary O'Sullivan
PRODUCT DESIGNER	Tom Kulesa
MARKETING MANAGER	John LaVacca III
SENIOR PROGRAM ASSISTANT	Michael O'Neal
CREATIVE DIRECTOR	Harry Nolan
COVER DESIGN	Tom Nery
SENIOR CONTENT MANAGER	Elle Wagner
SENIOR PRODUCTION EDITOR	John Curley
COMPOSITOR	Sandeep Srivastava for Aptara
COPY EDITOR	Betty Pessagno

ISBN: 978-1-118-98929-6

Printed in the United States of America

V10014854_101719

ADVANCED ENGINEERING MATHEMATICS: APPLICATIONS GUIDE

PREFACE

In a traditional undergraduate engineering curriculum, mathematics and engineering courses tend to follow independent paths, and as a result these courses are only loosely connected to each other. As engineering educators, we wanted to address this issue by bridging the gap between formal and abstract mathematics, and applied engineering in a meaningful way. This led us to a guide, different in approach when compared to standard textbooks in the field, which will aid and motivate engineering students to learn how advanced mathematics is of practical importance in engineering. The strength of this guide lies in modeling applied engineering problems, which are the stimuli, while advanced mathematics is the means to describe, model, and ultimately find solutions to these problems.

The guide is intended for undergraduates who have a good working knowledge of calculus and linear algebra and are ready to use Computer Algebra Systems (CAS) to find solutions expeditiously. It is our recommendation—though not a strict requirement—that as students solve problems in this guide, they use CAS when appropriate to carry out the tedious mathematical steps that do not add new knowledge.

This guide can be used as stand-alone for a course in Applied Engineering Mathematics, as a complement to Kreyszig's *Advanced Engineering Mathematics* or to any other standard text. The use of classical mathematics or CAS to solve problems varies throughout the guide; for example, at one end of the spectrum, it is possible to solve first-order and second-order ordinary differential equations (ODEs) automatically using the Dsolve function in Mathematica and the dsolve function in Maple and MATLAB. However, in this guide we approach ODEs in a classical sense so that students understand key parameters in ODEs and their effect on system behavior, for example, the time constant in first-order ODEs, and the natural frequency and damping coefficient in second-order ODEs. At the other end of the spectrum, in Chapter 8, "Introduction to Computational Techniques," we show that after two or three iterations the process becomes tedious without adding knowledge; thus, CAS is utilized to complete the solution. The optional usage of CAS is also embedded in other chapters for expeditious solutions, such as the solution of a system of ODEs in Chapter 5, "Laplace Transform," and the carrying out of discrete Fourier transform algebra in Chapter 7, "Discrete Fourier Transform."

The following are the key features in this guide:

- Chapters start with motivational engineering problems in order to show the student how mathematics relates to engineering.

- Problems from different disciplines, including civil, electrical, and mechanical engineering, are dispersed throughout the guide and solved in detail for effective learning.

- Clear explanations and step-by-step mathematical formulations.

- A balance between manual approach to problem solutions and CAS.

- Clear pathways to learning topics that may be perceived as difficult for undergraduates, namely, the continuous and discrete Fourier transforms.

- Comprehensive problems at the end of each chapter in various engineering fields.

The guide starts with **Chapter 1, First-Order Ordinary Differential Equations**. The concept of an engineering system is introduced. Students learn how to model and find the solution of a first-order linear engineering system in the time domain. The significance of the time constant in first-order ODEs and its effect on the response of an engineering system to various forcing functions are discussed in detail.

Chapter 2, Second-Order Initial Value Ordinary Differential Equations, has a similar structure to Chapter 1 in terms of engineering form and important engineering parameters, but addresses systems that are governed by second-order ODEs. Second-order oscillating systems with natural frequency and damping factor are covered in detail.

In **Chapter 3, Boundary Value Ordinary Differential Equations**, students are introduced to modeling second-order linear systems in the space domain as a boundary value problem with boundary conditions that are encountered quite often in engineering, namely, Dirichlet, Neumann, and Robin types. The chapter culminates with an introduction to eigenvalue problems.

Chapter 4, Systems of Ordinary Differential Equations, covers the solution of coupled linear ordinary differential equations. The concept of state-space variables is introduced, which is used widely in process control and automation. The emphasis in the chapter is on stability of systems of ODEs via eigenvalues and phase portrait. In addition, the student is introduced to a numerical technique to find the orbit in the phase plane and deduce information about stability.

In **Chapter 5, Laplace Transform**, students are introduced to forcing functions that are encountered quite often in engineering, namely, the unit step function, ramp function, and Dirac delta function. Students also learn how to use the Laplace transform method to solve first- and second-order linear ODEs, as well as first-order coupled ODEs, subject to the aforementioned forcing functions.

In **Chapter 6, Fourier Series and Continuous Fourier Transform**, students learn how to express periodic functions encountered in engineering using the Fourier series. Students are also introduced to the continuous Fourier transform and the concepts of frequency spectrum and bandwidth.

Chapter 7, Discrete Fourier Transform, is a continuation of Chapter 6 where the discrete Fourier transform is presented and applied to engineering problems. This chapter is written from a practical point of view by presenting real case studies and analyzing features such as dominant frequencies in random signals, and their importance in the case studies, for example, vis-à-vis safety.

In **Chapter 8, Introduction to Computational Techniques**, students are introduced to two widely used techniques in engineering: the finite difference and finite element methods. The methods are restricted to problems with one dependent variable. Engineering examples from several disciplines are presented in order to show the wide applications of both methods.

This guide would not have materialized without the guidance of David Dietz, Wiley's Senior Acquisitions Editor. David's advisory efforts and strategic approach are praiseworthy. Our gratitude is equally extended to Mary O'Sullivan, Sponsoring Editor, Product Solutions Group, John Curley, Senior Production Editor, and Michael O'Neal, Market Solutions Assistant, whose organizational guidance was exemplary. Special thanks to Betty Pessagno, Copy Editor, for her scrupulous editing. We also acknowledge the following reviewers who provided valuable advice on the development of this guide: Steven J. Desjardins, University of Ottawa, Paul Fahey, University of Scranton, Chiu Law, University of Wisconsin, Milwaukee, Wai Lau, Seattle Pacific University, Paul J. Laumakis, Rowan University, Laura L. Pauley, Pennsylvania State University, and Wen Shen, Pennsylvania State University. Last but not the least, we express our heartfelt gratitude to Ana Gonzalez for her support during the early development of this guide.

TABLE OF CONTENTS

CHAPTER ONE

FIRST-ORDER ORDINARY DIFFERENTIAL EQUATIONS

1.1 Introduction

In this chapter, engineering systems are modeled by a first-order linear ordinary differential equation (ODE). As an introduction to first-order systems in engineering, consider the soft-drink mixing system, as shown in **Figure 1.1**, where water, from the left tank, and syrup, from the right tank, are mixed with a ratio of ten to one. In order to meet production demand, the water in the left tank is maintained at a given water level h. This is achieved by controlling the volume flow rate of water $\dot{Q}_i$ with an inlet valve, located at the top of the tank. The inlet valve is motorized in order to control angle θ by computer. At the tank bottom, a second valve controls the water flow at a volume flow rate $\dot{Q}_o$, prior to the mixing device where it combines with syrup to produce the end product.

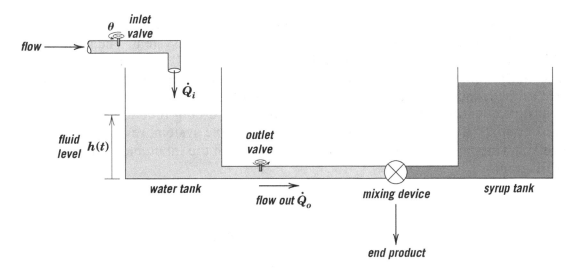

Figure 1.1 Soft-drink mixing system.

In this scenario, the fluid level depends on how fast water enters and leaves the tank; that is, h is a function of the variables $\dot{Q}_i$ and $\dot{Q}_o$, written as $h = h\left(\dot{Q}_i, \dot{Q}_o\right)$. The inlet volume flow rate is controlled by the valve angle setting. In functional form, $\dot{Q}_i = \dot{Q}_i\left(\theta\right)$, so that in lieu of $h = h\left(\dot{Q}_i, \dot{Q}_o\right)$, h is described as $h = h\left(\theta, \dot{Q}_o\right)$. The dependency of h on θ and $\dot{Q}_o$ will guide the engineer to predict how the valve angle setting, which may vary in time, $\theta(t)$, also affects the tank fluid level as a function of time, $h(t)$. In Section 1.3, you will learn how to reduce the mathematical model from $h = h\left(\dot{Q}_i, \dot{Q}_o\right)$ or $h = h\left(\theta, \dot{Q}_o\right)$ to $h = h(t, \theta)$ so that $h(t)$ is governed by the first-order equation

$$\frac{A}{k_o}\frac{dh}{dt}+h=\frac{k_v}{k_o}\theta(t) \tag{1.1}$$

where A is the tank cross-sectional area, and k_o and k_v are constants known as the outlet and inlet valve coefficients, respectively.

1.2 Basic Concepts

In this section, we define how variables of an engineering system are related to each other.

System: Forcing function, response, and block diagram

A system is an entity that performs an engineering task. It is isolated from everything else for the purpose of finding relationships between variables of interest. Associated with each system are two variables, forcing function and response.

The forcing function is a quantity applied to a system, which causes one or more effect to take place. The forcing function is defined as the input to the system. The system input could be physical quantity or a mathematical function that will have an effect on the solution of the equation that describes the system. The effect caused by the system input is defined as the response or output of the system. Similarly, the response of the system could be a physical quantity or a mathematical function. The forcing function (input) and response (output) are depicted by arrows in and out of the system, respectively, as shown in **Figure 1.2**. This representation of the system along with the input and output is called a block diagram.

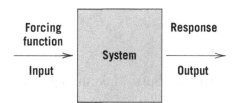

Figure 1.2 A block diagram of a system with a forcing function and a response.

For example, if the interest is in how the water tank height varies in time, then choose the fluid tank as the system enclosed in the dashed box, as shown in **Figure 1.3a**. From Eq. (1.1), the height in time, $h(t)$ depends on the valve angle, $\theta(t)$. Thus, $\theta(t)$ is selected as the forcing function or system input, and $h(t)$ as the response or system output, as shown in **Figure 1.3b**.

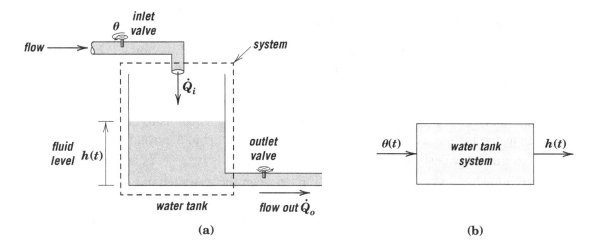

Figure 1.3 The water tank isolated as the system: (a) choosing a system confined by a dashed box; and (b) block diagram of the system with $\theta(t)$ as the input and $h(t)$ as the output.

First-order linear ODE

A general first-order linear ODE is defined as

$$\frac{dy}{dt} + a_0(t)y = r(t) \tag{1.2}$$

The coefficient $a_0(t)$ is continuous and bounded over some interval $t_a \leq t \leq t_b$, whereas $r(t)$ is assumed to be a piecewise continuous function, so that a unique solution of $y(t)$ exists over this interval, subject to an initial condition (IC)

$$y(t_0) = y_0 \tag{1.3}$$

The solution of $y(t)$ is found for two cases: $a_0(t) = 0$ and $a_0(t) \neq 0$.

When $a_0(t) = 0$, Eq. (1.2) reduces to

$$\frac{dy}{dt} = r(t) \tag{1.4}$$

with the solution

$$y(t) = \int r(t)dt + C \tag{1.5}$$

where C is a constant of integration found by applying the IC, $y(t_0) = y_0$. $r(t)$ is integrable by its piecewise continuous nature.

Example 1.1 Hydraulic Actuator Without Friction

Consider the piston in the hydraulic actuator, as shown in **Figure 1.4**, which is connected to a stamping machine. The stamping operation is done as the piston moves from left to right under the high pressure of oil supplied from a reservoir through the inlet port. Behind the cylinder, the low-pressure oil is returned to the reservoir through the outlet port. The operation requires a stamping time interval of 0.25 s for 1 stroke. Thus, the engineer wants to determine the piston stroke length as the piston accelerates from rest at the inlet port to a certain velocity at the outlet port. In this preliminary design, the net force of oil and load is $F_{net} = 8N$, and the piston mass is $m = 4kg$. No information is given about friction between the cylinder and the piston.

1. Use the free body diagram of the piston, and develop a mathematical model for the velocity as a function of time.
2. Use kinematics to find the distance the piston travels over one stroke as a function of time.
3. Find the piston stroke length if one stroke takes 0.25 s.

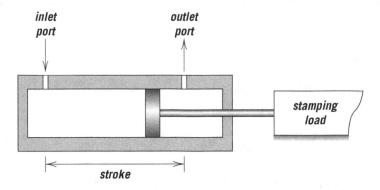

Figure 1.4 Accelerating piston in a hydraulic actuator.

Solution:

1. Use the free body diagram of the piston and develop a mathematical model for the velocity as a function of time.

The motion of the piston is modeled without friction by considering its free body diagram, as shown in **Figure 1.5**, where only the horizontal forces needed for accelerating the pistons are indicated: F_s is the oil supply force on the piston, F_r is the force exerted by the returning oil on the cylinder, and F_{load} is the load reaction force.

Figure 1.5 Free body diagram of the accelerating piston due to several forces.

Newton's second law of motion in the x-direction requires that $F_s - F_r - F_{load} = ma$. By definition, the net force is $F_{net} = F_s - F_r - F_{load}$. Therefore, $F_{net} = ma$. From kinematics, the acceleration a is the rate of change of the velocity v; that is, $a = \dfrac{dv}{dt}$, where t is time. Combining the last two equations yields

$$\frac{dv}{dt} = \frac{F_{net}}{m} \tag{1.6}$$

In comparison with Eq. (1.4), $\dfrac{F_{net}}{m}$ and $v(t)$ are equivalent to $r(t)$ and $y(t)$, respectively. The solution of Eq. (1.6) is given by Eq. (1.5), $v(t) = \int \left(\dfrac{F_{net}}{m} \right) dt + C$, which yields

$$v(t) = \frac{F_{net}}{m} t + C \tag{1.7}$$

for constant $\dfrac{F_{net}}{m}$. The constant of integration C is found by applying the IC $v(0)$. Given that the piston starts to move from rest, then $v(0) = 0$. Therefore, Eq. (1.7) yields $C = 0$, and

$$v(t) = \frac{F_{net}}{m} t \tag{1.8}$$

2. Use kinematics to find the distance the piston travels over one stroke as a function of time.

From kinematics, the velocity v is the rate of change of distance x. $v(t) = \dfrac{dx}{dt}$. Substituting this equation into Eq. (1.8) yields

$$\frac{dx}{dt} = \frac{F_{net}}{m} t \tag{1.9}$$

In comparison with Eq. (1.4), $\left(\dfrac{F_{net}}{m} \right) t$ and $x(t)$ are equivalent to $r(t)$ and $y(t)$, respectively. The solution of Eq. (1.9) is given by Eq. (1.5), $x(t) = \int \left(\dfrac{F_{net}}{m} \right) t \, dt + C$ that yields

$$x(t) = \frac{1}{2} \left(\frac{F_{net}}{m} \right) t^2 + C \tag{1.10}$$

The constant of integration C is found by applying the IC $x(0)$. Setting the coordinate $x = 0$ at the left port, where the piston is initially positioned, then $x(0) = 0$. Therefore, Eq. (1.10) yields $C = 0$. Thus,

$$x(t) = \frac{1}{2}\left(\frac{F_{net}}{m}\right)t^2 \qquad (1.11)$$

3. Find the piston stroke length if one stroke takes 0.25 s.

Substituting $F_{net} = 8N$, $m = 4kg$, and $t = 0.25\, s$ in Eq. (1.11) yields $x(0.25) = 0.0625\, m$.

This problem illustrates how to model a system (piston) by a first-order ODE, where in one part the dependent variable is the velocity $v(t)$ and the independent variable is time t, and in a second part the dependent variable is $x(t)$. Furthermore, Eq. (1.6) describes the piston as a system with $\dfrac{F_{net}}{m}$ as the input or forcing function, and $v(t)$ is the output or response. On the other hand, Eq. (1.9) describes the piston as a system with $\dfrac{F_{net}}{m}t$ as the input or forcing function, and $x(t)$ is the output or response.

■

Example 1.2 Hydraulic Actuator With Friction

What if in Example 1.1 the piston experiences a friction force F_f opposite the motion, which has a magnitude proportional to the piston speed v, that is,

$$|F_f| = kv \qquad (1.12)$$

where k is a constant?

1. How would you model the piston velocity $v(t)$ in this case?
2. How does the ODE compare with Eq. (1.4)?

Solution:

1. In the presence of friction, the motion of the piston is modeled by considering a free body diagram, as shown in **Figure 1.6**.

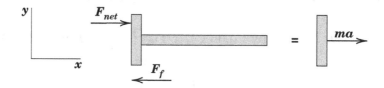

Figure 1.6 Free body diagram of an accelerating piston due to a net force F_{net} and an opposing friction force F_f.

Using Newton's second law of motion in the x-direction yields

$$F_{net} - \left| F_f \right| = ma \tag{1.13}$$

Combining Eq. (1.13) with Eq. (1.12) and $a = \dfrac{dv}{dt}$ yields $F_{net} - kv = m\dfrac{dv}{dt}$ that can be rearranged as

$$\frac{dv}{dt} + \left(\frac{k}{m} \right) v = \frac{F_{net}}{m} \tag{1.14}$$

2. The first-order ODE, Eq. (1.14), no longer compares with Eq. (1.4), which is based on $a_0(t) = 0$. Here $a_0 = \dfrac{k}{m} \neq 0$.

When the coefficient $a_0 \neq 0$, the solution of the linear ODE, Eq. (1.2), is the sum of a homogeneous solution, $y_h(t)$, and a particular solution $y_p(t)$ or $y(t) = y_h(t) + y_p(t)$.

Homogeneous solution

The homogeneous solution, $y_h(t)$, satisfies the homogeneous ODE $\dfrac{dy_h}{dt} + a_0(t) y_h = 0$ whose solution is given by

$$y_h(t) = Ce^{-g(t)} \tag{1.15}$$

where

$$g(t) = \int a_0(t) dt \tag{1.16}$$

For $g(t) > 0$, the homogeneous solution decays exponentially to zero, that is, $y_h(t) = Ce^{-g(t)} \to 0$ as t increases. Typical homogeneous solutions are shown in **Figure 1.7a**. In **Figure 1.7b** typical homogeneous solutions are shown for $g(t) < 0$, which grow exponentially as t increases.

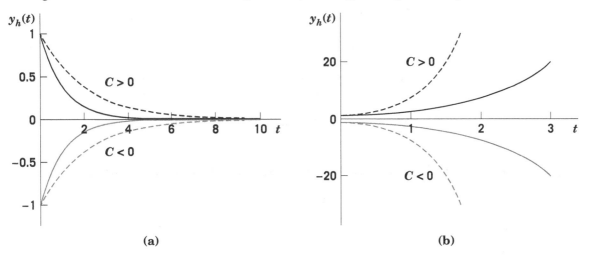

(a)   (b)

Figure 1.7 Homogeneous solution: (a) $y_h(t)$ decays to zero for $g(t) > 0$; and (b) $y_h(t)$ grows exponentially for $g(t) < 0$.

In this chapter, we are concerned mainly with the decaying solutions, which is the case in most first-order engineering systems. Applications do exist in other fields where $g(t) < 0$ such as modeling the exponential growth of interest compounded continuously in the finance and investment fields, or modeling the exponential growth of the human population in the field of geography.

Particular solution

The particular solution, $y_p(t)$, satisfies the nonhomogeneous ODE $\dfrac{dy_p}{dt} + a_0(t) y_p = r(t)$ whose solution is given by

$$y_p(t) = e^{-g(t)} \int r(t) \, e^{g(t)} dt \qquad (1.17)$$

The complete solution of Eq. (1.2) is the sum of Eqs. (1.15) and (1.17),

$$\underbrace{y(t)}_{\text{solution}} = \underbrace{Ce^{-g(t)}}_{\substack{\text{homogeneous} \\ \text{solution}}} + \underbrace{e^{-g(t)} \int r(t) e^{g(t)} dt}_{\text{particular solution}} \qquad (1.18)$$

Table 1.1 is a summary of solutions for first-order linear ODEs in two forms.

$$\frac{dy}{dt} + a_0(t) y = r(t); \text{ IC } y(t_0) = y_0$$

Coefficient, $a_0(t)$	Solution
$a_0(t) = 0$	$y(t) = \int r(t) dt + C$
$a_0(t) \neq 0$	$y(t) = y_h(t) + y_p(t)$ Homogeneous part: $y_h(t) = Ce^{-g(t)}$; $g(t) = \int a_0(t) dt$ Particular part: $y_p(t) = e^{-g(t)} \int r(t) \, e^{g(t)} dt$ Total solution: $y(t) = Ce^{-g(t)} + e^{-g(t)} \int r(t) e^{g(t)} dt$

Table 1.1 Solutions of the first-order linear ODE.

1.3 Engineering Form of The First-Order ODE and The Time Constant τ

In many engineering systems modeled by first-order ODEs, the coefficient a_0 in Eq. (1.2) is constant. In this case, another way to write the ODE so that the coefficient of $y(t)$ is one is by dividing Eq. (1.2) by $a_0 \neq 0$,

$$\frac{1}{a_0} \frac{dy}{dt} + y = \frac{r(t)}{a_0}$$

and write it as

$$\tau \frac{dy}{dt} + y = f(t) \tag{1.19}$$

where $\tau = \frac{1}{a_0}$ and $f(t) = \frac{r(t)}{a_0}$. An engineering system governed by Eq. (1.19) has $f(t)$ as its forcing function or system input, and $y(t)$ as its response or system output. Equation (1.19) is the engineering form of the first-order ODE. If t is time, then τ has the unit of time, which is called the time constant of the system described by the ODE. You can verify the unit of τ using dimensional analysis. Let the brackets [ ] represent the units of any physical quantity. Equation (1.19) must be dimensionally homogeneous so that $\tau \frac{dy}{dt}$ and $y(t)$ have the same units, that is,

$\frac{[\tau][dy]}{[dt]} = [y]$ or $[\tau][dy] = [y][dt]$. Moreover, $y(t)$ and its differential dy have the same units,

that is, $[dy] = [y]$. Thus, $[\tau] = [dt]$ so that τ has the unit of time.

Using $a_0 = \frac{1}{\tau}$ in **Table 1.1** yields $g(t) = \int \frac{1}{\tau} \, dt = \frac{t}{\tau}$. Furthermore, $r(t) = a_0 f(t) = \frac{1}{\tau} f(t)$, so that the general solution of the ODE is

$$y(t) = \underbrace{Ce^{-\frac{t}{\tau}}}_{\text{transient response}} + \underbrace{\frac{e^{-\frac{t}{\tau}}}{\tau} \int f(t) e^{\frac{t}{\tau}} \, dt}_{\text{forced response}} \tag{1.20}$$

In engineering, the time constant τ is positive and finite so that the homogeneous term, $Ce^{-\frac{t}{\tau}}$, in Eq. (1.20) decays in time. In this case, the response $y(t)$ approaches the particular solution given by the second term on the right side of Eq. (1.20). The decaying part is called the transient response so that $y(t)$ is forced to follow the particular solution where the forcing function $f(t)$ is embedded. For this reason, the particular solution is often called the forced response. Thus, think of the complete solution in Eq. (1.20) as the sum of a transient response given by the homogeneous solution, and a forced response given by the particular solution. **Table 1.2** is a summary of the first-order ODE and its general solution in engineering form.

$$\tau \frac{dy}{dt} + y = f(t); \text{ IC } y(t_0) = y_0$$

$$y(t) = y_h(t) + y_p(t)$$

Transient response (homogeneous solution): $y_h(t) = Ce^{-\frac{t}{\tau}}$

Forced response (particular solution): $y_p(t) = \dfrac{e^{-\frac{t}{\tau}}}{\tau} \int f(t) e^{\frac{t}{\tau}} \, dt$

Total solution: $y(t) = \underbrace{Ce^{-\frac{t}{\tau}}}_{\text{transient response}} + \underbrace{\dfrac{e^{-\frac{t}{\tau}}}{\tau} \int f(t) e^{\frac{t}{\tau}} \, dt}_{\text{forced response}}$

Table 1.2 Engineering form of the first-order linear ODE.

Example 1.3

1. Put the first-order ODE $\dfrac{dy}{dt} + 2y = 4t$ in engineering form.
2. Find the solution of the ODE in the time interval $t > 0$. The IC is $y(0) = 0$.
3. Show that the response is forced to behave like the ODE forcing function as the transient term decays in time.

Solution:

1. Put the ODE in engineering form.

Dividing the ODE by 2 yields

$$\frac{1}{2}\frac{dy}{dt} + y = 2t \tag{1.21}$$

A comparison of Eq. (1.21) with Eq. (1.19) yields $\tau = \dfrac{1}{2}$ and $f(t) = 2t$.

2. Find the solution of the ODE.

The general solution is given by Eq. (1.20),

$$y(t) = Ce^{-2t} + 2e^{-2t} \int e^{2t} (2t) \, dt \tag{1.22}$$

Integration by parts yields

$$y(t) = \underbrace{Ce^{-2t}}_{\text{transient response}} + \underbrace{2t - 1}_{\text{forced response}} \tag{1.23}$$

Applying the IC $y(0) = 0$ to Eq. (1.23) yields $C = 1$. Thus,

$$y(t) = e^{-2t} + 2t - 1 \tag{1.24}$$

3. Show that the solution is forced to behave like the ODE, forcing function as the transient term decays in time.

The solution $y(t)$ and the forcing function $f(t)$ are shown in **Figure 1.8**. The curvature in the solution $y(t)$ is clearly visible in the interval $0 \le t \le 0.5$. Beyond that point, the solution fairly follows a straight line. Thus, the homogeneous or transient solution e^{-2t} influences the solution at the initial stage. Subsequently, the particular solution $2t - 1$, which is a straight line, dominates the solution.

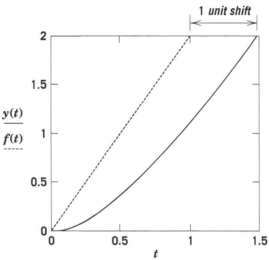

Figure 1.8 The complete solution $y(t)$ and the forcing function $f(t) = 2t$.

Thus, for large t, for example, as $t \to \infty$ Eq. (1.24) shows that $y(t)$ approaches the particular solution $y_p(t)$ as $y_h(\infty) = \lim_{t \to \infty} e^{-2t} = 0$, that is, $y(\infty) \approx 2t - 1$ so that $y(t)$ is forced to behave like $2t - 1$. In fact, the term $2t$ is the forcing function $f(t)$ in Eq. (1.22). For that reason, the particular solution is called a forced solution. In **Figure 1.8** you can see the forcing function of Eq. (1.22), $f(t) = 2t$, preceding the forced response $2t - 1$. In other words, the forced response resembles the forced function $f(t) = 2t$ in shape except that it is shifted rightward by -1. This shift in the response is due to the time lag caused by the transient that naturally occurs in temporal first-order ODEs.

The previous example raises the question about whether there exists a finite value for t beyond which the transient solution is no longer influential, instead of invoking the limiting process $t \to \infty$. The answer to this question lies in the time constant τ, which is discussed next.

Physical meaning of the time constant τ

For simplicity, consider a system governed by Eq. (1.19) with a forcing function
$$f(t) = \begin{cases} 0 & t < 0 \\ F & t \ge 0 \end{cases}$$
and IC $y(0) = 0$, where F is a constant. The general solution is given by

Eq. (1.20), $y(t) = Ce^{-\frac{t}{\tau}} + \frac{e^{-\frac{t}{\tau}}}{\tau} \int Fe^{\frac{t}{\tau}} dt$. Using the IC, the solution reduces to

$$y(t) = \underbrace{-Fe^{-\frac{t}{\tau}}}_{\text{transient response}} + \underbrace{F}_{\text{forced response}} \qquad (1.25)$$

Let $y_1(t)$ and $y_2(t)$ represent two responses that correspond to two time constants, τ_1 and τ_2, respectively, that is,

$$y_1(t) = \underbrace{-Fe^{-\frac{t}{\tau_1}}}_{\text{transient response}} + \underbrace{F}_{\text{forced response}}$$

and

$$y_2(t) = \underbrace{-Fe^{-\frac{t}{\tau_2}}}_{\text{transient response}} + \underbrace{F}_{\text{forced response}}$$

It is clear that as $t \to \infty$, $y_1(\infty) \to F$ and $y_2(\infty) \to F$. For definiteness, consider $\tau_2 < \tau_1$. This inequality implies that $e^{-\frac{t}{\tau_2}}$ decays faster than $e^{-\frac{t}{\tau_1}}$ so that the response $y_2(t)$ reaches F asymptotically faster than the response $y_1(t)$, as shown in **Figure 1.9**.

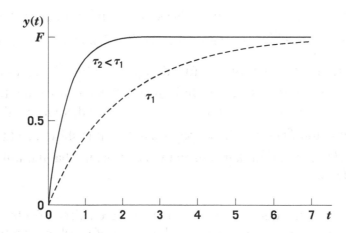

Figure 1.9 Response speed for $y(t)$ as a function of the time constant τ.

Thus, you can interpret the physical meaning of τ as a measure of the transient time lag. That is, τ determines how fast the transient response decays to zero, or equivalently how fast the system response approaches asymptotically the constant forcing function. In order to define a finite value for t beyond which the transient solution is no longer influential, write Eq. (1.25) in a dimensionless form as

$$\frac{y(t)}{F} = 1 - e^{-\frac{t}{\tau}} \qquad (1.26)$$

where $\dfrac{t}{\tau}$ and $\dfrac{y(t)}{F}$ are called the normalized time and normalized response, respectively. In this form, $\dfrac{y(t)}{F}$ versus $\dfrac{t}{\tau}$ degenerates to a single curve, as shown in **Figure 1.10**.

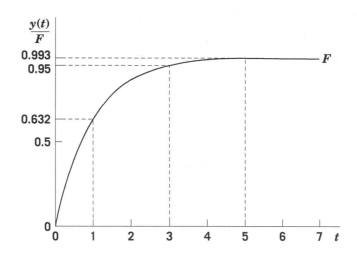

Figure 1.10 Normalized response of a first-order ODE to a constant forcing function.

Equation (1.26) and **Figure 1.10** show that it is possible to define a time threshold beyond which the transient solution is no longer influential, so that in lieu of $t \to \infty$, which is not practical in engineering, you can assign a finite value for t. **Table 1.3** lists values of $\dfrac{y(t)}{F}$ for different values of $\dfrac{t}{\tau}$.

$\dfrac{t}{\tau}$	0	1	2	3	4	5	6	7
$\dfrac{y(t)}{F}$	0	0.632	0.865	0.950	0.982	0.993	0.998	0.999

Table 1.3 Normalized response values as a function of normalized time.

At $\dfrac{t}{\tau} = 7$, $\dfrac{y}{F} = 0.999$, i.e., the response is 99.9% near the constant forcing function. In engineering, the transient time lag is taken conventionally at $t = 5\tau$ where $\dfrac{y}{F} = 0.993$, as shown in **Figure 1.10**.

Example 1.4

1. Identify the time constant τ and the forcing function of a system governed by the ODE

$$\frac{dy}{dt} + \frac{1}{2}y = 2 \qquad (1.27)$$

2. Find the solution of Eq. (1.27) in the time interval $t > 0$. The IC is $y(0) = 0$. Also estimate how long it takes the transient response to decay to zero.
3. Plot the solution and show the transient time lag on the plot.

Solution:

1. Identify the time constant and forcing function.

Equation (1.27) can be written in engineering form as

$$2\frac{dy}{dt} + y = 4 \qquad (1.28)$$

that yields $\tau = 2$. The forcing function is $f(t) = 4$.

2. Find the solution of Eq. (1.27) and estimate the time lag.

The general solution given by Eq. (1.20) yields $y(t) = Ce^{-\frac{t}{2}} + 4$. Applying the IC yields the complete solution $y(t) = -4e^{-\frac{t}{2}} + 4$. The transient time lag is conventionally given as 5τ, hence $5\tau = 10$.

3. Plot the solution and time lag.

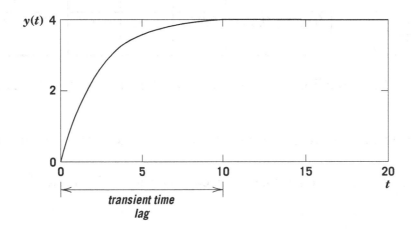

Figure 1.11 Response $y(t)$ to a constant forcing function, $f(t) = 4$.

At time $t=10$, as shown in **Figure 1.11**, $y(10)=3.973$ or 99.3% of the forcing function, $f(t)=4$ and at this point in time the response is considered to be operating at steady state.

Example 1.5

1. Identify the time constant and the forcing function of a system governed by the ODE

$$\frac{1}{2}\frac{dy}{dt}+y=2t \tag{1.29}$$

that was analyzed in Example 1.3. The solution is $y(t)=e^{-2t}+2t-1$.

2. Estimate how long it takes the transient response to decay to zero.

Solution:

1. Equation (1.29) is already in the engineering form given by Eq. (1.19). Therefore, the time constant is $\tau=0.5$ and the forcing function is $f(t)=2t$.

2. According to the convention $t=5\tau$ the transient response is expected to decay to zero in $t=5(0.5)=2.5$, as shown by $y_h(t)$ in **Table 1.4**. Beyond this point, you can safely state that the solution $y(t)$ is given by the forced solution $y_p(t)=2t-1$, that is,

$$y(t)\simeq y_p(t)=2t-1 \tag{1.30}$$

so that $y(t)$ is forced to follow the forcing function $f(t)=2t$ for $t\geq 2.5$ at the same rate of climb, or slope that is equal to 2.

t	$y_h(t)=e^{-2t}$	$y_p(t)=2t-1$	$y(t)=y_h(t)+y_p(t)$
0	1.000	−1.000	0
0.5	0.3679	0.000	0.3679
1	0.1353	1.000	1.135
1.5	0.04979	2.000	2.050
2	0.01831	3.000	3.018
2.5	0.006738	4.000	4.007

Table 1.4 Values of the homogeneous, particular, and complete solution as t increases.

The last two examples illustrated the time constant for a constant forcing function and a linear forcing function. Both times constants were measured from the initial time $t = 0$. In both examples, the forcing functions do not change over the interval $t \geq 0$. What if the forcing function $f(t)$ changes from one time interval to the next? In other words, if $f(t)$ is a piecewise continuous function, then how is 5τ evaluated? In this case 5τ is evaluated relative to the initial time of the interval under consideration, that is, for an ODE whose solution is found over a time interval $t_a \leq t \leq t_b$, 5τ is measured from $t = t_a$, as illustrated in the next example.

Example 1.6

Consider the following first-order ODE

$$5\frac{dy}{dt} + 5y = r(t) \tag{1.31}$$

where

$$r(t) = \begin{cases} 10 & 0 \leq t < 2 \\ 0 & t \geq 2 \end{cases}$$

The IC is $y(0) = 0$.

1. Put the ODE in engineering form, and determine the forcing function $f(t)$ and the time constant τ.
2. Find the solution of the ODE in the time interval $t > 0$.
3. Plot the forcing function $f(t)$ and the response $y(t)$ in the time interval $t \geq 0$.
4. Estimate the transient time lag of the system.

Solution:

1. Put the ODE in engineering form, and determine the forcing function $f(t)$ and time constant τ.

In engineering form, the coefficient of $y(t)$ is one. Therefore, divide Eq. (1.31) by 5

$$\frac{dy}{dt} + y = f(t) \tag{1.32}$$

where $f(t) = \dfrac{r(t)}{5}$, or

$$f(t) = \begin{cases} 2 & 0 \leq t < 2 \\ 0 & t \geq 2 \end{cases} \tag{1.33}$$

A comparison of Eq. (1.32) with Eq. (1.19) yields a time constant $\tau = 1$.

2. Find the solution of the ODE in the interval $t > 0$.

Step 1: Find the general solution in the interval $0 \leq t < 2$.

The general solution is given by Eq. (1.20), $y(t) = \underbrace{Ce^{-\frac{t}{\tau}}}_{\text{transient response}} + \underbrace{\frac{e^{-\frac{t}{\tau}}}{\tau} \int f(t) e^{\frac{t}{\tau}} dt}_{\text{forced response}}$.

for $\tau = 1$, $\qquad\qquad\qquad y(t) = Ce^{-t} + e^{-t} \int f(t) e^{t} dt \qquad\qquad$ (1.34)

In the interval $0 \leq t < 2$, $f(t) = 2$, thus, Eq. (1.34) yields

$$y(t) = \underbrace{Ce^{-t}}_{\text{transient response}} + \underbrace{2}_{\text{forced response}} \qquad\qquad\qquad (1.35)$$

Using the IC $y(0) = 0$ in Eq. (1.35) yields $C = -2$. Thus, the solution in the interval $0 \leq t < 2$ is

$$y(t) = -2e^{-t} + 2 \qquad\qquad\qquad (1.36)$$

Step 2: find the general solution in the interval $t \geq 2$.

From Eq. (1.33), $f(t) = 0$ in the interval $t \geq 2$, thus Eq. (1.20) yields

$$y(t) = Ce^{-t} \qquad\qquad\qquad (1.37)$$

The IC is $y(2)$ at the beginning of the interval, $t_0 = 2$, which is found from Eq. (1.36) at the end of the first time interval $y(2) = -2e^{-2} + 2 = 1.729$. Substituting $t = 2$ and $y(2) = 1.729$ in Eq. (1.37) yields $C = 12.78$. Thus, the solution in the interval $t \geq 2$ is $y(t) = 12.78e^{-t}$. The complete solution in the interval $t \geq 0$ is given by

$$y(t) = \begin{cases} -2e^{-t} + 2 & 0 \leq t < 2 \\ 12.78e^{-t} & t \geq 2 \end{cases} \qquad\qquad (1.38)$$

3. Plot the forcing function $f(t)$ and the response $y(t)$.

The forcing function and the response given by Eqs. (1.33) and (1.38), respectively, are shown in **Figure 1.12** as part of a block diagram.

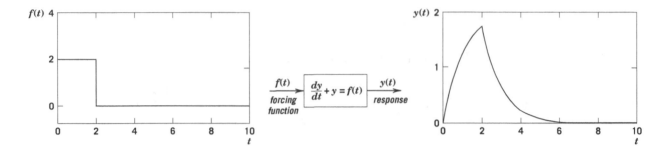

Figure 1.12 First-order ODE block diagram with forcing function $f(t)$ and response $y(t)$.

4. Estimate the transient time lag of the system.

In the time interval $t \geq 2$, the system reaches steady state at $t = 5\tau$ relative to $t = 2$. Therefore, relative to $t = 0$, the system reaches steady state at $t = 2 + 5\tau = 2 + 5(1) = 7$.

Next, you will learn how to model engineering systems as first-order ODEs.

First-order fluid system: Fluid level in a tank

The water tank system you will analyze is enclosed by the dashed line as shown in **Figure 1.3**. It has an inlet valve to control the fluid entering the tank at a volume flow rate $\dot{Q}_i(t)$ and an outlet valve to control the fluid leaving the tank at a volume flow rate $\dot{Q}_o(t)$. The following tasks are carried out in order to determine the fluid level in the tank:

*   Applying conservation of mass.

*   Expressing the mass flow rate in terms of volume flow rate.

*   Expressing the rate of volume inside the tank in terms of fluid level.

*   Expressing the outlet valve volume flow rate in terms of fluid level.

Conservation of mass

Conservation of mass states that the mass flow rate into the tank, $\dot{m}_i$, must be equal to the mass flow rate out of the tank, $\dot{m}_o$, plus the rate of change of mass inside the tank $\left(\dfrac{dm}{dt} \right)_{tank}$. In other words,

$$\dot{m}_i = \dot{m}_o + \left(\frac{dm}{dt} \right)_{tank} \tag{1.39}$$

1-18

$\dot{m}_i$ and $\dot{m}_o$ are always positive, whereas $\left(\dfrac{dm}{dt}\right)_{\text{tank}}$ may be negative, zero, or positive, depending on whether the fluid level is decreasing, not changing, or increasing, respectively.

Mass flow rate-volume flow rate relationship

Mass is the product of density ρ and volume Q, $m = \rho Q$; thus, the mass flow rate $\dot{m}$ is related to the volume flow rate $\dot{Q}$ by

$$\dot{m} = \rho\dot{Q} \tag{1.40}$$

For an incompressible fluid, that is, ρ is constant; Eqs. (1.39) and (1.40) yield

$$\dot{Q}_i = \dot{Q}_o + \left(\frac{dQ}{dt}\right)_{\text{tank}} \tag{1.41}$$

where $\dot{Q}_i$ and $\dot{Q}_o$ are the inlet and outlet volume flow rates, respectively, and $\left(\dfrac{dQ}{dt}\right)_{\text{tank}}$ is the rate of change of volume inside the tank.

Relationship between volume flow rate in the tank and fluid level

The quantity $\left(\dfrac{dQ}{dt}\right)_{\text{tank}}$ is by definition the limit as $\Delta t \to 0$ of the volume change ΔQ in the tank over a time interval Δt, as shown in **Figure 1.13**.

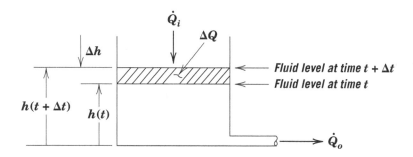

Figure 1.13 Volume change ΔQ over a time interval Δt.

The volume change is $\Delta Q = A\big[h(t+\Delta t) - h(t)\big]$, where A is the tank cross-sectional area and Δh is the fluid-level change. As $\Delta t \to 0$,

$$\left(\frac{dQ}{dt}\right)_{\text{tank}} = \lim_{\Delta t \to 0}\frac{\Delta Q}{\Delta t} = \lim_{\Delta t \to 0}A\frac{\Delta h}{\Delta t} = A\frac{dh}{dt} \tag{1.42}$$

Substituting Eq.(1.42) into Eq. (1.41) yields

$$A\frac{dh}{dt} + \dot{Q}_o = \dot{Q}_i \tag{1.43}$$

The prediction of the fluid level $h(t)$ is of interest in this case study. Thus, you will consider $h(t)$ as the response of the system (tank), which necessitates the left-hand side of Eq. (1.43) to contain only $h(t)$ and its derivative. This condition implies that a relationship between $\dot{Q}_o$ and $h(t)$ must be established in order to accomplish the required form of the ODE.

Outlet volume flow rate-fluid level relationship

Using theories well established in fluid mechanics, the outlet volume flow rate, $\dot{Q}_o$, can be expressed as either

$$\dot{Q}_o = k_o h(t) \tag{1.44}$$

or

$$\dot{Q}_o = k_o \sqrt{h(t)} \tag{1.45}$$

depending on fluid flow conditions at the pipe outlet. k_o is commonly called the output gain in engineering. It depends on several parameters, such as the geometry of the outlet pipe, and the fluid properties such as density and viscosity or the coefficient of resistance to flow. k_o is not the same constant in Eqs. (1.44) and (1.45). Substituting Eq. (1.44) in Eq. (1.43) yields a first-order linear nonhomogeneous ODE for the fluid level in the tank, given by

$$A\frac{dh}{dt} + k_o h = \dot{Q}_i(t) \tag{1.46}$$

whereas substituting Eq. (1.45) in Eq. (1.43) yields a first-order nonlinear ODE for the fluid level in the tank, given by

$$A\frac{dh}{dt} + k_o\sqrt{h} = \dot{Q}_i(t) \tag{1.47}$$

The solution of the nonlinear ODE, Eq. (1.47), is beyond the scope of this chapter. Notice, however, the desired ODE form obtained in either Eq. (1.44) or Eq. (1.45) by replacing $\dot{Q}_o$ with a function of $h(t)$. Dividing Eq. (1.46) by k_o yields the engineering form

$$\frac{A}{k_o}\frac{dh}{dt} + h = \frac{\dot{Q}_i(t)}{k_o} \tag{1.48}$$

From Eq. (1.48), the fluid level $h(t)$ is predicted as a function of time for a specified inlet flow rate function, $\dot{Q}_i$, and IC, $h(0)$. Moreover, if the valve time lag is neglected, the inlet valve flow rate $\dot{Q}_i(t)$ can be modeled by a linear equation as a function of the valve angular displacement $\theta(t)$,

$$\dot{Q}_i(t) = k_v \theta(t) \tag{1.49}$$

where k_v is called the gain of the valve. k_v is determined empirically by a calibration process. Substituting Eq. (1.49) into Eq. (1.48) yields

$$\frac{A}{k_o}\frac{dh}{dt} + h = \frac{k_v}{k_o}\theta(t) \tag{1.50}$$

If you compare Eq. (1.50) with the engineering form of the ODE, Eq. (1.19), you conclude that the time constant of the system is $\tau = \dfrac{A}{k_o}$. You should note that the mathematical input and output of an ODE may differ from the physical input and output of the system. In this case study, the system is the tank, and the physical input and output are the volumetric flow rates, $\dot{Q}_i$ and $\dot{Q}_0$, respectively. However, the mathematical input and output of the ODE, Eq. (1.50), are the forcing function $f(t) = \dfrac{k_v}{k_i}\theta(t)$ and the fluid level $h(t)$, respectively. The pertinent equations are summarized in **Table 1.5**.

ODE model	$\tau\dfrac{dh}{dt} + h = f(t)\,;\tau = \dfrac{A}{k_o}$	$h(t)$, fluid height; A, tank cross-sectional area; k_v, valve gain; k_o, output gain.
Forcing function	$f(t) = \dfrac{k_v}{k_o}\theta(t)$	$\theta(t)$, valve angular displacement.

Table 1.5 Summary of first-order tank model, Eq. (1.50).

Example 1.7 The Water Tank

Consider a water tank with uniform circular cross section. The water level $h(t)$ is modeled by the equations summarized in **Table 1.5**, where h is in (m) and t is in (s). Given data:

IC: $h(0) = 0$ (empty tank).

Tank diameter: $D = 2\,m$.

Valve gain: $k_v = 0.0103\ \left(\frac{m^3}{s}\right)/\text{degree}$.

Output gain: $k_o = 0.05236\ \left(\frac{m^3}{s}\right)/m$.

1. What is the water level $h(t)$ in the interval $t > 0$, for valve angles described in the following graph by a piecewise function consisting of a linear function and a constant function, as shown in **Figure 1.14**, where θ is in degrees?

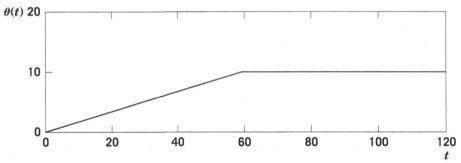

Figure 1.14 Valve angle as a piecewise function.

2. Plot the forcing function $f(t)$ and the response $h(t)$ in the time interval $t \geq 0$.
3. Estimate the transient time lag of the system, that is, the time for the water level to reach a steady state.

Solution:

1. What is the water level $h(t)$?

Step 1: Find the time constant and the forcing function $f(t)$.

The time constant is given by $\tau = \dfrac{A}{k_o} = \dfrac{\pi D^2}{4} \dfrac{1}{k_o} = \pi \dfrac{2^2}{4} m^2 \dfrac{1}{0.05236 \dfrac{m^2}{s}} = 60 \ s$.

The ratio $\dfrac{k_v}{k_o} = \dfrac{0.01030 \ m^3 s^{-1} / \text{degree}}{0.05236 \ m^2 s^{-1}} = 0.197 \ m/\text{degree}$.

The angular displacement is described as a piecewise function, $\theta(t) = \begin{cases} \dfrac{t}{6} & 0 \leq t < 60 \\ 10 & t \geq 60 \end{cases}$.

Thus, the forcing function is $f(t) = \left(\dfrac{k_v}{k_o}\right)\theta = 0.197 \begin{cases} \dfrac{t}{6} & 0 \leq t < 60 \\ 10 & t \geq 60 \end{cases}$.

Step 2: find the general solution in the interval $0 \leq t < 60$.

The ODE general solution is given by Eq. (1.20)

$$h(t) = \underbrace{Ce^{-\frac{t}{\tau}}}_{\text{transient response}} + \underbrace{\frac{e^{-\frac{t}{\tau}}}{\tau} \int f(t) e^{\frac{t}{\tau}} \, dt}_{\text{forced response}}.$$

For $\tau = 60$
$$h(t) = Ce^{-\frac{t}{60}} + \frac{e^{-\frac{t}{60}}}{60} \int f(t) e^{\frac{t}{60}} \, dt \tag{1.51}$$

In the interval $0 \leq t < 60$, $f(t) = 0.197 \frac{t}{6}$. Substituting $f(t)$ in Eq. (1.51), and integrating by parts or using a Computer Algebra System (CAS) yields

$$h(t) = \underbrace{Ce^{-\frac{t}{60}}}_{\text{transient response}} + \underbrace{0.0328t - 1.97}_{\text{forced response}} \tag{1.52}$$

The constant C is found using the IC $h(0) = 0$ in Eq. (1.52), $0 = C - 1.97$ or $C = 1.97$. Thus, the solution in the interval $0 \leq t < 60$ is

$$h(t) = 1.97 e^{-\frac{t}{60}} + 0.0328t - 1.97 \, (m) \tag{1.53}$$

Equation (1.53) describes the liquid-level response for a linear forcing function.

Step 3: find the general solution in the interval $t \geq 60$.

$f(t) = 1.97$ in the interval $t \geq 60$. Substituting $f(t)$ in the general solution yields

$$h(t) = Ce^{-\frac{t}{60}} + \frac{e^{-\frac{t}{60}}}{60} \int 1.97 e^{\frac{t}{60}} dt,$$

or

$$h(t) = Ce^{-\frac{t}{60}} + 1.97 \tag{1.54}$$

The IC for Eq. (1.54) is found from Eq. (1.53) at $t = 60s$,

$$h(60) = 1.97 e^{-\frac{60}{60}} + 0.0328(60) - 1.97 = 0.725m.$$

Substituting $t = 60s$ and $h(60) = 0.725m$ in Eq. (1.54) yields $C = -3.38$ so that the solution in the interval $t \geq 60$ is

$$h(t) = -3.38e^{-\frac{t}{60}} + 1.97 \, (m) \tag{1.55}$$

The complete solution in the interval $t \geq 0$ is given by

$$h(t) = \begin{cases} 1.97\left(e^{-\frac{t}{60}} - 1\right) + 0.0328t & 0 \leq t < 60 \\ \\ -3.38e^{-\frac{t}{60}} + 1.97 & t \geq 60 \end{cases}$$

2. Plot the forcing function $f(t)$ and the response $h(t)$ in the time interval $t \geq 0$.

The forcing function $f(t)$ and response $h(t)$ are shown in **Figure 1.15** using a block diagram.

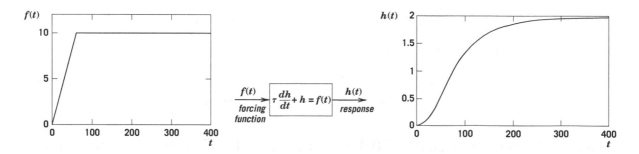

Figure 1.15 Tank system block diagram with forcing function $f(t)$ and response $h(t)$.

3. Estimate the transient time lag, that is, the time for the water level to reach a steady state.

The transient time lag interval of $h(t)$ in the interval $t \geq 60$ is 5τ relative to $t = 60$. Therefore, from $t = 0$, the time lag is $t = 60 + 5\tau = 60 + 5(60) = 360s$. At $t = 360s$, the steady water level is found from Eq. (1.55) to be $h(t) = 1.96m$. A steady water level means also that the inlet and outlet volumetric flow rates are equal to each other. Note that when $t \to \infty$, Eq. (1.55) shows that $h(\infty)$ approaches $1.97m$, which is only 0.4% greater than that estimated using the 5τ criterion.

In the next section, you will solve a first-order ODE to find the response caused by a sinusoid forcing function.

1.4 Response of a System to a Sinusoid Forcing Function

This section covers several important aspects of the sinusoid forcing function. First, you will find the solution of a first-order ODE with a sinusoid forcing function. Second, you will learn that, in addition to the transient time lag, which is inherent to first-order systems, there is another steady time lag between the response and the forcing function. Third, you will analyze the effect of the forcing function frequency on the amplitude of the response. This effect is important for understanding a specific engineering application in this section, the low-pass filter.

Basic concepts for a sinusoid function

A sinusoid function $r(t)$ is a continuous sine or cosine function in the open interval $-\infty < t < +\infty$. A sine function, is defined as

$$r(t) = R\sin(\omega t) \tag{1.56}$$

as shown in **Figure 1.16a**, where R is the amplitude, t is time and ω is the circular frequency.

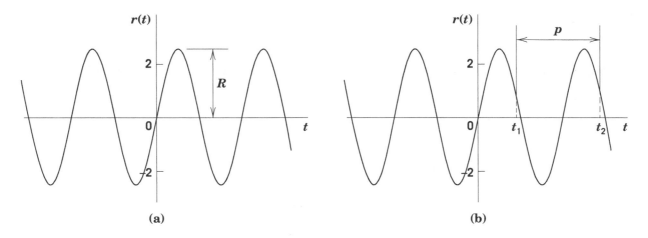

Figure 1.16 A sinusoid with: (a) amplitude R; and (b) period p.

The function $r(t)$ is said to be periodic or cyclic, with a period p, if $r(t)$ and its derivatives repeat their values at two instants, t_1 and t_2, as shown in **Figure 1.16b**, so that the period is $p = t_2 - t_1$. In other words, changing t in Eq. (1.56) by p causes $r(t)$ and all its derivatives to repeat their values. This cyclic property can be expressed as

$$\sin(\omega t) = \sin\left[\omega(t \pm p)\right] \tag{1.57}$$

Equation (1.57) allows you to find a relationship between the circular frequency ω and the period p as follows. From basic trigonometry, a sinusoid is cyclic if the argument ωt is changed by $\pm 2\pi$ radians, that is, $\sin(\omega t) = \sin(\omega t \pm 2\pi)$. Thus, a comparison of this cyclic property with Eq. (1.57) yields $\omega p = 2\pi \; rad$, or

$$p = \frac{2\pi}{\omega} \tag{1.58}$$

If p is in s, then ω is in $\dfrac{rad}{s}$. The frequency f is defined as the inverse of the period p,

$$f = \frac{1}{p} \tag{1.59}$$

in $\dfrac{1}{s}$ defined as *hertz* $\left(Hz = \dfrac{1}{s}\right)$. The hertz is also known as one cycle per second. Eliminating p from Eqs. (1.58) and (1.59) yields a relationship between f and ω as

$$\omega = 2\pi f \qquad (1.60)$$

Thus, the sine function $r(t)$ can be expressed alternatively in terms of the frequency f as $r(t) = R\sin(2\pi ft)$.

Phase angle

In many engineering problems, sinusoids may shift in time. If a sine function $r_1(t)$ of amplitude R_1 does not pass through $t = 0$, as shown in **Figure 1.17**, the sinusoid is said to be shifted by a phase angle ϕ. In this case $r_1(t)$ is written as

$$r_1(t) = R_1\sin(\omega t + \phi) \qquad (1.61)$$

The angle ϕ has the same units as ωt, which is *rad*.

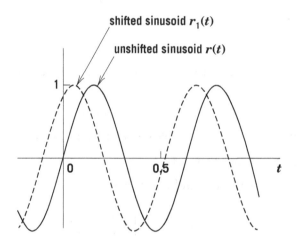

Figure 1.17 Shifted and unshifted sinusoids.

For $\phi \neq 0$ we often describe the sinusoid that is ahead of the other sinusoid. Thus, we use the modifier leading or lagging. Since you will compare two sinusoids you need a reference sinusoid, which will be chosen as $r(t) = R\sin(\omega t)$, or $r(t) = R\cos(\omega t)$.

Leading sinusoid

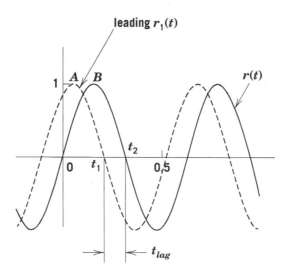

Figure 1.18 Two sinusoids with $r_1(t)$ leading $r(t)$ in time.

Suppose the second sinusoid is $r_1(t)$ given by Eq. (1.61) for $0 < \phi < \pi$. The sinusoid $r_1(t)$ is said to be leading $r(t)$, which means the following. If you consider the two closet maxima, or minima of $r(t)$ and $r_1(t)$, then the maximum of $r_1(t)$, point A, takes place earlier in time than the maximum of $r(t)$, point B, as shown in **Figure 1.18**.

Lagging sinusoid

On the other hand, for $-\pi < \phi < 0$ the sinusoid $r_1(t)$ is said to be lagging $r(t)$, which means the following. If you consider the two closest maxima, or minima of $r(t)$ and $r_1(t)$, the maximum of $r_1(t)$, point A, takes place later in time than the maximum of $r(t)$, point B. as shown in **Figure 1.19**.

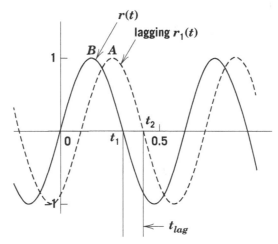

Figure 1.19 Two sinusoids with $r_1(t)$ lagging $r(t)$ in time.

Whether $r_1(t)$ is leading or lagging $r(t)$ we describe the delay in time between the two sinusoids as a time lag, t_{lag}, as shown in **Figures 1.18** and **1.19**. The time lag is defined as $t_{lag} = t_2 - t_1$ found as follows. When $r(t)$ crosses the t-axis, as shown in **Figure 1.18**, then $r(t) = R\sin(\omega t) = 0$. This identity is satisfied if ωt is an integer multiple of π,

$$\omega t = \pi n \tag{1.62}$$

for $n = 0, 1, 2, 3,$ and so on.

For $n = 0$ then $t = 0$, and $r(t)$ crosses the origin. For $n = 1$, $r(t)$ crosses the t-axis at $t = t_2$, so that $t_2 = \dfrac{\pi}{\omega}$ from Eq. (1.62). Similarly, when $r_1(t)$ crosses the t-axis, then $r_1(t) = R_1\sin(\omega t + \phi) = 0$, which is satisfied if $\omega t + \phi$ is an integer multiple of π,

$$\omega t + \phi = \pi n \tag{1.63}$$

for $n = 0, 1, 2, 3,$ and so on.

For $n = 1$, $r_1(t)$ crosses the t-axis at $t = t_1$ so that $t_1 = \dfrac{\pi - \phi}{\omega}$ from Eq. (1.63). Thus, the time lag is

$$t_{lag} = t_2 - t_1 = \frac{\pi}{\omega} - \frac{\pi - \phi}{\omega}, \text{ or}$$

$$t_{lag} = t_2 - t_1 = \frac{\phi}{\omega} \tag{1.64}$$

For $0 < \phi < \pi$ then t_{lag} is positive as required. When $r_1(t)$ is lagging $r(t)$ as shown in **Figure 1.19**, then $r(t)$ crosses the t axis at $t_1 = \dfrac{\pi}{\omega}$ and $r_1(t)$ crosses the t axis at $t_2 = \dfrac{\pi - \phi}{\omega}$. In this case, the time lag is given by

$$t_{lag} = t_2 - t_1 = \frac{\pi - \phi}{\omega} - \frac{\pi}{\omega} = -\frac{\phi}{\omega} \tag{1.65}$$

For $-\pi < \phi < 0$ then t_{lag} is positive as required.

$t_{lag} > 0$; thus Eqs. (1.64) and (1.65) can be written compactly as

$$t_{lag} = t_2 - t_1 = \frac{|\phi|}{\omega} \tag{1.66}$$

Useful trigonometric relationships

Quite often, in the course of an engineering problem, you may encounter a function such as

$$f(\theta) = a\cos(\theta) + b\sin(\theta) \tag{1.67}$$

where $\theta = \omega t$ is the argument of the sinusoid in *rad*. In this form, you may not recognize that $f(\theta)$ represents a shifted sinusoid with a phase angle. However, you can identify a phase angle if Eq. (1.67) is converted to a single sinusoid of amplitude c and phase angle ϕ. For example, if you write $f(\theta)$ as a cosine function,

$$f(\theta) = c\cos(\theta + \phi) \tag{1.68}$$

then

$$c = \sqrt{a^2 + b^2} \tag{1.69}$$

and ϕ is given by one of the following formulas depending on the signs of a and b:

- $a > 0$ and $(b > 0 \text{ or } b < 0)$ $\qquad \phi = -\tan^{-1}\left(\dfrac{b}{a}\right)$ $\qquad$ (1.70)

- $a < 0$ and $b > 0$, $\qquad \phi = -\tan^{-1}\left(\dfrac{b}{a}\right) - \pi$ $\qquad$ (1.71)

- $a < 0$ and $b < 0$, $\qquad \phi = \pi - \tan^{-1}\left(\dfrac{b}{a}\right)$ $\qquad$ (1.72)

A sine function $\sin(\theta)$ can be written in terms of a shifted cosine function as

$$\sin(\theta) = \cos\left(\theta - \frac{\pi}{2}\right) \tag{1.73}$$

On the other hand, a cosine function $\cos(\theta)$ can be written in terms of a shifted sine function as

$$\cos(\theta) = \sin\left(\theta + \frac{\pi}{2}\right) \tag{1.74}$$

The general solution of a first-order ODE with a sinusoid forcing function

To find the response to a sinusoid forcing function, we begin with the engineering form of the first-order ODE, Eq. (1.19), $\tau\dfrac{dy}{dt} + y = f(t)$, with IC $y(0) = y_0$, which is subject to a sinusoid forcing function,

$$f(t) = F\cos(\omega t) \tag{1.75}$$

or

$$f(t) = F\sin(\omega t) \tag{1.76}$$

where F is the amplitude and ω is the angular frequency of the forcing function.

As far as the theory is concerned, it is immaterial at this point whether you find the solution using a sine or cosine function as long as you are consistent when you compare the response with the forcing function. That is, if you wish to compare graphically the response $y(t)$ with the forcing function given by Eq. (1.75) or Eq. (1.76), then express $y(t)$ as either a cosine or sine function, respectively. For definiteness, consider the solution of the ODE that corresponds to the forcing function $f(t)$ given by Eq. (1.75). The general solution is given by Eq. (1.20),

$$y(t) = \underbrace{Ce^{-\frac{t}{\tau}}}_{\text{transient response}} + \underbrace{\frac{1}{\tau}e^{-\frac{t}{\tau}}\int F\cos(\omega t)\, e^{\frac{t}{\tau}}dt}_{\text{forced response}}$$

The integral is found by integration by parts or using CAS, which yields

$$y(t) = \underbrace{Ce^{-\frac{t}{\tau}}}_{\text{transient}} + \underbrace{\frac{F}{1+\omega^2\tau^2}\left[\cos(\omega t) + \omega\tau\sin(\omega t)\right]}_{\text{forced response}} \tag{1.77}$$

Express the trigonometric term in Eq. (1.77) as

$$\cos(\omega t) + \omega\tau\sin(\omega t) = \sqrt{1+\omega^2\tau^2}\,\cos(\omega t + \phi) \tag{1.78}$$

where the phase angle ϕ is given by Eq. (1.70)

$$\phi = -\tan^{-1}(\omega\tau) \tag{1.79}$$

by virtue of $\omega\tau > 0$. Substituting Eq. (1.78) in (1.77) yields

$$y(t) = Ce^{-\frac{t}{\tau}} + \frac{F}{\sqrt{1+\omega^2\tau^2}}\cos(\omega t + \phi) \tag{1.80}$$

The constant of integration is found using the IC $y(0) = y_0$ from either Eq. (1.77) or Eq. (1.80). If you try it, you will find it is easier to use Eq. (1.77),

$$y_o = C + \frac{F}{1+\omega^2\tau^2}\left[\underbrace{\cos(0)}_{=1} + \omega\tau\underbrace{\sin(0)}_{=0}\right]$$

or $C = y_o - \dfrac{F}{1+\omega^2\tau^2}$. Thus Eq. (1.77) becomes

$$y(t) = \underbrace{\left(y_o - \frac{F}{1+\omega^2\tau^2} \right) e^{-\frac{t}{\tau}}}_{\text{transient response}} + \underbrace{\frac{F}{\sqrt{1+\omega^2\tau^2}} \cos\left(\omega t + \phi\right)}_{\text{forced response}} \qquad (1.81)$$

Equation (1.81) is the complete solution for a cosine forcing function, with IC $y(0) = y_0$. The phase angle represents the shift between two sinusoids. As in any first-order ODE in time, there is a transient time lag τ, which describes the duration of the decaying transient response. When the exponential term decays to values close to zero, Eq. (1.81) converges to the forced response so that

$$y(t) \simeq y_p(t) = \underbrace{\frac{F}{\sqrt{1+\omega^2\tau^2}} \cos\left(\omega\, t + \phi\right)}_{\text{forced response}} \qquad (1.82)$$

By virtue of the phase angle ϕ, the forced response has a steady time lag relative to the forcing function $f(t) = F\cos(\omega t)$, which is given by Eq. (1.66). The pertinent equations are summarized in **Table 1.6**.

Transient response: $y_h(t) = \left(y_o - \dfrac{F}{1+\omega^2\tau^2} \right) e^{-\frac{t}{\tau}}$ Forced response: $y_p(t) = \dfrac{F}{\sqrt{1+\omega^2\tau^2}} \cos\left(\omega\, t + \phi\right)$ Response: $y(t) = y_h(t) + y_p(t)$	Transient time, 5τ; Steady time lag, $t_{lag} = \dfrac{\|\phi\|}{\omega}$; Eq. (1.66)

Table 1.6 Solution summary of the first-order ODE $\tau\dfrac{dy}{dt} + y = f(t) = F\cos(\omega t)$.

Example 1.8

1. Find the solution of the first-order ODE

$$0.1\frac{dy}{dt} + y = f(t) \qquad (1.83)$$

where $f(t) = \sin(\omega t)$, $\omega = 3\dfrac{rad}{s}$, and t is in seconds. The IC is $y(0) = 0$.

2. Determine the transient time lag, τ, and verify your answer graphically by plotting the total response and forced response in the time interval $t \geq 0$.

3. Compare the forced response with the sinusoid forcing function, and determine whether it is in-phase with, lagging, or leading the forcing function.
4. Determine the steady time lag.

Solution:

1. Find the solution of the first-order ODE.

Given the forcing function $f(t) = \sin(3t)$, the general solution of Eq. (1.83) is given by Eq. (1.20),

$$y(t) = Ce^{-\frac{t}{\tau}} + \frac{1}{\tau}e^{-\frac{t}{\tau}}\int \sin(3t)e^{\frac{t}{\tau}}\, dt$$

The integral is found by integration by parts or using CAS, $y(t) = Ce^{-\frac{t}{\tau}} + \dfrac{-3\tau\cos(3t) + \sin(3t)}{1 + 9\tau^2}$.

For $\tau = 0.1$,

$$y(t) = Ce^{-10t} + \frac{-0.3\cos(3t) + \sin(3t)}{1.09} \tag{1.84}$$

The term $-0.3\cos(3t) + \sin(3t)$ in Eq. (1.84) can be written as $\sqrt{1.09}\sin(3t - 0.2915)$. Thus,

$$y(t) = Ce^{-10t} + \frac{\sqrt{1.09}\sin(3t - 0.2915)}{1.09}$$

or

$$y(t) = Ce^{-10t} + 0.9578\sin(3t - 0.2915) \tag{1.85}$$

The constant C is found using the IC $y(0) = 0$ in Eq. (1.84) that yields $C = 0.2752$, so that the complete response is

$$y(t) = \underbrace{0.2752e^{-10t}}_{\text{transient response}} + \underbrace{0.9578\sin(3t - 0.2915)}_{\text{forced response}} \tag{1.86}$$

2. Determine the transient time lag, τ, and verify your answer graphically by plotting the complete solution and the forced response in the time interval $t \geq 0$.

Express Eq. (1.86) as $y(t) = y_h(t) + y_p(t)$, where the homogeneous solution $y_h(t)$ corresponds to the transient term, and the particular solution $y_p(t)$ corresponds to the forced term. The transient response is given by

$$y_h(t) = 0.2752\,e^{-10t} \qquad\qquad (1.87)$$

and the forced response is given by

$$y_p(t) = 0.9578\sin\left(3t - 0.2915\right) \qquad\qquad (1.88)$$

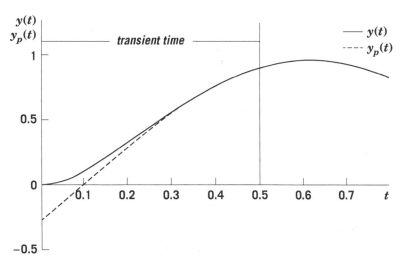

Figure 1.20 Comparison of the complete response with the steady response near the transient time of 0.5 s.

Using the conventional setting of 5τ shows that it takes $5(0.1) = 0.5\ s$ for the transient response, Eq. (1.87), to decay exponentially to zero. After that time, the response $y(t)$ converges to the forced response, Eq. (1.88). This can be verified graphically, as shown in **Figure 1.20**, where $y(t)$ matches closely $y_p(t)$ for $t \geq 0.5\ s$, that is, $y(t) \simeq y_p(t) = 0.9578\sin\left(3t - 0.2915\right)$.

3. Compare the forced response with the sinusoid forcing function, and determine whether the response is in-phase with, lagging, or leading the forcing function.

The sinusoid forcing function $f(t) = \sin(3t)$ and the forced response $y_p(t) = 0.9578\sin\left(3t - 0.2915\right)$ are shown in **Figure 1.21**. The phase angle $\phi = -0.2915\,rad$ is negative, meaning that the forced response lags the forcing function by $0.2915\,rad$. This is clearly seen where the forcing function peaks earlier than the response.

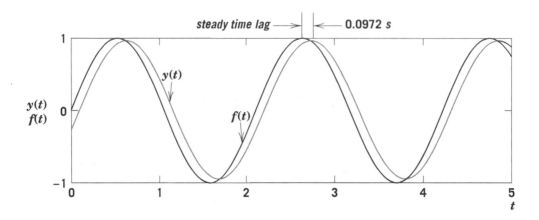

Figure 1.21 A comparison of the response and the forcing function with a time lag of $0.0972\ s$.

4. Determine the steady time lag.

The steady time lag is given by $t_{\text{lag}} = \dfrac{|\phi|}{\omega} = \dfrac{.2915}{3} = 0.0972\ s$, which is shown in **Figure 1.21**.

Notice that the amplitude of the forced response is 0.9578, which is less than that of the forcing function, $F = 1$. In the next section, you will learn that the forcing function frequency ω has a significant influence on the amplitude of the forced response.

■

Effect of the forcing function frequency ω on the forced response amplitude

By design, the transient response decays rapidly in engineering systems so that the response converges to the forced response. For a forcing function given by $f(t) = F\cos(\omega t)$, the response is given by Eq. (1.82), $y_p(t) = \dfrac{F}{\sqrt{1 + \omega^2 \tau^2}}\cos(\omega t + \phi)$, so that the forced response and the forcing function oscillate with the same frequency ω except that one function lags the other one. However, it is important to notice that while the amplitude of the forcing function F is constant, the forced response amplitude, $\dfrac{F}{\sqrt{1 + \omega^2 \tau^2}}$, depends on the frequency ω. In order to describe analytically and graphically how the forced response amplitude varies with frequency in the range $0 < \omega < \infty$, write $y_p(t)$ in Eq. (1.82) as $y_p(t) = Y_p\cos(\omega t + \phi)$, where the amplitude Y_p is given by $Y_p = \dfrac{F}{\sqrt{1 + \omega^2 \tau^2}}$ that is normalized as

$$\frac{Y_p}{F} = \frac{1}{\sqrt{1 + \omega^2 \tau^2}} \tag{1.89}$$

The dimensionless terms $\omega\tau$ and $\dfrac{Y_p}{F}$ are known as the normalized frequency and normalized amplitude, respectively.

Low frequency

For very small frequencies, that is, as $\omega \to 0$, the normalized amplitude approaches 1,

$$\frac{Y_p}{F} = \lim_{\omega \to 0} \frac{1}{\sqrt{1 + \omega^2 \tau^2}} = 1$$

High frequency

For very high frequencies, that is, as $\omega \to \infty$, the normalized amplitude approaches zero,

$$\frac{Y_p}{F} = \lim_{\omega \to \infty} \frac{1}{\sqrt{1 + \omega^2 \tau^2}} = 0$$

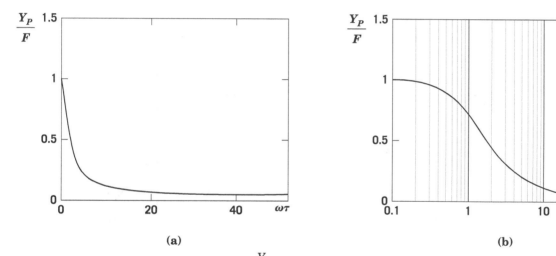

(a)

(b)

Figure 1.22 Normalized amplitude, $\dfrac{Y_p}{F}$ versus normalized frequency $\omega\tau$ on: (a) linear-linear scale; and (b) linear-log scale.

Thus, as ω increases from very low frequencies to very high frequencies, $\dfrac{Y_p}{F}$ starts at 1 and decays to zero, as shown in **Figure 1.22**. In **Figure 1.22a**, $\dfrac{Y_p}{F}$ is plotted versus $\omega\tau$ using a linear-linear scale. In **Figure 1.22b**, $\dfrac{Y_p}{F}$ is plotted versus $\omega\tau$ using a linear-log scale: linear scale for $\dfrac{Y_p}{F}$, and logarithm for $\omega\tau$. Plotting $\omega\tau$ on a $\log_{10}$ scale, as in **Figure 1.22b**, is equivalent to stretching the linear scale of $\omega\tau$ in **Figure 1.22a**. As a result, a better resolution of the curve is gained, especially in the low-frequency range.

Cutoff frequency

If $\dfrac{Y_p}{F}$ is plotted versus $\omega\tau$ on a log-log scale, as shown in **Figure 1.23a**, $\dfrac{Y_p}{F}$ drops off almost as a straight line for $\omega\tau > 1$. Thus, $\dfrac{Y_p}{F}$ can be enclosed by an envelope of two straight lines, as shown in **Figure 1.23b**. The two lines intersect at $\omega\tau = 1$. The frequency ω where this condition exists is called the cutoff frequency, ω_c, which is defined as $\omega_c\tau = 1$, or

$$\omega_c = \frac{1}{\tau} \tag{1.90}$$

So that Eq. (1.89) is expressed as

$$\frac{Y_p}{F} = \frac{1}{\sqrt{1 + \left(\dfrac{\omega}{\omega_c}\right)^2}} \tag{1.91}$$

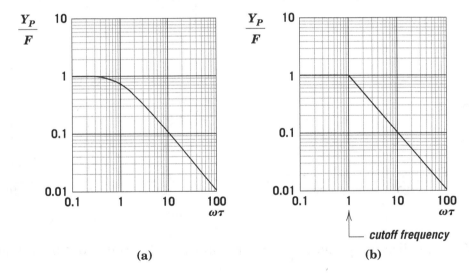

Figure 1.23 $\dfrac{Y_p}{F}$ versus $\omega\tau$ on a log-log scale for a low-pass filter: (a) actual curve; and (b) approximate envelope.

For $\omega \ll \omega_c$, the system is characterized as a low-pass filter where $\dfrac{Y_p}{F} \approx 1$ from Eq. (1.91). For $\omega \gg \omega_c$, Eq. (1.91) shows that $\dfrac{Y_p}{F}$ drops off to zero rapidly. When $\omega = \omega_c$, Eq. (1.91) yields an output magnitude that is $\sqrt{2}$ of the input amplitude, $\dfrac{Y_p}{F} = \dfrac{1}{\sqrt{2}}$.

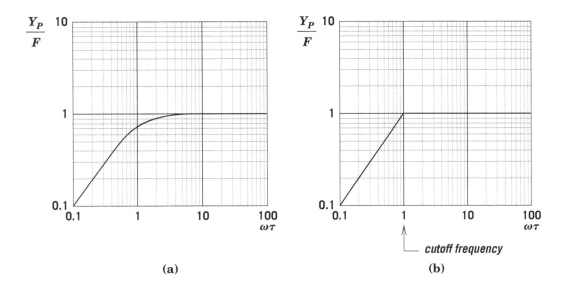

(a)   **(b)**

Figure 1.24 $\dfrac{Y_p}{F}$ versus $\omega\tau$ on a log-log scale for a high-pass filter: (a) actual curve; and (b) approximate envelope.

In a homework problem you may analyze a first-order system ODE described by

$\tau\dfrac{dy}{dt} + y(t) = \tau g(t)$, where $g(t)$ is the derivative of the forcing function $f(t) = F\cos(\omega t)$. In

other words, $\tau\dfrac{dy}{dt} + y(t) = \tau\dfrac{df}{dt} = -\tau\omega F\sin(\omega t)$. In this case, the system described by this ODE

operates as a high-pass filter, as shown in **Figure 1.24**, so that for $\omega \ll \omega_c$, $\dfrac{Y_p}{F}$ is nearly zero,

and for $\omega \gg \omega_c$, $\dfrac{Y_p}{F} \approx 1$.

Example 1.9

Revisit Example 1.8 and indicate whether or not the problem describes a low-pass filter system.

Solution:

From Example 1.8, the time constant is $\tau = 0.1$. Therefore, the cutoff frequency, given by

Eq. (1.90), is $\omega_c = \dfrac{1}{\tau} = \dfrac{1}{0.1} = 10\dfrac{rad}{s}$. The forcing frequency is $\omega = 3\dfrac{rad}{s}$, which is below ω_c.

Furthermore, the steady solution that was found to be $y_p(t) = 0.9578\sin(3t - 0.2915)$ has an

amplitude $y_p(t) = 0.9578$. Therefore, $\dfrac{Y_p}{F} = \dfrac{0.9578}{1} = 0.9578$. Considering that $\omega < \omega_c$ and

$\dfrac{Y_p}{F} \approx 1$, the ODE and the forcing function in Example 1.8 describe a low-pass filter system.

First-order electrical system: RC low-pass filter

A low-pass filter is an electrical circuit that consists of a resistor, R, a capacitor, C, and an input voltage $V(t)$ applied between ground and the resistor, as shown in **Figure 1.25**. The circuit output $V_o(t)$ is measured across the capacitor.

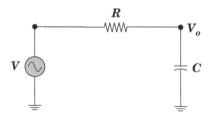

Figure 1.25 Schematic of a low-pass filter.

When this RC circuit is modeled as a block diagram, the system is governed by a first-order ODE, with $V(t)$ as the system input or forcing function and $V_o(t)$ as the system output or response.
The relationship between $V(t)$ and $V_o(t)$ is developed using:

- Kirchhoff's voltage loop law.

- A current–voltage relationship across the capacitor.

Kirchhoff's voltage law

Kirchhoff's voltage loop law requires the input voltage $V(t)$ to balance the sum of voltages across the resistor $\left(V_R = iR\right)$ and capacitor, $V_o(t)$, that is,

$$V(t) = iR + V_o(t) \tag{1.92}$$

where i is the current through the circuit. The current i is a time variable and should be removed from the right-hand side in favor of $V_o(t)$. This is accomplished by using a current–voltage relationship across the capacitor.

Current–voltage relationship across the capacitor

The current across a capacitor is given by

$$i = C\frac{dV_o}{dt} \tag{1.93}$$

Substituting Eq. (1.93) into Eq. (1.92) yields a first-order linear nonhomogeneous ODE

$$RC\frac{dV_o}{dt}+V_o=V(t) \tag{1.94}$$

which is equivalent to

$$\tau\frac{dV_o}{dt}+V_o=V\left(t\right) \tag{1.95}$$

where $\tau=RC$ in s when R is in *ohm* and C in *farad*. The time constant fixes the cutoff frequency, $\omega_c=\dfrac{1}{\tau}$. For a small τ that causes the transient to decay rapidly, the solution of Eq. (1.95) for a sinusoid input voltage, $V\left(t\right)=V_{in}\cos\left(\omega t\right)$, is given by Eq. (1.82),

$V_o\left(t\right)=\dfrac{V_{in}}{\sqrt{1+\omega^2\tau^2}}\cos\left(\omega t+\phi\right)$, or

$$V_o\left(t\right)=V_{out}\cos\left(\omega t+\phi\right) \tag{1.96}$$

where $V_{out}=\dfrac{V_{in}}{\sqrt{1+\omega^2\tau^2}}$, or

$$V_{out}=\frac{V_{in}}{\sqrt{1+\left(\dfrac{\omega}{\omega_c}\right)^2}} \tag{1.97}$$

This solution satisfies the concepts of a low-pass filter as described in **Figure 1.23**; that is, for an input voltage with a frequency $\omega \ll \omega_c$, Eq. (1.97) yields $V_{out}\simeq V_{in}$, as shown in **Figure 1.26**.

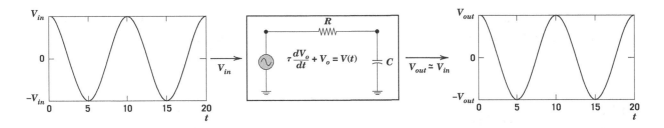

Figure 1.26 A low-frequency input sinusoid passing through the low-pass filter.

For an input voltage with a frequency $\omega \gg \omega_c$, Eq. (1.97) yields $V_{out} \simeq 0$, as shown in **Figure 1.27**.

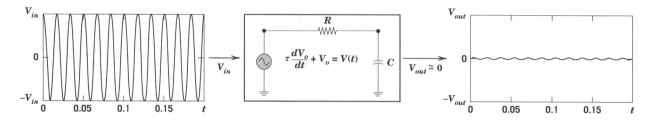

Figure 1.27 A high-frequency input sinusoid not passing through the low-pass filter.

What if the input voltage consists of two sinusoids, one sinusoid with a frequency ω_L such that $\omega_L \ll \omega_c$ and a second sinusoid with a frequency ω_H such that $\omega_H \gg \omega_c$, for example, $V(t) = V_{in,L} \cos(\omega_L t) + V_{in,H} \cos(\omega_H t)$? In this case, and assuming rapid decay of the transient, the solution of the ODE, Eq. (1.95), is

$$V_o(t) = \frac{V_{in,L}}{\sqrt{1 + \left(\dfrac{\omega_L}{\omega_c}\right)^2}} \cos(\omega_L t + \phi_L) + \frac{V_{in,H}}{\sqrt{1 + \left(\dfrac{\omega_H}{\omega_c}\right)^2}} \cos(\omega_H t + \phi_H)$$

For $\omega_L \ll \omega_c$, the first term on the right-hand side, $\dfrac{V_{in,L}}{\sqrt{1 + \left(\dfrac{\omega_L}{\omega_c}\right)^2}}$ approaches $V_{in,L}$, whereas the

second term $\dfrac{V_{in,H}}{\sqrt{1 + \left(\dfrac{\omega_H}{\omega_c}\right)^2}}$ approaches zero for $\omega_H \gg \omega_c$, so that $V_o(t) \simeq V_{in,L} \cos(\omega_L t + \phi_L)$.

The point is that a low-pass filter always outputs a very low-frequency input with almost the same amplitude while eliminating a very high-frequency input, as shown in **Figure 1.28**.

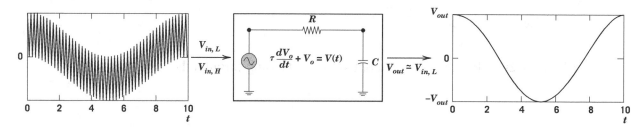

Figure 1.28 High and low frequencies input sinusoid through the low-pass filter.

Typical applications of this simple low-pass filter circuit are encountered in sensors, for example, a thermocouple as a temperature measuring device, which outputs a signal in the form of a voltage, for example, $V_{in,L} \cos(\omega_L t)$. Quite often, the measurements are taken in environments with high levels of electrical noise, for example, 60-Hz line in the proximity to the thermocouple, which is modeled by $V_{in,H} \cos(\omega_H t)$. There could be other electrical noises, for example, those that are generated by running machinery. Thus, the low-pass filter is essential in filtering the unwanted high-frequency noises.

Example 1.10 The Low-Pass Filter

In the design of a homemade low-pass filter used with a thermocouple, an engineer runs a preliminary simulation with a function generator under controlled conditions, as shown in **Figure 1.29**. A function generator is a device that generates various waveforms over a wide range of frequencies. A sinusoid is an example of a waveform.

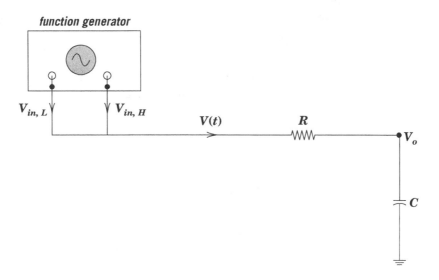

Figure 1.29 A preliminary simulation with controlled inputs $V_{in,L}(t)$ and $V_{in,H}(t)$ using a function generator.

In the particular application where the thermocouple is used, the sinusoid has a frequency $f_L = 0.25 Hz$ and amplitude $V_{in,L} = 2 volts$. The engineer also knows that the electrical noise from machinery in the proximity to the application has a frequency $f_H = 60 Hz$ and an amplitude $V_{in,H} = 0.5 volts$.

1. Plot the input voltage to the low-pass filter.
2. Choose a cutoff frequency in Hz above which the output amplitude of the noise is significantly reduced so that the output voltage matches closely the low-frequency sinusoid.
3. What should be the resistance R and capacitance C of the low-pass filter?

4. Compare the output of the low-pass filter with the low-frequency input sinusoid and discuss the results.

Solution:

1. Plot the input voltage to the low-pass filter.

The input voltage is given by $V(t) = V_{in,L}\cos(\omega_L t) + V_{in,H}\cos(\omega_H t)$. Using the given data,

$$V(t) = 2\cos(0.5\pi t) + 0.5\cos(120\pi t) \qquad (1.98)$$

$V(t)$ is seen in **Figure 1.30**, which shows the dominating high-frequency noise at 60 *Hz*.

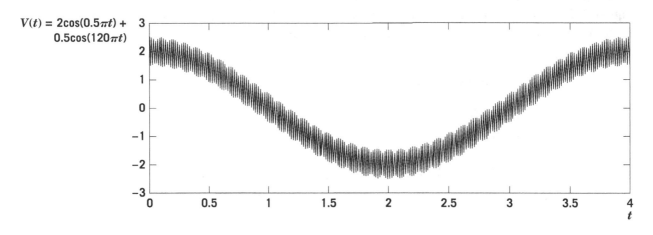

Figure 1.30 High and low frequencies input sinusoid prior to going through the low-pass filter.

2. Choose a cutoff frequency in *Hz* above which the output amplitude of the noise is significantly reduced so that the output voltage matches closely the low-frequency sinusoid.

The cutoff frequency f_c should be chosen so that $f_L \ll f_c$ and $f_H \gg f_c$. Considering the infinite choices, you need to make an educated guess and examine the output signal based on that choice. For example, try $f_c = 3Hz$. This implies that the cutoff frequency is 12 times larger than the low frequency and 20 times smaller than the high-frequency noise.

3. What should be the resistance R and capacitance C of the low-pass filter?

The choice of $f_c = 3Hz$, the frequency $\omega_c = 2\pi f_c = 6\pi \dfrac{rad}{s}$, so that $\tau = \dfrac{1}{\omega_c} = \dfrac{1}{6\pi} = 0.05305s$.

From the low-pass filter analysis, $\tau = RC$. Thus, choose R and C such that $RC = 0.05305s$. Here also you have infinite choices. For definiteness, choose $R = 1000 ohm$ that yields a capacitance $C = 0.05305 \times 10^{-3}$ *farad* or $53.05 \mu farad$.

4. Compare the output of the low-pass filter with the low-frequency input sinusoid and discuss the results.

Assuming rapid decay of the transient, we find that the output voltage is given by

$$V_o(t) = \frac{V_{in,L}}{\sqrt{1 + \left(\dfrac{\omega_L}{\omega_c}\right)^2}} \cos(\omega_L t + \phi_L) + \frac{V_{in,H}}{\sqrt{1 + \left(\dfrac{\omega_H}{\omega_c}\right)^2}} \cos(\omega_H t + \phi_H)$$

or

$$V_o(t) = \frac{2}{\sqrt{1 + \left(\dfrac{0.25}{3}\right)^2}} \cos(2\pi t + \phi_L) + \frac{0.5}{\sqrt{1 + \left(\dfrac{60}{3}\right)^2}} \cos(120\pi t + \phi_H) \qquad (1.99)$$

where $\phi_L = -\tan^{-1}(\omega_L \tau) = -\tan^{-1}\left(\dfrac{\omega_L}{\omega_c}\right) = -\tan^{-1}\left(\dfrac{1}{12}\right) = -0.08314 rad$, and

$\phi_H = -\tan^{-1}(\omega_H \tau) = -\tan^{-1}\left(\dfrac{\omega_H}{\omega_c}\right) = -\tan^{-1}(20) = -1.521 rad$.

Substituting these values in Eq. (1.99) yields

$$V_o(t) = 1.993\cos(0.5\pi t - 0.08314) + 0.0250\cos(120\pi t - 1.521) \qquad (1.100)$$

In comparison to the desired low-frequency input sinusoid given as $2\cos(0.5\pi t)$, the output voltage $V_o(t)$ reproduces this sinusoid quite well, as shown by the solid curve in **Figure 1.31**. Obviously, a steady time lag is expected between the input and output by virtue of the phase angles. However, what is important to notice is that the low-pass filter circuit filtered the noise shown in **Figure 1.30** and produced an output signal with an average amplitude of 1.993 *volts*, as indicated by the first term on the right-hand side of Eq. (1.100), with little noise on top of the curve with an amplitude of 0.025 *volt*.

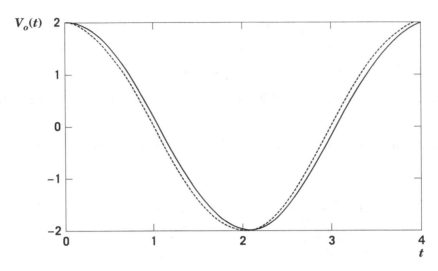

Figure 1.31 The low-pass filter system performance with voltage output $V_o(t)$ in comparison to the low-frequency input sinusoid given by $2\cos(0.5\pi t)$.

In reality, for example, in a thermocouple application, the low-frequency input is not known, but rather inferred from the first component of the measured output voltage, $V_o(t)$. Say, for example, that the measured output frequency and average amplitude are 0.25 Hz and 3 $volts$, respectively. For this circuit with a cutoff frequency of 3 Hz, the input voltage is inferred from

$$\frac{V_{in,L}}{\sqrt{1+\left(\dfrac{\omega_L}{\omega_c}\right)^2}} = 3 \quad \text{as} \quad V_{in,L} = 3.01 volts.$$

HOMEWORK PROBLEMS FOR CHAPTER ONE

Problems for Section 1.2

1. In an assembly manufacturing plant, rod A rotates continuously causing shaft B to oscillate back and forth, as shown in **Figure 1.32**. The shaft is designed to reciprocate with an acceleration $a = -0.02\pi^2 \sin(\pi t)\frac{m}{s^2}$.

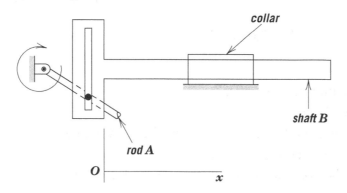

Figure 1.32 Reciprocating shaft by virtue of an oscillating rod.

a. Find an expression for the shaft displacement $x(t)$. The initial displacement and velocity are $x(0) = 0$ and $v(0) = 0.02\pi\frac{m}{s}$, respectively.

2. An airplane begins to move from rest in a straight line before taking off with a piecewise acceleration in $\frac{m}{s^2}$ given by $a = \begin{cases} 2 + 0.3t & 0 \le t < 10 \\ 5 & t \ge 10 \end{cases}$.

a. What is the airplane initial velocity, $v(0)$?
b. Find the velocity $v(t)$.
c. Find the distance $x(t)$.
d. Plot $a(t)$, $v(t)$, and $x(t)$ versus t in the interval $t \ge 0$.

3. A speedboat moving at $15\frac{m}{s}$ suddenly shuts off its engines. The magnitude of the hydrodynamic drag F_f is proportional to v^2, where v is the boat velocity, that is, $\left|F_f\right| = kv^2$ where k is a constant that depends on the fluid properties and the boat shape. Usually, k is determined experimentally. According to Newton's second law of motion, $-\left|F_f\right| = Ma = M\frac{dv}{dt}$, as shown in **Figure 1.33**, where a is the boat acceleration (deceleration in this case) and M is the boat mass.

Figure 1.33 A speedboat decelerating by virtue of drag.

Combining the hydrodynamic force with Newton's second law of motion yields $\dfrac{dv}{dt} = -k_1 v^2$,

where $k_1 = \dfrac{k}{M}$. For this boat $k_1 = 0.025 m^{-1}$.

a. What is the boat initial velocity, $v(0)$?
b. Find the boat velocity $v(t)$.
c. Find the boat distance $x(t)$ using $x(0) = 0$ as the IC when the engines shut off.
d. Plot $v(t)$ and $x(t)$ versus t in the interval $t \geq 0$.
e. Estimate how far the boat travels when its velocity decreases by 99% from the initial value.

4. When the shutter of a camera is depressed, the energy stored in the camera capacitor is converted to an intense light through the flash tube. This discharge is accompanied by a drop in the capacitor voltage over a very short period of time. You can simulate mathematically the voltage drop across a capacitor by considering the following simple circuits.

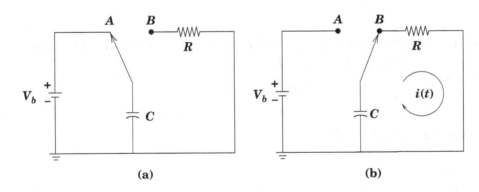

Figure 1.34 Camera shutter simplified circuit: (a) charging a capacitor; and (b) discharging the capacitor.

When the switch is in position A, as shown in **Figure 1.34a**, the battery with voltage V_b charges the capacitor. When the switch is thrown to position B at $t = 0$, as shown in **Figure 1.34b**, the capacitor whose initial voltage $V_C(0) = V_b$ discharges its energy through resistor R.

a. Use Kirchhoff's voltage loop law in **Figure 1.34b**,

$$V_C + V_R = 0 \qquad (1.101)$$

where $V_R = iR$ is the voltage across the resistor R, i is the current, and $V_C = \dfrac{1}{C}\int i\, dt$ is the voltage across the capacitor in order to derive the first-order ODE

$$RC\frac{dV_C}{dt} = -V_C \qquad (1.102)$$

b. Solve Eq. (1.102) with $V_C(0) = 24\,Volts$, $R = 10\ ohm$ and $C = 100\left(10^{-6}\right) farad$. Note that $1 ohm \cdot farad = 1s$.

c. Plot $V_C(t)$ and $V_R(t)$ versus t in the interval $t \geq 0$. Estimate how long it takes the capacitor voltage to drop to 0.1% of its initial value.

5. You are charged in a food processing plant to document the time to discharge various tanks with different geometrics. Consider such a tank as shown **Figure 1.35**, which is cylindrical.

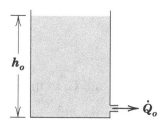

Figure 1.35 A cylindrical tank discharging at flow rate $\dot{Q}_o$.

a. Use the conservation of mass, Eq. (1.41), where $\dot{Q}_i = 0$ in this case, and Eq. (1.44) to derive a first-order ODE for the fluid height at any time $h(t)$.

b. Solve the ODE with IC $h(0) = h_0$.

c. Plot $h(t)$ versus time in the interval $t \geq 0$, and determine how long it takes to empty the tank for $A = \dfrac{\pi}{4} m^2$, $h_0 = 2m$, and $k_o = 0.01 \left(\dfrac{m^3}{s}\right)\dfrac{1}{\sqrt{m}}$.

6. Follow a similar approach to that in Problem 5 for the conical tank, as shown **Figure 1.36**.

The radius of the tank varies linearly from zero at the bottom of the tank to R_o at the top of the tank at height H_o. For a right circular cone, the local volume $Q = \dfrac{1}{3}\pi r^2 h$.

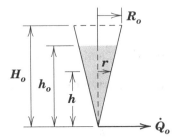

Figure 1.36 A conical tank discharging at flow rate $\dot{Q}_o$.

a. Show that for this geometry Eqs. (1.41) and (1.44) lead to the first-order ODE in $h(t)$ given

   by $\dfrac{dh}{dt} = -\dfrac{k_o}{\pi}\left(\dfrac{H_o}{R_o}\right)^2 h^{-\frac{3}{2}}$.

b. Solve the ODE with IC $h(0) = h_0$.

c. Plot $h(t)$ versus time in the interval $t \geq 0$, and determine how long it takes to empty the tank

   for $H_o = 2m$, $R_o = 2m$, $h_0 = 1m$, and $k_o = 0.01 \left(\dfrac{m^3}{s}\right)\dfrac{1}{\sqrt{m}}$.

Problems for Section 1.3

7. Consider three first-order systems, A, B, and C, in the time interval $t \geq 0$, which are

   governed by $\dfrac{dy}{dt} + \dfrac{1}{60}y = \dfrac{2}{60}$; $\dfrac{dy}{dt} + \dfrac{1}{30}y = \dfrac{2}{30}$; and $\dfrac{dy}{dt} + \dfrac{1}{15}y = \dfrac{2}{15}$, respectively.

a. Which system reaches a steady state the fastest?
b. How long does it take the fastest system to reach a steady state?
c. Using the time found in part b, find $y(t)$ of each system at that time. The IC of each system
   is $y(0) = 0$. Normalize your answer for each system as $y(t)/F$ when F is the constant
   forcing function of a particular system.

8. Consider two first-order systems, A and B, which are governed by $\dfrac{1}{4}\dfrac{dy}{dt} + y = f(t)$ in the

   time interval $t \geq 0$. For system A, $f(t) = t$ for $0 < t < 1$, whereas for system B,

$$f(t) = \begin{cases} 0 & t < 1 \\ t-1 & t \geq 1 \end{cases}.$$

a. Which system reaches a steady state the fastest?
b. What is the time lag between the two systems?
c. How long does it take each system to follow its forcing function steadily?

d. Find the solution of each system using the IC $y(0)=0$. Plot both solutions versus t on the same graph.

e. If $y(5)$ is the output of system A at $t=5$, estimate when system B will have the same output as system A.

For Problems 9–13 refer to Example 1.7. Use the data given in the example except for the IC $h(0)$; use $h(0)=2m$.

a. Find the water level $h(t)$ for $t\geq0$.

b. Plot $f(t)$ and $h(t)$ in the time interval $t\geq0$.

c. Estimate the transient time lag of the system, that is, the time for the water level to reach a steady state.

d. Determine the steady water level.

9.

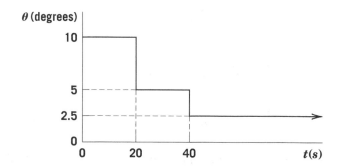

Figure 1.37 Valve angle as a piecewise function for Problem 9.

10.

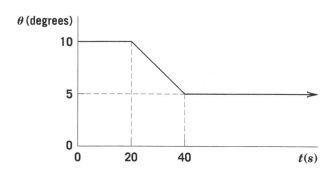

Figure 1.38 Valve angle as a piecewise function for Problem 10.

11.

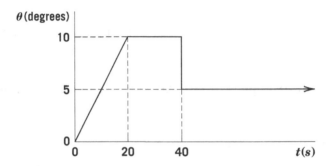

Figure 1.39 Valve angle as a piecewise function for Problem 11.

12.

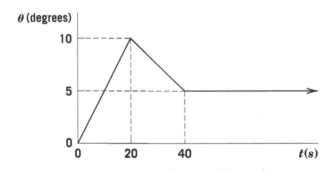

Figure 1.40 Valve angle as a piecewise function for Problem 12.

13.

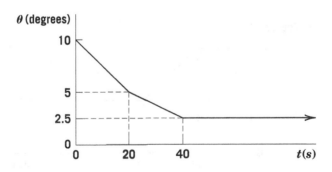

Figure 1.41 Valve angle as a piecewise function for Problem 13.

14. Consider a warm object, as shown in **Figure 1.42**, which is cooled by a surrounding fluid, such as air. By the first law of thermodynamics, conservation of energy, the energy decrease of the object is transformed to energy by virtue of the moving fluid, known as convective heat transfer.

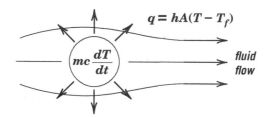

Figure 1.42 A warm object cooled by a surrounding fluid.

The energy decrease of the object is given by $-mc\dfrac{dT}{dt}$, where m is the mass (kg), c is specific heat $\left(\dfrac{kJ}{kg\,^{o}C}\right)$, T is instantaneous temperature $\left(^{o}C\right)$ of the object, and t is time (s). The convective heat transfer is given by Newton's cooling law , $q = hA\left(T - T_f\right)$ where q is the rate of heat transfer to the fluid $\left(\dfrac{kJ}{s}\right)$, A is the surface area $\left(m^2\right)$ of the object, $h\left(\dfrac{kJ}{sm^{2\,o}C}\right)$ is a constant known as the heat transfer coefficient, which depends on the fluid properties, and T_f is the fluid temperature $\left(^{o}C\right)$ in the vicinity of the object

a. Using the first law of thermodynamics, show that the instantaneous temperature of the object is a first-order ODE as $\tau\dfrac{dT}{dt} + T = f(t)$.

b. Define the time constant τ in terms of the object and fluid properties and show that it has the units of time by direct substitution of properties units.

c. Define $f(t)$.

d. If the object is the tip of a thermocouple with a spherical iron bead, estimate how long it will take the thermocouple to reach the steady-state temperature of fluid. The volume and surface area of a sphere of radius r are $\dfrac{4}{3}\pi r^3$ and $4\pi r^2$, respectively. Use $r = 0.5mm$, $\rho = 7800\dfrac{kg}{m^3}$, $c = 0.5\dfrac{kJ}{kg\,^{o}C}$, and $h = 10\dfrac{kJ}{sm^{2\,o}C}$.

e. Find the solution of the ODE using the IC $T(0) = 100^{o}C$. The fluid temperature is $T_f = 15^{o}C$.

f. Plot the solution in the interval $t \geq 0$.

15. Consider a cold object, as shown in **Figure 1.43**, which is heated by a surrounding fluid, such as water.

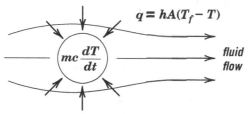

Figure 1.43 A cold object heated by a surrounding fluid.

By the first law of thermodynamics, conservation of energy, the energy by virtue of the moving fluid, known as convective heat transfer, is transformed to energy increase of the object. The convective heat transfer is given by $q = hA\left(T_f - T\right)$. The energy increase of the object is given by $mc\dfrac{dT}{dt}$. See Problem 14 for nomenclature and units. The density, specific heat, and heat transfer coefficients are $\rho = 7800\dfrac{kg}{m^3}$, $c = 0.5\dfrac{kJ}{kg\,^oC}$, and $h = 50\dfrac{kJ}{sm^{2o}C}$, respectively.

a. Using the first law of thermodynamics, show that the instantaneous temperature of the object can be written as a first-order ODE as $\tau\dfrac{dT}{dt} + T = f(t)$.

b. Find the solution of the ODE using the IC $T(0) = 15^oC$ for a thermocouple, with a 1-*mm*-diameter spherical bead. The volume and surface area of a sphere of radius r are $\dfrac{4}{3}\pi r^3$ and $4\pi r^2$, respectively. The fluid temperature is $T_f = 100^oC$

16. Consider as part of a chemical process a tank that initially contains a mixture of water and alcohol. The mixture's initial volume is $Q_{mix,tank}(0) = 1000L$. During the process, water and alcohol mixture is introduced into the tank at a rate of $\dot{Q}_{mix,i} = 1\dfrac{L}{s}$ while $\dot{Q}_{mix,o} = 1.5\dfrac{L}{s}$ of the mixture is discharged.

a. Integrate Eq. (1.41) over the time interval $t > 0$ and use the foregoing data to show that the instantaneous mixture volume in the tank is given by $Q_{mix,tank}(t) = (-0.5t + 1000)L$.

b. Initially, the tank contains 75% water and 25% alcohol by volume. Determine the initial volume of alcohol in the tank $\dot{Q}_A(0)$ in *L*, where the subscript A stands for alcohol.

c. If the mixture that enters the tank is 50% alcohol and 50% water by volume, infer from $\dot{Q}_{mix,i} = 1\dfrac{L}{s}$ the rate of alcohol entering the system, $\dot{Q}_{A,i}$ in *L/s*,

Define $Q_A(t)$ as the volume of alcohol in the tank at any time so that the fraction of alcohol in the tank is $\dfrac{Q_A(t)}{Q_{mix,tank}(t)}$, which is dimensionless. The foregoing expression yields the rate of alcohol leaving the tank as $\dot{Q}_{A,o} = \dot{Q}_{mix,o}\dfrac{Q_A(t)}{Q_{mix,tank}(t)}$.

d. Use Eq. (1.41) along with $\dot{Q}_{A,i}$ and $\dot{Q}_{A,o}$ to show that the alcohol volume in the tank, $Q_A(t)$, is governed by the first-order ODE $\dfrac{dQ_A}{dt} + \dfrac{3}{2000 - t}Q_A = \dfrac{1}{2}$.

e. Solve the ODE to find an expression for $Q_A(t)$ subject to the IC $\dot{Q}_A(0)$.

f. Plot $Q_A(t)$ in the interval $t \geq 0$.

g. Find the percentage of alcohol volume in the tank at $t = 15s$.

Problems for Section 1.4

17. Consider the first-order ODE $0.2\dfrac{dy}{dt} + y = f(t)$, where t is in s and $f(t) = \cos(5t)$. The IC is $y(0) = 0$.

a. Find the solution $y(t)$.

b. What is the transient time lag? Verify your answer graphically by plotting the complete solution and the steady response in the time interval $0 \leq t \leq 2s$.

c. Compare the forced response with the sinusoid forcing function, and determine whether it is in phase with, lagging or leading the forcing function.

d. Determine the steady time lag and the amplitude of the steady response.

18. A function generator outputs two voltages, $V_{in,L}(t)$ and $V_{in,H}(t)$ to a low-pass filter (see **Figure 1.29** in Example 1.10). The voltage $V_{in,L}(t)$ has an amplitude of 2 *volts* and a frequency of 2 *Hz*. The voltage $V_{in,H}(t)$ has an amplitude of 1 *volt* and a frequency of 120 *Hz*.

a. Find an expression for $V(t)$, the input voltage to the low-pass filter.

b. Select the resistance R and the capacitance C in the low-pass filter for a cutoff frequency $f_c = 15Hz$.

c. Find $V_o(t)$, the output of the low-pass filter by ignoring the transient voltage.

d. Compare $V_o(t)$ with $V_{in,L}(t)$ graphically in order to describe the effect of the filter. Discuss your results.

19. Refer to Example 1.10. Find $V_o(t)$ if the input voltage is expressed as $V(t) = V_{in,L}\sin(\omega_L t) + V_{in,H}\sin(\omega_H t)$. Ignore the transient voltage.

20. Consider the first-order ODE $\tau\dfrac{dy}{dt} + y(t) = \tau g(t)$, where $g(t)$ is the first derivative of the forcing function $f(t) = F\cos(\omega t)$. In this problem you will demonstrate that this ODE describes a high-pass filter in an electrical system.

a. Find a general solution $y(t)$ expressed in terms of a phase angle. Ignore the transient term.

b. Designate the output amplitude as Y_P and show that $\dfrac{Y_P}{F} = \dfrac{\tau\omega}{\sqrt{1 + \omega^2\tau^2}}$.

c. Show that for very low frequencies, that is, as $\omega \to 0$, the normalized amplitude of the forced response approaches zero.

d. Show that for very high frequencies, that is, as $\omega \rightarrow \infty$, the normalized amplitude of the forced response approaches 1.

e. Plot on a log-log scale $\dfrac{Y_P}{F}$ versus $\omega\tau$ to validate your result; see **Figure 1.24**.

21. In reference to **Figure 1.25**, you learned that the RC circuit acted as a low-pass filter when the output voltage is measured across the capacitor. In this problem you will show that the RC circuit acts as a high-pass filter when the output voltage is measured across the resistor, as shown in **Figure 1.44**.

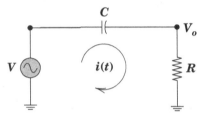

Figure 1.44 An RC circuit for a high-pass filter.

From physics, the voltage across R is $V_o(t) = iR$. The voltage V_C across C is related to the current $i(t)$ by $i = C\dfrac{dV}{dt}$, where $V(t)$ is the input voltage to the circuit.

a. Apply Kirchhoff's voltage loop law to this circuit and show that the voltage $V_o(t)$ across the resistor R is governed by the first-order ODE $\tau \dfrac{dV_o}{dt} + V_o = \tau g(t)$, where $g(t)$ is the first derivative of $V(t)$. See Eq. (1.92).

b. Define the time constant τ and cutoff frequency.

c. Solve the ODE with a zero IC and use the output amplitude to demonstrate analytically that this circuit describes a high-pass filter. Ignore the transient term.

22. A function generator outputs two voltages, $V_{in,L}$ and $V_{in,H}$ to a high-pass filter, as shown in **Figure 1.45**.

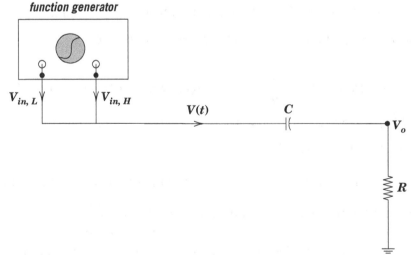

Figure 1.45 A function generator with controlled inputs $V_{in,L}(t)$ and $V_{in,H}(t)$ to a high-pass filter.

$V_{in,L}$ has an amplitude of 2 *volts* and a frequency of 2 *Hz*, whereas $V_{in,H}$ has an amplitude of 1 *volt* and a frequency of 120 *Hz*.

a. Find an expression for $V(t)$, the input voltage to the high-pass filter.
b. Select the resistance R and the capacitance C in the high-pass filter for a cutoff frequency $f_C = 15\ Hz$.
c. Find $V_o(t)$, the steady output of the high-pass filter. Ignore the transient term.
d. Compare $V_o(t)$ and $V_{in,H}(t)$ to describe the effect of the filter. Discuss your results.

23. Electrical filters can be developed using different components. For example, in the circuit shown in **Figure 1.46**, an inductor L is used with a resistor R instead of a capacitor. From physics, the voltage across R is $V_o(t) = iR$. The voltage V_L across L is related to the current $i(t)$ by $V_L = L\dfrac{di}{dt}$.

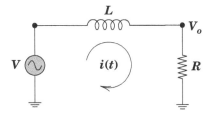

Figure 1.46 An RL circuit for a low-pass filter.

a. Apply Kirchhoff's voltage loop law to this circuit and show that $V_o(t)$ is governed by the first-order ODE $\tau\dfrac{dV_o}{dt} + V_o = V(t)$, where $V(t)$ is the input voltage to the circuit.
b. Define the time constant τ in terms of L and R.
c. Solve the ODE with a zero IC and use the output amplitude to demonstrate analytically that this circuit describes a low-pass filter. Ignore the transient term.

24. If in Problem 23 the output voltage is measured across the inductor L instead of the resistor R, as shown in **Figure 1.47**, then the system acts as a high-pass filter. In this circuit, the voltage V_R across R is $V_R = iR$. The voltage $V_o(t)$ across L is related to the current $i(t)$ by $V_o(t) = L\dfrac{di}{dt}$.

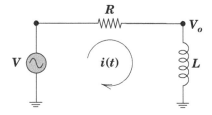

Figure 1.47 An RL circuit a high-pass filter.

a. Apply Kirchhoff's voltage law to this circuit and show that $V_o(t)$ is governed by the first-order ODE $\tau \dfrac{dV_o}{dt} + V_o = \tau g(t)$, where $g(t)$ is the first derivative of $V(t)$, the input voltage to the circuit.

b. Define the time constant τ.

c. Solve the ODE with a zero IC and use the output amplitude to demonstrate analytically that this circuit describes a high-pass filter. Ignore the transient term.

CHAPTER TWO

SECOND-ORDER INITIAL VALUE ORDINARY DIFFERENTIAL EQUATIONS

2.1 Introduction

In this chapter, we will describe engineering systems that are modeled by a second-order linear ordinary differential equation, also known as second-order systems. As an introduction to second-order systems, consider the suspension of a car system, as shown in **Figure 2.1**.

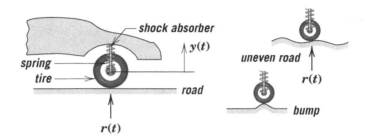

Figure 2.1 Suspension of a car under different road conditions.

Second-order ODEs are extremely important in many engineering fields. They appear in vibration, control systems, chemical and hydraulic systems, and electrical systems, to name a few applications. As you well know from daily experience, driving comfort depends on how quickly the spring-shock system dampens oscillations on uneven roads. From an engineering point of view, the control of oscillations will require an understanding of how the spring-shock system interacts with various road conditions. In this chapter, you will learn that by invoking several laws and physical relationships the displacement of the car $y(t)$, as shown in **Figure 2.1**,

is governed by a second-order ordinary differential equation given by $\dfrac{d^2y}{dt^2} + a_1 \dfrac{dy}{dt} + a_0 y = r(t)$,

where a_0 and a_1 are coefficients that depend on the car mass, spring stiffness, tire elasticity, and shock absorber resistance to motion, to name a few parameters. This ODE, which describes the car frame displacement $y(t)$ as a function of time t, depends on the road condition described by $r(t)$ and the initial conditions of the system.

As you will learn in this chapter, oscillating second-order systems display interesting characteristics. One of these characteristics is resonance. You will learn about the conditions that lead to this phenomenon that exhibits a large-amplitude response for a certain frequency of the forcing function. Physically, when a system resonates, it absorbs the maximum amount of energy from its forcing function. In some engineering systems resonance is destructive and should be avoided, while in other engineering systems resonance is desirable and should be exploited. In mechanical systems, for example, resonance leading to very large amplitudes may be catastrophic and should be avoided. Consider the case of the car suspension system in

Figure 2.2. If the shock absorber in the system is not working properly so that in effect damping is negligible, the system is essentially a spring and the car mass, as shown in **Figure 2.2a**. Also, if the uneven road is configured such that the tires displace sinusoidally, as shown in **Figure 2.2b**, then it is conceivable that the car frame oscillates at a particular frequency that causes the oscillation amplitude to grow so large to the point that the passengers could be in danger. This specific frequency at which the amplitude of oscillation grows unboundedly is called the resonance frequency.

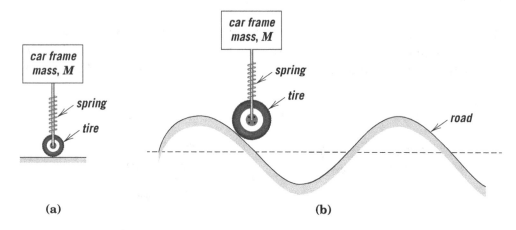

Figure 2.2 Suspension system model: (a) spring model and no shock absorber; and (b) system interaction with a sinusoidally shaped road.

Conversely, in electrical systems resonance may be useful. An example is a radio receiver tuner circuit, as shown in **Figure 2.3a**.

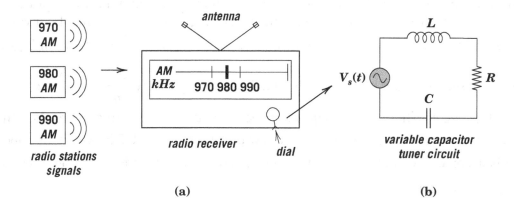

Figure 2.3 Electrical system resonance: (a) radio receiver; and (b) oscillating tuner circuit.

As the radio is dialed, mechanically or digitally, the turner circuit inside the receiver, as shown in **Figure 2.3b**, detects the signal from a particular radio station. The circuit is designed to resonate at the frequency of a signal from a radio station. In this case, the radio signal forces the turner circuit current to oscillate with a growing magnitude until the radio signal is clearly detected.

Another characteristic of second-order oscillating systems is damping due to friction between the system and its environment. In mechanical systems, damping is usually described by a force that is proportional to velocity. The direction of a positive damping force is opposite to the motion, and therefore positive damping impedes motion. Positive damping also removes energy from the system. In the case of the suspension system, positive damping is provided by the shock absorber, which mitigates the oscillation of the car frame. In the case of the radio tuner circuit, positive damping is provided by the resistor of the tuning circuit, which removes energy from the system in the form of heat. Damping is shown in this chapter as the first-order term in the ODE, $a_1 \dfrac{dy}{dt}$. Thus, when we speak of positive, we imply that $a_1 > 0$. For example, you will learn that the spring-shock system displacement $y(t)$, as shown in **Figure 2.1**, is governed by

$\dfrac{d^2 y}{dt^2} + \dfrac{c_d}{m}\dfrac{dy}{dt} + \dfrac{k}{m}y = f(t)$. In this equation, the coefficient of the first derivative, $a_1 = \dfrac{c_d}{m}$ turns out to be positive and is the cause for dampening oscillations as you will learn later on. On the other hand, when $a_1 < 0$, we speak of negative damping that may exist in some mechanical systems. An example of a system that may experience negative damping is the wing of an aircraft as shown in **Figure 2.4**.

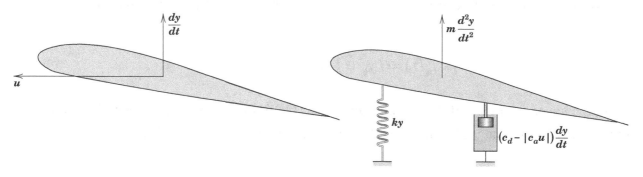

Figure 2.4 Cross section of an airplane wing flying at speed u while lifting at a rate $\dfrac{dy}{dt}$.

In this case, the equation of motion of the wing is given by $m\dfrac{d^2 y}{dt^2} + \left(c_d - |c_a u|\right)\dfrac{dy}{dt} + ky = 0$,

where m is wing mass; c_d is a positive damping coefficient that is a function of the fluid property and wing geometry; k is a spring-like stiffness of the wing; and c_a is an aerodynamic coefficient that depends on the wing geometry and the inclination of the wing relative to the air (angle of attack). In this equation $a_1 = c_d - |c_a u|$.

If $|c_a u| > c_d$, then a_1 is negative, and a_1 is called negative damping. When $a_1 < 0$, the oscillations grow in time. In other words, negative damping feeds energy into the oscillating system, which leads to a catastrophic state.

2.2 Basic Concepts

A general second-order linear ODE is defined as

$$\frac{d^2 y}{dt^2} + a_1(t)\frac{dy}{dt} + a_0(t)y = r(t) \tag{2.1}$$

The coefficients $a_0(t)$ and $a_1(t)$ are at most function of the independent variable, t, in order to classify the ODE as linear. Otherwise, if either a_0 or a_1 is also a function of the dependent variable, y, that is, $a_0(t,y)$ or $a_1(t,y)$, then the ODE is classified as nonlinear. Equation (2.1) is homogeneous if $r(t)=0$. Otherwise, it is a nonhomogeneous ODE.

Homogeneous and particular solution

Equation (2.1) possesses two homogeneous solutions $y_{h_1}(t)$ and $y_{h_2}(t)$, which are combined linearly with two arbitrary constants C_1 and C_2 by the principle of linear superposition as follows

$$y_h(t) = C_1 y_{h_1}(t) + C_2 y_{h_2}(t) \tag{2.2}$$

Thus, the general solution of the second-order linear ODE, Equation (2.1), is found by adding the homogeneous solution $y_h(t)$ to the particular solution $y(t) = y_h(t) + y_p(t)$, or

$$y_h(t) = C_1 y_{h_1}(t) + C_2 y_{h_2}(t) + y_p(t) \tag{2.3}$$

C_1 and C_2 are found using the initial conditions (IC),

$$y(t_0) = y_0 \tag{2.4}$$

and

$$\frac{dy}{dt}(t_0) = y_0' \tag{2.5}$$

at one point in time, t_0. The solutions of the homogeneous second-order linear ODE are summarized in **Table 2.1**.

$$\frac{d^2 y_h}{dt^2} + a_1 \frac{dy_h}{dt} + a_0 y_h = 0 \;;\; \Delta = a_1^2 - 4a_0$$

Discriminant	Homogeneous solution $y_h(t)$	Oscillation
$\Delta < 0$	$e^{-\frac{a_1}{2}t}\left[a\cos\left(\frac{\sqrt{\lvert\Delta\rvert}}{2}t\right)+b\sin\left(\frac{\sqrt{\lvert\Delta\rvert}}{2}t\right)\right]$ or $ce^{-\frac{a_1}{2}t}\cos\left(\frac{\sqrt{\lvert\Delta\rvert}}{2}t+\phi\right)$	Yes
$\Delta = 0$	$e^{-\frac{a_1}{2}t}\left(a+bt\right)$	No
$\Delta > 0$	$e^{-\frac{a_1}{2}t}\left(C_1 e^{\frac{\sqrt{\Delta}}{2}t}+C_2 e^{-\frac{\sqrt{\Delta}}{2}t}\right)$ or $e^{-\frac{a_1}{2}t}\left(a\sinh\frac{\sqrt{\Delta}}{2}t+b\cosh\frac{\sqrt{\Delta}}{2}t\right)$	No

Table 2.1 Solutions of the homogeneous second-order linear ODE.

The next step is to cast the second-order ODE, Eq. (2.1), into the engineering form so that the coefficients a_0 and a_1 are related to important physical parameters widely used in engineering, such as the frequency of vibration and damping coefficient.

2.3 Engineering Form of the Second-Order ODE

Many engineering systems can be modeled in the time domain by a second-order ODE where the coefficients a_0 and a_1 are constant. In this case, it is advantageous to put Eq. (2.1) in a form widely used by engineers, so that the coefficients in the modified equation have physical meaning. For a second-order system, the engineering form for a general second-order linear nonhomogeneous ODE is given by

$$\frac{d^2y}{dt^2}+2\zeta\omega_n\frac{dy}{dt}+\omega_n^2 y=f(t) \tag{2.6}$$

where $f(t)$ is defined as the system forcing function, $y(t)$ is the system response, ω_n is the undamped natural circular frequency of the system, and ζ is the damping factor. The factor 2 in

Eq. (2.6) is introduced for mathematical convenience. A comparison of Eq. (2.6) with Eq. (2.1) yields $a_0 = \omega_n^2$ and $a_1 = 2\zeta\omega_n$. From **Table 2.1**, $\Delta = 4\omega_n^2\left(\zeta^2 - 1\right)$.

Units of ζ and ω_n

The unit of the natural frequency ω_n is *rad/s*, whereas the damping factor ζ is dimensionless so that Eq. (2.6) is dimensionally consistent. The natural frequency ω_n is related to the frequency f_n in *Hz* or $\dfrac{1}{s}$ by the relationship $\omega_n = 2\pi f_n$.

Example 2.1

Given the second-order linear ODE

$$0.04\frac{d^2 y}{dt^2} + 0.8\frac{dy}{dt} + y = \sin(t) \tag{2.7}$$

1. Put the ODE in the engineering form of Eq. (2.6).
2. Determine the forcing function $f(t)$ and the values of the natural frequency ω_n and the damping factor ζ.

Solution:

1. Notice that the coefficient of $\dfrac{d^2 y}{dt^2}$ in Eq. (2.7) is 0.04. Therefore, to obtain the same form as in Eq. (2.6) divide Eq. (2.7) by 0.04

$$\frac{d^2 y}{dt^2} + 20\frac{dy}{dt} + 25y = 25\sin(t) \tag{2.8}$$

2. Comparing Eq. (2.8) with Eq. (2.6), you find $f(t) = 25\sin(t)$, $\omega_n = 5$, and $2\zeta\omega_n = 20$ that yields $\zeta = 2$.

∎

Next, cast the homogeneous solution in **Table 2.1** into the engineering form, that is, with $f(t) = 0$. An engineering system governed by Eq. (2.6) with no forcing function $f(t)$ is said to be in free vibration. In this case, four different forms are discerned depending on the value of the damping factor:

- $\zeta = 0$, known as the undamped solution or undamped free vibration.

- $0 < \zeta < 1$, known as the underdamped solution or underdamped free vibration.

- $\zeta = 1$, known as the critically damped solution.

- $\zeta > 1$, known as the overdamped solution.

The undamped homogeneous solution: Undamped free vibration

When $\zeta = 0$, the system is said to be undamped. In this case, the homogeneous ODE can be inferred from Eq. (2.6) with $f(t) = 0$ as

$$\frac{d^2 y_h}{dt^2} + \omega_n^2 y_h = 0 \tag{2.9}$$

Equation (2.9) describes an undamped system in free vibration, where $a_1 = 0$. In this case, $\Delta = -4\omega_n^2$, which corresponds to the first row of **Table 2.1** that yields

$$y_h(t) = a\cos\omega_n t + b\sin\omega_n t \tag{2.10}$$

or

$$y_h(t) = c\cos(\omega_n t + \phi) \tag{2.11}$$

Equation (2.11) is plotted in **Figure 2.5**. With no forcing function and damping, that is, $f(t) = 0$ and $\zeta = 0$, respectively, the system oscillates with a constant amplitude c. Moreover, the system oscillates with a constant frequency ω_n that is solely determined by the natural properties of the system as you will learn in the next example. For that reason ω_n is called the natural frequency of the system in free vibration. To emphasize that there is no damping in the system, ω_n is called the undamped natural frequency. The period of oscillation given by $p = \dfrac{1}{f_n} = \dfrac{2\pi}{\omega_n}$ is shown in **Figure 2.5**.

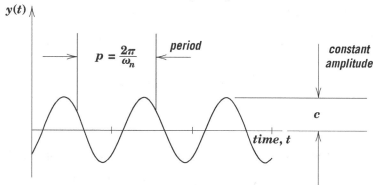

Figure 2.5 Solution of an undamped homogeneous second-order linear ODE.

Example 2.2 Undamped Elevator System

An elevator is moving downward at constant velocity, as shown in **Figure 2.6**. As part of engineering safety, the engineer must envision all possible scenarios leading to failure and take the necessary precautions in the design process. One of those scenarios would be a failure in the motor drive causing the elevator and cable system to suddenly stop. Although the cable is made out of strong material, such as steel, the sudden stop will cause the elevator to vibrate about an equilibrium point at the instance it stopped due to cable elasticity. The oscillation means an initial increase in cable length, which will result in an increase in cable stress (internal tension force per unit area). Thus, if the cable is not designed to handle this maximum stress, it may break.

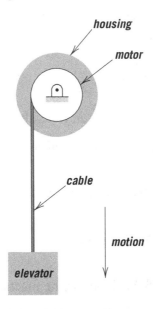

Figure 2.6 Elevator and cable system in downward motion.

1. Model the cable as a spring with a stiffness k and show that the displacement at the moment the elevator stopped is governed by an undamped homogeneous second-order linear ODE.
2. Show that the natural frequency ω_n is determined by the natural properties of the system, namely, the elevator mass m and the cable stiffness k.
3. Solve for the elevator displacement if the elevator velocity is $1\dfrac{m}{s}$ when it suddenly stopped.

   The cable stiffness is $k = 4\left(10^6\right)\dfrac{N}{m}$, and the elevator mass is $m = 2000 kg$.

Solution:

1. Without the elevator, the cable free length is L_f, as shown in **Figure 2.7a**. When attached to the cable, the elevator weight causes the cable to stretch by a distance y_0, as shown in **Figure 2.7b**. Since the elevator was moving at constant velocity, just before it stopped, then Newton's second law of motion requires

$$W - \left| F_{s,b} \right| = 0 \qquad (2.12)$$

where $\left| F_{s,b} \right|$ is the magnitude of the spring restoring force as shown in **Figure 2.7b**, which is given by Hooke's law as

$$\left| F_{s,b} \right| = k y_0 \qquad (2.13)$$

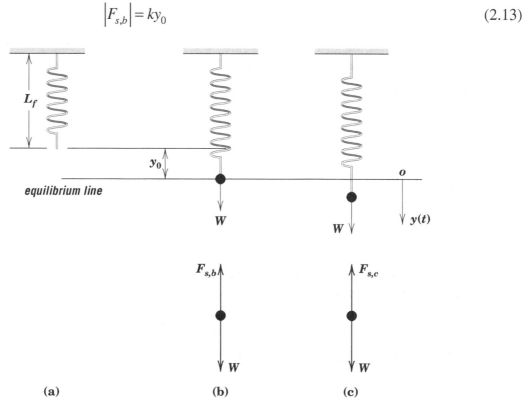

Figure 2.7 Cable-elevator system: (a) cable length without the elevator; (b) cable elongation with the attached elevator; and (c) oscillation of the elevator due to the sudden stop.

Equations (2.12) and (2.13) yield

$$W = k y_0 \qquad (2.14)$$

When the elevator suddenly stops, it tends to oscillate due to the inertial effect about a new equilibrium line as shown in **Figure 2.7c**.

In this case, there is an acceleration due to oscillation. Thus, Newton's second law requires

$$W - \left| F_{s,c} \right| = ma \qquad (2.15)$$

where $\left| F_{s,c} \right|$ is the spring restoring force, as shown in **Figure 2.7c**, which is given by

$$\left| F_{s,c} \right| = k \left(y_0 + y \right) \qquad (2.16)$$

Substituting Eqs. (2.14) and (2.15) in Eq. (2.16) yields

$$ky_0 - k\left(y_0 + y\right) = ma \tag{2.17}$$

which simplifies to

$$-ky = ma \tag{2.18}$$

Note how the elevator weight W is eliminated from the equation when the displacement $y(t)$ is measured from the new equilibrium line. From kinematics, the acceleration a is the second derivative of the instantaneous displacement $y(t)$,

$$a = \frac{d^2 y}{dt^2} \tag{2.19}$$

Substituting Eq. (2.19) in Eq. (2.18) yields $-ky = m\dfrac{d^2 y}{dt^2}$, or

$$m\frac{d^2 y}{dt^2} + ky = 0 \tag{2.20}$$

Equation (2.20) is a linear homogeneous second-oder ODE, which governs the instantaneous displacement of the elevator with respect to the equilibrium point. One IC for Eq. (2.20) is the initial displacement $y(0) = 0$. The second IC is the initial velocity $\dfrac{dy(0)}{dt} = 1\dfrac{m}{s}$. Equation (2.20) is cast into the engineering form, Eq. (2.6) by dividing by m,

$$\frac{d^2 y}{dt^2} + \frac{k}{m} y = 0 \tag{2.21}$$

Equation (2.21) implies that $\zeta = 0$. Hence, the system is undamped.

2. From Eq. (2.21) you can infer that $\omega_n^2 = \dfrac{k}{m}$. Thus, the natural frequency of the system in free vibration is

$$\omega_n = \sqrt{\frac{k}{m}} \tag{2.22}$$

which is solely dependent on the intrinsic or natural properties of the system, namely, the elevator mass m and the spring stiffness k.

3. The solution of Eq. (2.21) is given by Eq. (2.10)

$$y_h(t) = a\cos\omega_n t + b\sin\omega_n t \tag{2.23}$$

2-10

Using the IC $y(0)=0$ in Eq. (2.23) yields $a=0$. Therefore,

$$y_h(t)=b\sin\omega_n t \qquad (2.24)$$

The derivative of $y_h(t)$ is

$$\frac{dy_h}{dt}=b\omega_n\cos\omega_n t \qquad (2.25)$$

Using the IC $\frac{dy(0)}{dt}=1$ yields $b=\frac{1}{\omega_n}$. Therefore, the solution is $y_h(t)=\frac{1}{\omega_n}\sin\omega_n t$.

Using the given data yields $\omega_n=\sqrt{\frac{4(10^6)}{2000}}=44.72\frac{rad}{s}$ so that the solution is

$$y_h(t)=0.02236\sin(44.72t) \qquad (2.26)$$

For safety, the engineer will calculate the maximum elongation of the cable from its free length and use that information to determine the maximum stress in the cable. This is a specialty in engineering mechanics, which is beyond the scope of this chapter. Nevertheless, the maximum elongation is given by the sum of the static displacement y_0, as shown in **Figure 2.7b**, and the amplitude of oscillation from Eq. (2.26), which is equal to $0.02236m$. From Eq. (2.14)

$$y_0=\frac{W}{k}=\frac{mg}{k}=\frac{2000(9.81)}{4(10^6)}=0.00491m$$

Therefore, the maximum elongation is equal to $0.0273m$.

This example illustrates how to model an undamped engineering problem as a homogeneous linear second-order ODE where the dependent variable is the elevator displacement $y(t)$ and the independent variable is time t. Furthermore, the engineering form clearly reveals the undamped natural frequency of the system ω_n and the damping factor that is equal to zero in this case. The engineering form is suitable because it provides information such as ω_n and ζ without solving the ODE. In this case, however, the solution of the ODE yielded other useful information such as the cable maximum elongation under oscillation.

Later on, you will revisit this example by including friction and examine its effect on the results.

Physical analogy for the undamped solution

To clarify the physical meaning of no forcing function and no damping in a system, you may consider a simple system familiar to you, that is, the swing system as shown in **Figure 2.8**.

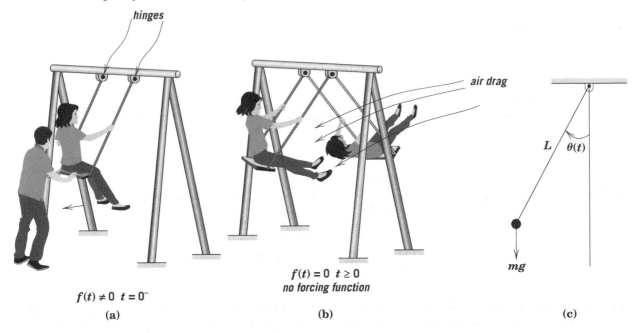

Figure 2.8 Swing analogy: (a) initial pull at $t = 0^-$; (b) releasing the swing at $t = 0$; and (c) ideal pendulum model.

Initially, you pull the swing with a force $f(t) \neq 0$ at $t = 0^-$, the time just before releasing the swing, as shown in **Figure 2.8a**. Then you release the swing at $t = 0$. Without ever pushing the swing again, as shown in **Figure 2.8b**, the forcing function, $f(t)$, is equal to zero. In this case, the motion of the swing can be modeled, for small angular displacements, by the equation of motion of an ideal oscillating pendulum

$$\frac{d^2\theta}{dt^2} + \frac{g}{L}\theta = 0 \tag{2.27}$$

where θ is the pendulum angular displacement, L is the length of the cable, and g is the acceleration of gravity, as shown in **Figure 2.8c**. Equation (2.27) is derived later on. If you

compare Eq. (2.27) with Eq. (2.9), you find that the natural frequency is $\omega_n = \sqrt{\dfrac{g}{L}}$.

Notice that without solving Eq. (2.27), you are able to obtain important information about the system, for example, its natural frequency of oscillation, simply by inspecting the equation. Notice also that the damping factor is $\zeta = 0$ because in an ideal pendulum friction at the hinges and any energy dissipation due to air drag against the swing system are neglected. Under this condition, the angular displacement of the swing is governed by Eq. (2.11)

$$\theta(t) = c\cos(\omega_n t + \phi) = c\cos\left(\sqrt{\frac{g}{L}}t + \phi\right)$$

Thus, the swing oscillates at a constant amplitude similar to the curve shown in **Figure 2.5**.

The constants c and ϕ are fixed by the initial conditions, which are the initial displacement $\theta(0) = \theta_o$, and the initial angular velocity $\dfrac{d\theta(0)}{dt} = 0$.

Next you will consider the underdamped homogeneous solution where the damping factor, ζ, is less than one.

The underdamped homogeneous solution: Underdamped free vibration

When $0 < \zeta < 1$, the system is said to be underdamped. In this case, the homogeneous ODE is given by

$$\frac{d^2 y_h}{dt^2} + 2\zeta\omega_n \frac{dy_h}{dt} + \omega_n^2 y_h = 0 \tag{2.28}$$

For $0 < \zeta < 1$, $\Delta = 4\omega_n^2\left(\zeta^2 - 1\right)$ implies that $\Delta < 0$. Therefore, from row 1 of **Table 2.1** the solution is given by $y_h(t) = e^{-\zeta\omega_n t}\left(a\cos\omega_n\sqrt{\left|\zeta^2 - 1\right|}t + b\sin\sqrt{\left|\zeta^2 - 1\right|}t\right)$

Define ω_d as

$$\omega_d = \omega_n\sqrt{1 - \zeta^2} \tag{2.29}$$

and write $y_h(t)$ as either

$$y_h(t) = e^{-\zeta\omega_n t}\left(a\cos\omega_d t + b\sin\omega_d t\right) \tag{2.30}$$

or

$$y_h(t) = ce^{-\zeta\omega_n t}\cos\left(\omega_d t + \phi\right) \tag{2.31}$$

ω_d has the same units as the natural frequency; it is called the damped natural frequency. A typical solution of an underdamped system that is given by either Eq. (2.30) or Eq. (2.31) is shown in **Figure 2.9**.

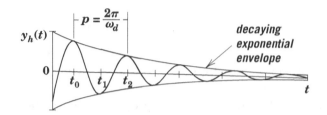

Figure 2.9 Underdamped homogeneous solution.

The system described by Eq. (2.31) oscillates at the damped natural frequency due to the cosine term, but it also decays exponentially and asymptotically due to the exponential term $e^{-\zeta \omega_n t}$ to a static equilibrium point, which is zero in this case, as shown in **Figure 2.9** in the time interval $0 \le t < +\infty$.

As $\zeta \to 0$ in the limit, the damped natural frequency, given by Eq.(2.29), approaches the natural frequency, that is, $\omega_d \to \omega_n$. The damping natural frequency, ω_d, can also be expressed in terms of a frequency f_d in Hz as $\omega_d = 2\pi f_d = \dfrac{2\pi}{p}$, where $f_d = \dfrac{1}{p}$ and p is the period, as shown in **Figure 2.9**.

Physical analogy of the underdamped solution

The concept of an underdamped system can be extended to the swing system because in reality there is damping in the system due to friction at the hinges and air resistance, or drag, which dissipates energy from the system. In this case a damping term proportional to the angular velocity $\dfrac{d\theta}{dt}$ is added to Eq. (2.27),

$$\frac{d^2\theta}{dt^2} + \underbrace{\frac{c_d}{mL^2}\frac{d\theta}{dt}}_{\text{damping term}} + \frac{g}{L}\theta = 0 \qquad (2.32)$$

Thus, for $0 < \zeta < 1$ the swing amplitude will decrease to a zero static equilibrium point, in the absence of an external force $f(t)$, as shown in **Figure 2.10**. This oscillating and decaying response is characteristic of the underdamped homogeneous solution.

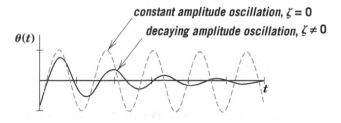

Figure 2.10 Ideal and real system responses with no forcing function.

The final two cases are the critically damped homogeneous solution where $\zeta = 1$ and the overdamped homogeneous solution for $\zeta > 1$. A physical analogy that can be used to clarify both of these conditions is a properly functioning suspension system in a car as discussed in the introduction (see **Figure 2.1**). Imagine the car frame to be forced by a sudden impulse such as a hole in the road. For comfort, it is desirable to design the suspension system with enough damping so that the car frame displaces and settles back to its normal position. A decaying response to zero without oscillation is characteristic of the critically damped or overdamped homogeneous solution.

The critically damped homogeneous solution: Critically damped free vibration

When $\zeta = 1$, the system is said to be critically damped. In this case, $\Delta = 4\omega_n^2\left(\zeta^2 - 1\right)$ yields $\Delta = 0$, which corresponds to a nonoscillatory solution given from the second row of **Table 2.1** as

$$y_h\left(t\right) = e^{-\omega_n t}\left(a + bt\right) \tag{2.33}$$

The critically damped case is the transition point from the underdamped case, $0 < \zeta < 1$, to the overdamped case, $\zeta > 1$, discussed next.

The overdamped homogeneous solution: Overdamped free vibration

When $\zeta > 1$ the system is said to be overdamped. In this case, $\Delta = 4\omega_n^2\left(\zeta^2 - 1\right)$ implies that $\Delta > 0$. Therefore, the nonoscillating general solution is given from the third row of **Table 2.1** as

$$y_h(t) = e^{-\zeta\omega_n t}\left[C_1 e^{\left(\omega_n\sqrt{\zeta^2-1}\right)t} + C_2 e^{-\left(\omega_n\sqrt{\zeta^2-1}\right)t}\right] \tag{2.34}$$

or in terms of hyperbolic sine and cosine functions as

$$y_h(t) = e^{-\zeta\omega_n t}\left[a\sinh\left(\omega_n\sqrt{\zeta^2-1}\right)t + b\cosh\left(\omega_n\sqrt{\zeta^2-1}\right)t\right] \tag{2.35}$$

Hyperbolic sine and cosine functions do not oscillate. Thus, the overdamped solution has a similar behavior as the critical damped solution, as shown in **Figure 2.11**. You may write Eqs. (2.34) and (2.35) in terms of ω_d as summarized in **Table 2.2**. In comparison to all responses, a critically damped response takes the shortest time to reach asymptotically a static equilibrium point.

$$\frac{d^2 y_h}{dt^2} + 2\zeta\omega_n \frac{dy_h}{dt} + \omega_n^2 y_h = 0$$

System	Homogeneous solution or response $y_h(t)$	Oscillation
Undamped $\zeta = 0$	$a\cos(\omega_n t) + b\sin(\omega_n t)$	Yes constant amplitude
Underdamped $0 < \zeta < 1$	$e^{-\zeta\omega_n t}\left[a\cos(\omega_d t) + b\sin(\omega_d t)\right]$ $\omega_d = \omega_n\sqrt{1-\zeta^2}$	Yes decaying amplitude
Critically damped $\zeta = 1$	$(a+bt)e^{-\omega_n t}$	No
Overdamped $\zeta > 1$	$e^{-\zeta\omega_n t}\left(ae^{\omega_d t} + be^{-\omega_d t}\right)$ or $e^{-\zeta\omega_n t}\left[a\sinh(\omega_d t) + b\cosh(\omega_d t)\right]$ $\omega_d = \omega_n\sqrt{\zeta^2 - 1}$	No

Table 2.2 Solutions of the homogeneous second-order linear ODE.

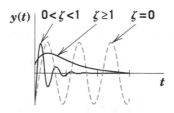

Figure 2.11 Homogeneous solutions of undamped and damped systems.

Example 2.3

Consider the second-order linear ODE

$$4\frac{d^2 y}{dt^2} + y = 0 \tag{2.36}$$

with ICs $y(0) = 0.1$ and $\frac{dy(0)}{dt} = 0$.

1. Classify the ODE as undamped, underdamped, critically damped, or overdamped.
2. Find the undamped natural frequency, the damping factor, and, where applicable, the damped natural frequency.

3. Find the solution $y(t)$.

4. Plot $y(t)$ in the time interval $t > 0$.

Solution:

1. To compare Eq. (2.36) with the general ODE, Eq. (2.28), divide Eq. (2.36) by 4:

$$\frac{d^2y}{dt^2} + \frac{1}{4}y = 0 \qquad (2.37)$$

Thus $\omega_n^2 = \frac{1}{4}$ or $\omega_n = \frac{1}{2}$, and $2\zeta\omega_n = 0$ that yields $\zeta = 0$. Therefore the ODE is undamped.

2. Since the ODE is undamped, then ω_n, the undamped natural frequency, is the only frequency.

3. For an undamped equation, the general solution is given by row 1 in **Table 2.2** as

$$y_h(t) = a\cos(\omega_n t) + b\sin(\omega_n t) \qquad (2.38)$$

The constants a and b are found using the ICs. Using $y(0) = 0.1$, Eq. (2.38) yields $0.1 = a\cos(0)$. Thus, $a = 0.1$.

The derivative of Eq. (2.38) is

$$\frac{dy_h(t)}{dt} = -a\,\omega_n\sin(\omega_n t) + b\,\omega_n\cos(\omega_n t) \qquad (2.39)$$

Using $\frac{dy(0)}{dt} = 0$ Eq. (2.39) yields $0 = b\omega_n\cos(0)$. Thus, $b = 0$. Substituting the values of a and b in Eq. (2.38) yields the solution

$$y_h(t) = 0.1\cos(0.5t) \qquad (2.40)$$

4. Equation (2.40) is plotted in **Figure 2.12** in the interval $t \geq 0$. The amplitude is 0.1.
The period is $p = \dfrac{2\pi}{\omega_n} = \dfrac{2\pi}{0.5} = 4\pi$.

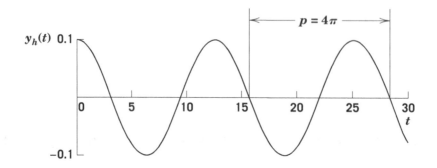

Figure 2.12 Undamped homogeneous solution.

Example 2.4

Answer the questions in Example 2.3 for the ODE

$$\frac{1}{2}\frac{d^2y}{dt^2}+\frac{3}{2}\frac{dy}{dt}+y=0 \tag{2.41}$$

with ICs $y(0)=0$ and $\frac{dy(0)}{dt}=1$.

Solution:

1. Write the ODE using the engineering form, Eq. (2.28), by multiplying Eq. (2.41) by 2:

$$\frac{d^2y}{dt^2}+3\frac{dy}{dt}+2y=0 \tag{2.42}$$

Thus, $\omega_n^2=2$ or $\omega_n=\sqrt{2}$, and $2\zeta\omega_n=3$ that yields $\zeta=\frac{3\sqrt{2}}{4}=1.061$. The ODE is overdamped since $\zeta>1$.

2. The damped frequency is given by by row 4 in **Table 2.2** as $\omega_d=\sqrt{2\left[(1.061)^2-1\right]}=0.5$.

3. For an overdamped equation, the homogeneous solution is given by row 4 in **Table 2.2** as

$$y_h(t)=e^{-\zeta\omega_n t}\left(ae^{\omega_d t}+be^{-\omega_d t}\right) \tag{2.43}$$

Using the values of ζ, ω_n, and ω_d, Eq. (2.43) becomes $y_h(t)=e^{-1.5t}\left(ae^{0.5t}+be^{-0.5t}\right)$ or

$$y_h(t) = ae^{-t} + be^{-2t} \tag{2.44}$$

Using the IC $y(0) = 0$ in Eq. (2.44) yields

$$a + b = 0 \tag{2.45}$$

The derivative of Eq. (2.44) is

$$\frac{dy_h}{dt} = -ae^{-t} - 2be^{-2t} \tag{2.46}$$

Using the IC $\dfrac{dy(0)}{dt} = 1$ in Eq. (2.46) yields

$$-a - 2b = 1 \tag{2.47}$$

Solving Eqs. (2.45) and (2.47) simultaneously yields $a = 1$ and $b = -1$. Therefore Eq. (2.44) becomes

$$y_h(t) = e^{-t} - e^{-2t} \tag{2.48}$$

4. Equation (2.48) is shown in **Figure 2.13** in the time interval $t \geq 0$.

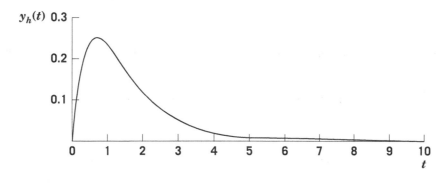

Figure 2.13 Overdamped homogeneous solution.

Example 2.5

Answer the questions in Example 2.3 for the ODE

$$\frac{d^2y}{dt^2} + 0.4\frac{dy}{dt} + 4y = 0 \tag{2.49}$$

with ICs $y(0)=1$ and $\dfrac{dy(0)}{dt}=1$.

Solution:

1. Compare the ODE, Eq. (2.49), with the engineering form, Eq. (2.28).

Thus, $\omega_n = 2$ and $2\zeta\omega_n = 0.4$ that yields $\zeta = 0.1$. The ODE is underdamped since $\zeta < 1$.

2. The damped natural frequency is given by by row 2 in **Table 2.2** as
$$\omega_d = \omega_n\sqrt{1-\zeta^2} = 2\sqrt{1-(0.1)^2} = 1.99.$$

3. For an underdamped equation, the homogeneous solution is given by row 2 in **Table 2.2** as

$$y_h(t) = e^{-\zeta\omega_n t}\left[a\cos(\omega_d t) + b\sin(\omega_d t)\right] \qquad (2.50)$$

where

$$\zeta\omega_n = (0.1)(2) = 0.2$$

Hence, Eq. (2.50) becomes

$$y_h(t) = e^{-0.2t}\left[a\cos(1.99t) + b\sin(1.99t)\right] \qquad (2.51)$$

Using the IC $y(0)=1$ in Eq. (2.51) yields $a=1$. Substituting the value of a and taking the derivative of Eq. (2.51) yield

$$\frac{dy_h}{dt} = -0.2e^{-0.2t}\left[\cos(1.99t) + b\sin(1.99t)\right] + 1.99e^{-0.2t}\left[b\cos(1.99t) - \sin(1.99t)\right] \quad (2.52)$$

Using the IC $\dfrac{dy(0)}{dt}=1$ in Eq. (2.52) yields $-0.2 + 1.99b = 1$ or $b = 0.603$.

Thus,

$$y_h(t) = e^{-0.2t}\left[\cos(1.99t) + 0.603\sin(1.99t)\right] \qquad (2.53)$$

4. Equation (2.53) is shown in **Figure 2.14** in the time interval $t \geq 0$.

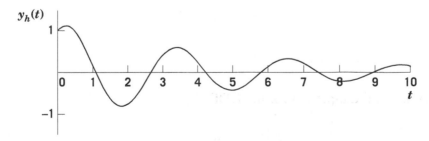

Figure 2.14 Underdamped homogeneous solution.

Next, the damped homogenous linear ODE is applied to two oscillating systems. The first ODE describes the angular motion of an unforced pendulum, and the second ODE describes the linear motion of an unforced spring-dashpot system.

2.4 Modeling of Second-Order Linear Homogeneous Systems: Free Vibration

Modeling of a system in rotation: The damped pendulum

You may be quite familiar with the analysis of the pendulum problem from introductory physics course. Consider the pendulum as shown in **Figure 2.15**, consisting of a rod of length L and negligible mass, attached to a body with mass m, and a hinge with a rotational friction coefficient c_d. In this model, the instantaneous angular displacement $\theta(t)$ describes the motion of the pendulum as a function of time t. Once the pendulum is displaced by an initial angular displacement $\theta(0) = \theta_0$, and released from rest, that is, $\dfrac{d\theta(0)}{dt} = 0$, the pendulum will oscillate about the equilibrium point, which is the vertical line.

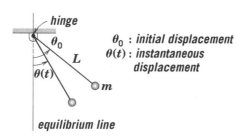

Figure 2.15 The pendulum geometry model.

First, to find the equation of motion of the pendulum, start by constructing a free body diagram (FBD) of the rod and mass as shown in **Figure 2.16**. On the left-hand side of the FBD the external forces and moments are shown, and on the right-hand side the tangential and centripetal inertia forces as required by Newton's second law of motion are shown.

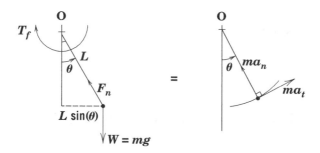

Figure 2.16 Forces and moments in the free body diagram of the rod and mass.

On the left-hand side of the FBD you find the following:

- the weight of the object $W = mg$; the rod mass is negligible.

- the tension in the rod F_n; and

- the resisting frictional moment in the hinge T_f.

On the right side of the FBD you find the inertia force resolved into two components:

- the tangential inertia force tangent to the circular path of the swing ma_t; and

- the centripetal inertia force directed toward the hinge ma_n.

Second, you equate the moment of the forces about the hinge end of the rod on the left-hand side of **Figure 2.16** to the moment of the inertia forces on the right-hand side. The weight W, whose line of action is at a distance $L\sin(\theta)$ from the hinge, will have a moment about point O given by

$$T_w = mg\left[L\sin(\theta)\right] \tag{2.54}$$

The line of action of F_n passes through O; thus F_n has no moment about O. The moment arm of the tangential inertial force ma_t is L, thus the moment of ma_t about O is

$$T_t = ma_t(L) \tag{2.55}$$

The moment of the centripetal force is zero because ma_n passes through O. At the instant shown in **Figure 2.16**, the weight W tends to rotate the pendulum in the clockwise direction, thus inducing a clockwise moment T_w. On the other hand, the inertial force ma_t tends to accelerate the mass in the counterclockwise direction, thus inducing a counterclockwise moment. If you assume counterclockwise moments to be positive, then

$$-T_f - T_w = T_t \tag{2.56}$$

The frictional moment, T_f is modeled as

$$T_f = c_d\omega \tag{2.57}$$

where c_d is the rotational friction coefficient of the hinge and ω is the angular velocity.

Substituting Eqs. (2.54), (2.55), and (2.57) in Eq. (2.56) yields

$$-c_d\omega - mgL\sin(\theta) = ma_t L \tag{2.58}$$

From kinematics, the tangential acceleration a_t is, by definition, the rate of change of the linear velocity v, that is,

$$a_t = \frac{dv}{dt} \tag{2.59}$$

In turn, the linear velocity v is related to the angular velocity, ω, by

$$v = L\omega \tag{2.60}$$

By definition, the angular velocity ω is the rate of change of the angular displacement $\theta(t)$, that is,

$$\omega = \frac{d\theta}{dt} \tag{2.61}$$

Substituting Eqs. (2.59)–(2.61) in Eq. (2.58) yields $-c_d \dfrac{d\theta}{dt} - mgL\sin(\theta) = mL\left(L\dfrac{d^2\theta}{dt^2}\right)$ or

$$\frac{d^2\theta}{dt^2} + \frac{c_d}{mL^2}\frac{d\theta}{dt} + \frac{g}{L}\sin(\theta) = 0 \tag{2.62}$$

Equation (2.62) is a homogeneous nonlinear second-order ODE. The nonlinearity arises from the function $\sin(\theta)$. For small angular displacement $\theta(t)$ in radians, $\sin(\theta) \cong \theta$. For example,

$\theta = 10^o\left(\dfrac{\pi rad}{180^o}\right) = 0.1745 rad$ and $\sin(10^o) = 0.1736 rad$ so that the error between $\sin(\theta)$ and θ

in radians is $\dfrac{|0.1736 - 0.1745|}{0.1736} = 0.52\%$, which is small. Thus, for small angles, substituting

$\sin(\theta) \cong \theta$ in Eq. (2.62) yields $\dfrac{d^2\theta}{dt^2} + \dfrac{c_d}{mL^2}\dfrac{d\theta}{dt} + \dfrac{g}{L}\theta = 0$, which is Eq. (2.32). This equation is a

homogeneous linear second-order ODE, which is valid for small angular displacements only. If you compare this ODE with Eq. (2.28), you conclude that the natural frequency is

$$\omega_n = \sqrt{\frac{g}{L}} \tag{2.63}$$

and $2\zeta\omega_n = \dfrac{c_d}{mL^2}$. Thus, the damping factor is

$$\zeta = \frac{c_d}{2mL^2\omega_n} = \frac{c_d}{2m}\sqrt{\frac{1}{gL^3}} \tag{2.64}$$

You should note that if the hinge friction is neglected, then $c_d = 0$, and the equation of motion reduces to the ideal pendulum described by Eq. (2.27).

Example 2.6 Hinge Design

In a toy manufacturing company, an engineer is faced with the problem of determining the friction coefficient of a hinge as part of a new child swing design. The engineer simulated a maximum weight of a child by hanging a mass $m = 30$ *kg* to a 1-*m* rod. The weight was lifted to an initial angle of 10° and released from rest, as shown in **Figure 2.17a**. The rod mass is negligible. The engineer recorded the motion with a video camera. Using special software, the engineer captured a series of frames on the computer, as shown in **Figure 2.17b**. Knowing the speed of the camera, the engineer measured 2.094 *s* between the first maximum (initial position), and the second maximum, through one cycle.

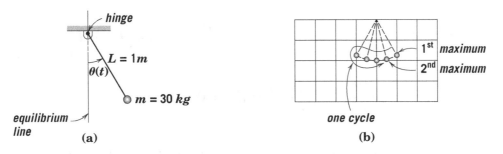

Figure 2.17 Swing model: (a) rod and mass displaced from the equilibrium line; and (b) digitized frames of the pendulum motion.

1. How should the engineer model the swing to describe the angular displacement as a function of time, if air drag is neglected? The weight of the rod is negligible.
2. What is the friction coefficient of the hinge?

3. Plot the solution in the time interval $t \geq 0$ for the following ICs: $\theta(0) = 10^o$ and $\dfrac{d\theta(0)}{dt} = 0$.

Solution:

1. How should the engineer model the swing to describe the angular displacement as a function of time, if air drag is neglected? The weight of the rod is negligible.

For a small initial angle of 10°, the swing can be modeled by Eq. (2.32),

$$\frac{d^2\theta}{dt^2} + \frac{c_d}{mL^2}\frac{d\theta}{dt} + \frac{g}{L}\theta = 0 \tag{2.65}$$

2. What is the friction coefficient of the hinge?

From Eq. (2.64) the friction coefficient of the hinge is given by

$$c_d = 2mL^2\omega_n\zeta \qquad (2.66)$$

The measured time, 2.094 s, between two maxima is the period p, which yields

$$\omega_d = \frac{2\pi}{p} = \frac{2\pi}{2.094} = 3\frac{rad}{s} \qquad (2.67)$$

From Eq. (2.63)

$$\omega_n = \sqrt{\frac{g}{L}} = \sqrt{\frac{9.81}{1}} = 3.132\frac{rad}{s} \qquad (2.68)$$

For an oscillating pendulum, $\zeta < 1$. Thus, from row 2 in **Table 2.2**, $\omega_d = \omega_n\sqrt{1-\zeta^2}$, Using Eqs. (2.67) and (2.68) yields $3 = 3.132\sqrt{1-\zeta^2}$ or $\zeta = 0.287$.

Finally, from Eq. (2.66) $c_d = 2mL^2\omega_n\zeta = 2(30)(1)^2(3.132)(0.287) = 53.93\frac{Nm}{s}$.

3. Plot the solution in the time interval $t \ge 0$ for the following ICs: $\theta(0) = 10^o$ and $\dfrac{d\theta(0)}{dt} = 0$.

The general solution of Eq. (2.65) is given by row 2 in **Table 2.2** for an underdamped system as

$$\theta(t) = e^{-\zeta\omega_n t}\left[a\cos(\omega_d t) + b\sin(\omega_d t)\right] \qquad (2.69)$$

Substituting the values of ω_n, ω_d, and ζ in Eq. (2.69) yields

$$\theta(t) = e^{-0.9t}\left[a\cos(3t) + b\sin(3t)\right] \qquad (2.70)$$

Using the IC $\theta(0) = 10^o = 0.1745rad$ in Eq. (2.70) yields $0.1745 = a\cos(0)$. Thus, $a = 0.1745$. The derivative of Eq. (2.70) is

$$\frac{d\theta(t)}{dt} = -0.9e^{-0.9t}\left[a\cos(3t) + b\sin(3t)\right] + 3e^{-0.9t}\left[-a\sin(3t) + b\cos(3t)\right] \quad (2.71)$$

Using the IC $\dfrac{d\theta(0)}{dt}=0$ in Eq. (2.71) yields $0=-0.9a+3b$. Using the found value of a yields $b=0.05235$. Thus, Eq. (2.70) becomes

$$\theta(t)=e^{-0.9t}\left[0.1745\cos(3t)+0.05235\sin(3t)\right] \tag{2.72}$$

Alternatively, you can express Eq. (2.72) by a single sinusoid as $\theta(t)=c\,e^{-0.9t}\cos(3t+\phi)$, where

$c=\sqrt{a^2+b^2}=\sqrt{0.1745^2+0.05235^2}=0.1822$, and the phase angle ϕ is given by

$\phi=-\tan^{-1}\left(\dfrac{b}{a}\right)=-\tan^{-1}\left(\dfrac{0.05235}{0.1745}\right)=-0.291rad$.

Thus,

$$\theta(t)=0.1822\,e^{-0.9t}\cos(3t-0.291) \tag{2.73}$$

Equation (2.73) is shown in **Figure 2.18**.

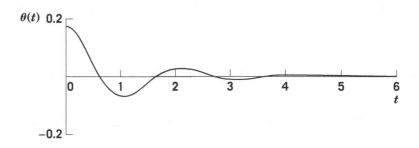

Figure 2.18 Swing underdamped motion.

In this example, you can see that the underdamped swing oscillates sinusoidally with a decaying amplitude. This is confirmed by the mathematical model of a second-order ODE with a natural damped frequency ω_d and a damping factor in the range $0<\zeta<1$. In this equation, the dependent variable is the angular displacement $\theta(t)$ and the independent variable is time t.

Next, you will learn about the spring-dashpot system.

Modeling of a system in translation: The spring-dashpot system

The spring-damper system is one of the most important second-order systems in engineering. The example that is most familiar to you is the car. For safety and comfort, the oscillation of the car suspension should dampen quickly when the car passes over potholes or bumps or any uneven rough roads. For this reason, the car suspension system has a shock absorber as shown in **Figure 2.19**.

Figure 2.19 Suspension system of a car consisting of a spring and a shock absorber.

The shock absorber is the unit that dampens oscillations. The common symbol used for a general-purpose damper is a piston inside a cylinder filled with a resisting fluid such as air or oil, as shown in **Figure 2.20**.

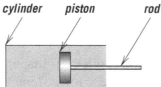

Figure 2.20 Symbol of a general-purpose damper or dashpot.

The general-purpose damper is commonly called the dashpot in engineering. If one end of the dashpot, for example, the cylinder is restrained as shown in **Figure 2.21**, and the piston-rod unit is connected to a body moving at velocity v, then the magnitude of the resisting fluid is modeled as

$$\left|F_d\right| = c_d v \tag{2.74}$$

where F_d is the resisting or damping force and c_d is a damping coefficient that depends on the geometry of the device and fluid properties. A dashpot that is modeled by Eq. (2.74) is known as a linear dashpot.

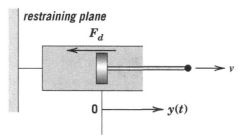

Figure 2.21 Model of a linear dashpot.

From kinematics, the velocity v is the rate of change of displacement,

$$v = \frac{dy}{dt} \tag{2.75}$$

Substituting Eq. (2.75) in Eq. (2.74) yields

$$\left|F_d\right| = c_d \frac{dy}{dt} \tag{2.76}$$

Mathematically speaking, Eq. (2.76) implies that a damping force appears through the first derivative term in a second-oder ODE. Using the spring and dashpot symbols you can now model the car suspension system in **Figure 2.19** as a spring-dashpot system as shown in **Figure 2.22a**.

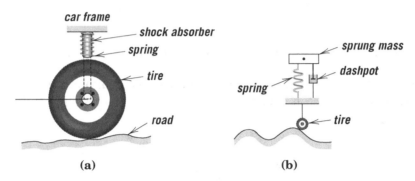

(a)   **(b)**

Figure 2.22 Car suspension model: (a) suspension main components; and (b) suspension system model.

By design, a car of mass M has four suspension units, two in the front and two in the rear. For simplicity, the motion of one suspension unit loaded with one quarter of the mass of the car $m = \dfrac{M}{4}$, can be analyzed, as shown in **Figure 2.22b**. The oscillating car frame is called the sprung mass.

The car suspension system does have two separate units, namely, the spring and shock absorber. However, there are systems in engineering that are modeled by a spring-dashpot system but do not appear as two separate units. For example, the rubber pads, as shown in **Figure 2.23a**, can be modeled as a spring and a damper, as shown in **Figure 2.23b**.

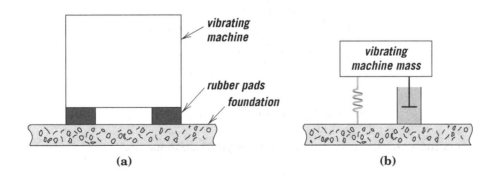

(a)   **(b)**

Figure 2.23 Vibrating machine: (a) isolation rubber pads between a vibrating machine and the foundation; and (b) spring-dashpot model of rubber pads.

The rubber pads act as vibration isolators by damping undesirable vibration transmitted from a machine to the foundation or the structure supporting the machine. These rubber pads with spring and damper characteristics are special types of elastomers. The mathematical spring-dashpot model of these elastomers is validated experimentally. An example of a vibrating machine on a foundation is the refrigerator unit in a kitchen, which vibrates when the compressor of the cooling system runs.

The simplest mathematical model of a spring-dashpot system is one that is connected to a body moving horizontally, as shown in **Figure 2.24**. If in this configuration the body is displaced a distance y_0 and released from rest, then the body tends to oscillate about the equilibrium point $y = 0$, due to the spring. The effect of the dashpot is to dampen the oscillations until the body eventually comes to rest.

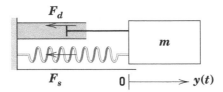

Figure 2.24 Body moving horizontally and restrained by a spring-dashpot system.

If the Coulomb friction force between the body and the horizontal surface is ignored, then the spring and dashpot forces are the only external forces acting on the body. In this case, Newton's second-law of motion requires

$$-\left|F_d\right| - \left|F_s\right| = ma \qquad (2.77)$$

The magnitude of F_s is given by Hooke's law, that is, $\left|F_s\right| = ky$. Substituting $\left|F_s\right|$, $\left|F_d\right|$ from Eq. (2.76) and $a = \dfrac{d^2y}{dt^2}$ in Eq. (2.77) yields $m\dfrac{d^2y}{dt^2} + c_d\dfrac{dy}{dt} + ky = 0$. If you cast this ODE into the engineering form, Eq. (2.28), by dividing by m,

$$\frac{d^2y}{dt^2} + \frac{c_d}{m}\frac{dy}{dt} + \frac{k}{m}y = 0 \qquad (2.78)$$

then $\omega_n = \sqrt{\dfrac{k}{m}}$ and $2\zeta\omega_n = \dfrac{c_d}{m}$. Thus,

$$\zeta = \frac{c_d}{2m\omega_n} \qquad (2.79)$$

Notice that the natural frequency is the same as if the body is connected to the spring alone. The solution of Eq. (2.78) will be underdamped, critically damped, or overdamped depending on whether $0 < \zeta < 1$, $\zeta = 1$, or $\zeta > 1$, respectively. In the configuration shown in **Figure 2.24** the initial displacement is $y(0) = y_0$ and the velocity is $\dfrac{dy(0)}{dt} = 0$ if the body is released from rest.

The next mathematical model for a spring-dashpot system is in a vertical configuration, as shown in **Figure 2.25**. Without a body attached to the spring-dashpot system, the spring will assume its natural length L_f, and the dashpot will be floating freely as shown in **Figure 2.25a**.

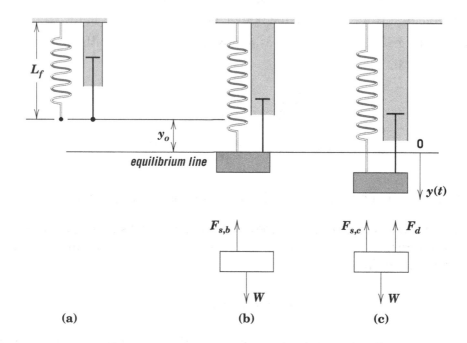

Figure 2.25 Spring-dashpot system: (a) free cable length; (b) cable elongation due to attached weight; and (c) oscillation due to some initial displacement from equilibrium.

When a body of weight W is attached to the spring-dashpot system, a new equilibrium position is established, as shown in **Figure 2.25b**. In this case, the only external forces acting on the body are its weight W and the spring force F_s, as shown on the free body diagram. The dashpot does not contribute a force in this equilibrium position because there is no motion. In other words, the dashpot will simply assume a new free-floating position. In this equilibrium position, Newton's first law requires $W - |F_{s,b}| = 0$. Using Hooke's law, $|F_{s,b}| = ky_0$ yields

$$W = ky_0 \tag{2.80}$$

where y_0 is the equilibrium stretched length of the spring from its free length. If now the body is displaced a distance y from the equilibrium line and released, as shown in **Figure 2.25c**, the damping force, spring force and weight are related by Newton's second law of motion

$$-|F_d| - |F_{s,c}| + W = ma$$

Substituting Hooke's law, $|F_{s,c}| = k[y_0 + y(t)]$ in the foregoing equation yields

$$-|F_d| - ky - ky_0 + W = ma \tag{2.81}$$

Substituting Eqs. (2.76) and (2.80) together with $a = \dfrac{d^2y}{dt^2}$ in Eq. (2.81) yields

$m\dfrac{d^2y}{dt^2} + c_d\dfrac{dy}{dt} + ky = 0$ which is the same as Eq. (2.78). This is because W is eliminated from the equation when the displacement $y(t)$ is measured from the equilibrium line.

Example 2.7 Damped Elevator System

Revisit Example 2.2 by including damping due to the friction between the elevator and the guiderails. The frictional effect can be modeled as a dashpot as shown in **Figure 2.25**.

1. For the same conditions as given in Example 2.2, assume a damping coefficient such that $\zeta = 0.05$, and solve for the elevator displacement $y(t)$.
2. Compare the maximum elongation with that found in Example 2.2, and comment.

Solution:

1. The equation of motion of the elevator is governed by Eq. (2.78), $\dfrac{d^2y}{dt^2} + \dfrac{c_d}{m}\dfrac{dy}{dt} + \dfrac{k}{m}y = 0$,

   which can be put in engineering form as $\dfrac{d^2y}{dt^2} + 2\zeta\omega_n\dfrac{dy}{dt} + \omega_n y = 0$. From Example 2.2

   $\omega_n = 44.72\dfrac{rad}{s}$. For $\zeta = 0.05$, the motion is underdamped. In this case the displacement is

   given by row 2 in **Table 2.2** as $y(t) = e^{-\zeta\omega_n t}(a\cos\omega_d t + b\sin\omega_d t)$, where $\omega_d = \omega_n\sqrt{1-\zeta^2}$.

The ICs are $y(0) = 0$ and $\dfrac{dy(0)}{dt} = 1\dfrac{m}{s}$. Using the IC $y(0) = 0$ yields $a = 0$. Thus,

$y(t) = be^{-\zeta\omega_n t}\sin\omega_d t$.

The first derivative of $y(t)$ is $\dfrac{dy}{dt} = -b\zeta\omega_n e^{-\zeta\omega_n t}\sin\omega_d t + b\omega_d e^{-\zeta\omega_n t}\cos\omega_d t$.

Using the IC $\dfrac{dy(0)}{dt}=1$ yields $b=\dfrac{1}{\omega_d}$ so that the displacement equation is

$y(t)=\dfrac{1}{\omega_d}e^{-\zeta\omega_n t}\sin\omega_d t$. The damped natural frequency is $\omega_d=44.72\sqrt{1-0.05^2}=44.66\dfrac{rad}{s}$.

Thus,

$$y(t)=0.02239e^{-2.236t}\sin\left(44.66t\right) \tag{2.82}$$

The undamped displacement found in Example 2.2 and the damped displacement given by Eq. (2.82) are plotted in **Figure 2.26** for comparison.

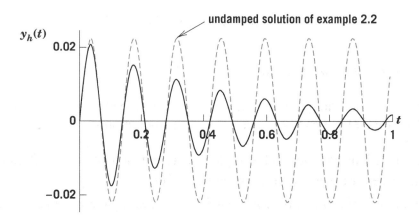

Figure 2.26 Damped displacement given by Eq. (2.82) and the undamped solution found in Example 2.2.

2. The maximum displacement from the equilibrium line occurs when the first derivative is zero and the second derivative is negative. From the first derivative derived above,

$$0=-\zeta\omega_n e^{-\zeta\omega_n t}\sin\omega_d t+\omega_d e^{-\zeta\omega_n t}\cos\omega_d t$$

Divide by $e^{-\zeta\omega_n t}$ since the exponential function is always positive, and rearrange the equation as

$$\zeta\omega_n\sin\omega_d t=\omega_d\cos\omega_d t$$

that yields $\tan\omega_d t=\dfrac{\omega_d}{\zeta\omega_n}$. From this equation you can solve for the time $t=\dfrac{1}{\omega_d}\tan^{-1}\left(\dfrac{\omega_d}{\zeta\omega_n}\right)$.

Therefore, the maximum displacement occurs at $t=\dfrac{1}{44.66}\tan^{-1}\left(\dfrac{44.66}{0.05\left(44.72\right)}\right)=0.034s$.

The maximum displacement is found by substituting this time value in Eq. (2.82); thus

$$y_{max} = 0.02239e^{-2.236(0.034)}\sin\left(44.66x0.034\right) = 0.0207m.$$

The maximum elongation is given by the sum of $y_0 = 0.00491m$ found in Example 2.2 and y_{max}, which comes out to be equal to $0.0256m$.

This elongation is only 6% less than the maximum elongation found in Example 2.2. For the worst case scenario, it is recommended that a conservative estimate be used for the maximum elongation as found in Example 2.2.

This example illustrates how to model a damped engineering problem using a spring-dashpot system, as a homogeneous linear second-order ODE where the dependent variable is the elevator displacement $y(t)$ and the independent variable is the time t.

Before you proceed to the next section, you should note the analogies between friction in translation and friction in rotation.

In this section, friction was modeled by a damper or dashpot where the damping force is modeled to be proportional to the linear velocity, as given by Eq. (2.76). In many engineering problems, there are also rotational dashpots whose purpose is to dampen oscillations of a body in rotation. In this case, the torque is modeled to be proportional to the angular speed. We took advantage of this relationship in modeling a damped pendulum as shown in **Figure 2.16.** Although there is no physical rotational dashpot in a pendulum, we modeled the resisting frictional moment in the hinge as a rotational dashpot by stating that the frictional moment T_f is proportional to the angular speed ω, as given by Eq. (2.57).

2.5 Solution of Nonhomogeneous Second-Order Linear ODE

At this point, you have learned about engineering systems that are governed by a homogeneous ODE. There are, however, many engineering problems that are governed by a nonhomogeneous linear second-order ODE, Eq. (2.1). We stated already that the complete solution of this equation is given by the sum of two linearly independent homogeneous solutions, y_{h_1} and y_{h_2}, and a particular solution y_p, as given by Eq. (2.3). The particular solution y_p is found by satisfying the ODE

$$\frac{d^2 y_p}{dt^2} + a_1 \frac{dy_p}{dt} + a_0 y_p = r(t)$$

For most of the forcing functions that are encountered in engineering, you may be able to find the particular solution using the undetermined coefficients method.

Undetermined coefficients method

The undetermined coefficients method is based on proposing solutions with unknown coefficients, which have the same general form as $r(t)$ in Eq. (2.1) or $f(t)$ in Eq. (2.6). The unknown or undetermined coefficients are fixed by having the particular solution satisfy either ODE. The forcing functions that are commonly encountered in engineering are shown in the left column of **Table 2.3**. For each forcing function, the corresponding proposed particular solution, $y_p(t)$, is shown in the right column.

$r(t)$ or $f(t)$	Particular solution $y_p(t)$
Step function A	B
$A\sin(\omega t)$ or $A\cos(\omega t)$	$B_1\cos(\omega t) + B_2\sin(\omega t)$
At^n; $n = 1, 2, 3, \dots$	$B_n t^n + B_{n-1} t^{n-1} + \dots + B_1 t + B_0$
$Ae^{\alpha t}$	$B_1 e^{\alpha t}$
$Ae^{\alpha t}\sin(\omega t)$ or $Ae^{\alpha t}\cos(\omega t)$	$e^{\alpha t}\left[B_1\cos(\omega t) + B_2\sin(\omega t)\right]$

Table 2.3 Undetermined coefficient method.

The constants that appear in the particular solution, for example, B, B_0, B_1, are the undetermined coefficients that you want to fix. To find the particular solution for a nonhomogeneous second-order linear ODE, use the following four steps:

1. Given $r(t)$ or $f(t)$ use **Table 2.3** to identify $y_p(t)$.

2. Check if $y_p(t)$ is linearly dependent on one of the homogeneous solutions. If it is, then the general solution will degenerate to the homogeneous solution and will not satisfy the ODE. In this case, multiply the particular solution by t. Check the new particular solution for linear dependency on either y_{h_1} and y_{h_2}. If it is, then repeat the previous step by multiplying by t^2. If necessary, repeat this procedure until $y_p(t)$ is linearly dependent on neither y_{h_1} nor y_{h_2}.

3. If $r(t)$ or $f(t)$ is a product of two functions, for example, $r(t) = te^{\alpha t}$, then the particular solution is given by the product of two suggested solutions, that is, $y_p(t) = (A_1 t + A_0)e^{\alpha t}$.

4. If $r(t)$ or $f(t)$ is the sum of several functions, for example, $r(t) = t + 3t^3$, then the particular solution is given by the sum of the corresponding suggested solutions. That is, for $r_1(t) = t$, $y_{p_1}(t) = A_1 t + A_0$, and for $r_2(t) = 3t^3$, $y_{p_2}(t) = B_3 t^3 + B_2 t^2 + B_1 t + B_0$. Since $y_{p_1}(t)$ appears in $y_{p_2}(t)$ through a smiliar term, namely $B_1 t + B_0$, then it is sufficient to suggest $y_p(t) = C_3 t^3 + C_2 t^2 + C_1 t + C_0$ as the particular solution for $r(t) = t + 3t^3$.

Transient and steady responses

As you have seen already, the solution $y(t)$ depends on the homogeneous solution and the particular solution. If you inspect **Table 2.2** and **Figure 2.11**, you can observe that for $\zeta > 0$ the homogeneous solution eventually decays asymptotically to zero, that is, $y_h \to 0$ as $t \to \infty$. In this case, the homogeneous solution $y_h(t)$ is called the transient response. Thus, from $y(t) = y_h(t) + y_p(t)$, you can see that $y(t) \to y_p(t)$ as $y_h(t) \to 0$; in other words, the solution $y(t)$ is forced to behave like the particular solution $y_p(t)$ as time increases. For that reason, you may label the solution and its components as follows,

$$
\begin{array}{ccccc}
y(t) & & y_h(t) & + & y_p(t) \\
\text{total solution} & = & \text{homogeneous solution} & & \text{particular solution} \\
\text{or} & & \text{or} & & \text{or} \\
\text{response} & & \text{transient response} & & \text{forced response}
\end{array}
$$

For $\zeta = 0$, the homogeneous solution is a sinusoid and bounded as you can see in **Table 2.2** and **Figure 2.11**. This means that $y_h(t)$ is not a transient and makes itself present in the total solution at any time. Thus, both solutions $y_h(t)$ and $y_p(t)$ are important in the total solution. Although $y_h(t)$ is bounded, the total solution $y(t)$ may or may not be bounded depending on the particular solution. For example, you will learn in the next section that for a sinusoid forcing function $f(t)$ with a frequency ω, the particular solution is also a sinusoid with the same frequency ω, based on **Table 2.3**. If this frequency ω happens to be equal to the natural frequency of the system ω_n, a phenomenon defined as resonance, the amplitude of $y(t)$ will grow unbounded.

For $\zeta < 0$, as is the case with systems with negative damping, the exponential function in the homogeneous solution will grow unbounded as $t \to \infty$.

Next, you will learn how to model engineering systems as nonhomogeneous linear second-order ODEs. Of all the forcing functions encountered in engineering, you will find that a sinusoid forcing function will lead to very interesting phenomena such as resonance and beats.

2.6 Modeling of Second-Order Linear Nonhomogeneous Systems: Forced Vibration

In this section you will learn about two systems that are modeled as a second-order nonhomogeneous ODE: the RLC circuit for a radio turner and the mass-spring-dashpot suspension system of a car.

Modeling of an electrical system: The RLC circuit

Consider the RLC circuit with an inductor L, a resistor R, a capacitor C, and an input voltage to the circuit $V_S(t)$, as shown in **Figure 2.3**. Radio stations transmit an electromagnetic field by oscillating electric charges in their very long antennas. When you tune your radio to a particular station, say 990AM, you cause a current $i(t)$ to oscillate up and down the receiver antenna that is now interacting with the electromagnetic field transmitted from a radio station at a frequency of 990 kHz. It is the oscillation of the current in the receiver circuit at the same frequency as that of the transmitter that allows you to detect a radio signal. Although the circuit of a receiver is quite complex, we will analyze a simple RLC circuit, which is one of the building blocks of a receiver circuit, as shown in **Figure 2.3b**. The electromagnetic field interacting with the antenna creates a voltage V_S. To achieve this frequency, the receiver is designed to vary either L or C in order to match the transmitter frequency. When the receiver frequency matches the transmitter forcing frequency, the circuit is said to be resonating.

You will also learn shortly that ideal resonance takes place in an LC circuit without a resistor. However, in a real circuit the coil or inductor possesses some resistance R that causes the system to dissipate energy in the form of heat.

Let us proceed next to develop a mathematical model for the oscillating current in the RLC circuit. From this model you should be able to define: (1) the forcing function, $f(t)$; (2) the forcing frequency, ω; (3) the natural frequency of the system; and (4) the damping factor of the system. The mathematical model of the oscillating current begins by applying Kirchhoff's voltage loop law, which requires that the sum of voltages across L, R, and C balances the source voltage $V_S(t)$. In other words,

$$V_L + V_C + V_R = V_S(t) \tag{2.83}$$

where the voltage across the inductor is

$$V_L = L\frac{di}{dt} \tag{2.84}$$

the voltage across the capacitor is

$$V_C = \frac{1}{C}\int i\,dt \tag{2.85}$$

the voltage across the resistor is

$$V_R = i(t)R \tag{2.86}$$

and $i(t)$ is the current in the circuit as a function of time t.

Substituting Eqs. (2.84)–(2.86) in Eq. (2.83) yields

$$L\frac{di}{dt} + \frac{1}{C}\int i\, dt + iR = V_s(t) \tag{2.87}$$

Taking the time derivative of Eq. (2.87) yields

$$L\frac{d^2i}{dt^2} + \frac{1}{C}i + R\frac{di}{dt} = \frac{dV_s(t)}{dt} \text{ or}$$

$$\frac{d^2i}{dt^2} + \frac{R}{L}\frac{di}{dt} + \frac{1}{LC}i = \frac{1}{L}\frac{dV_s(t)}{dt} \tag{2.88}$$

Equation (2.88) is a second-order linear nonhomogeneous ODE that governs an oscillating current $i(t)$ in an RLC system. A comparison of this ODE with the engineering form, Eq. (2.6) yields $\omega_n^2 = \frac{1}{LC}$ and $2\zeta\omega_n = \frac{R}{L}$ from which you can solve for the natural frequency and the damping factor,

$$\omega_n = \frac{1}{\sqrt{LC}} \tag{2.89}$$

and

$$\zeta = \frac{R}{2}\sqrt{\frac{C}{L}} \tag{2.90}$$

Furthermore, the forcing frequency is

$$f(t) = \frac{1}{L}\frac{dV_s}{dt} \tag{2.91}$$

As you can see, although $V_s(t)$ is the physical forcing function in the RLC circuit, $f(t)$ in Eq. (2.91) is the mathematical forcing function containing the derivative of $V_s(t)$. For a given receiver sinusoid with a frequency f in Hz that matches the transmitter frequency, the corresponding voltage $V_s(t)$ is given by

$$V_s = A\sin(\omega t) \tag{2.92}$$

where the circular frequency is $\omega = 2\pi f$ and A is the amplitude of the voltage V_s. Do not confuse the symbol of frequency in Hz with the symbol of the forcing function.

The RLC circuit will resonate when the natural frequency ω_n matches the transmitter frequency ω, that is, $\omega_n = \omega$, which yields with Eq. (2.89) $\omega = \frac{1}{\sqrt{LC}}$.

The pertinent equations are summarized in **Table 2.4**.

ODE model	$\dfrac{d^2 i}{dt^2} + \dfrac{R}{L}\dfrac{di}{dt} + \dfrac{1}{LC}i = f(t)$; Eq. (2.88)	$i(t)$, current; C, capacitance; L, inductance; R, resistance.
Forcing function	$f(t) = \dfrac{1}{L}\dfrac{dV_s}{dt}$	$V_s(t)$, source voltage.
Source voltage	$V_s(t) = A\sin(\omega t)$	ω, forcing frequency.

Table 2.4 Summary of second-order RLC circuit model.

Example 2.8 The Ideal LC Circuit

Consider a radio receiver tuned to a particular radio station, say 980 AM, as shown in **Figure 2.3**, which corresponds to $980 kHz$. The resistance in the circuit is small, which can be neglected. Thus, the tuner circuit model consists of a variable capacitor C and a fixed inductor L. The forcing sinusoid amplitude, A_s, is $\sqrt{2}$ volts.

1. What is the governing ODE of this system?
2. Select the capacitance C required for the receiver to resonate with the transmitter at $980\ kHz$, given an inductance $L = 5\left(10^{-4}\right) henry$.

Solution:

1. The receiver signal is modeled as a sinusoid given by $V_s(t) = A_s \sin(\omega t)$, where A_s is the voltage amplitude and ω is the forcing frequency, $\omega = 2\pi f$. For $f = 980\left(10^3\right) Hz$, $\omega = 6.16\left(10^6\right)\dfrac{rad}{s}$. The governing ODE for a LC circuit is found from Eq. (2.88) without the resistor term as $L\dfrac{d^2 i}{dt^2} + \dfrac{1}{C}i = \dfrac{dV_s(t)}{dt}$.

Using the engineering form, the foregoing ODE is written as $\dfrac{d^2 i}{dt^2} + \dfrac{1}{LC}i = \dfrac{1}{L}\dfrac{dV_s(t)}{dt}$, which in comparison with the engineering form, Eq. (2.6), yields $\omega_n^2 = \dfrac{1}{LC} = 37.915\left(10^{12}\right)\dfrac{rad}{s^2}$, and $f(t) = \dfrac{1}{L}\dfrac{dV_s}{dt} = \dfrac{A_s \omega}{L}\cos \omega t = 1.742\left(10^{10}\right)\cos\left(6.16\times10^6 t\right)$. Thus, the ODE takes the form

$$\dfrac{d^2 i}{dt^2} + 37.915\left(10^{12}\right)i(t) = 1.742\left(10^{10}\right)\cos\left(6.16\times10^6 t\right)$$

2-38

2. Resonance takes place when $\omega_n = \omega$, that is, when $\omega_n = 6.16\left(10^6\right)\dfrac{rad}{s}$.

For $L = 5\left(10^{-4}\right) henry$, solving for C using Eq. (2.89) yields

$$C = \frac{1}{L\omega^2} = \frac{1}{5\left(10^{-4}\right)(37.915)10^{12}} = 52.75\left(10^{-12}\right)(farad)$$

In this example, the radio circuit is modeled by a second-order nonhomogeneous ODE without damping, which oscillates at the natural frequency ω_n. The dependent variable is the current $i(t)$, and the independent variable is time t.

Modeling of a mechanical system: The mass and spring-dashpot system

The mass and spring-dashpot system model is based on an earlier discussion of a car suspension system as shown in **Figure 2.22**. In this oversimplified model, the weight of the car is modeled to be distributed equally on four suspension systems, that is, the sprung mass per suspension is taken as $m = \dfrac{M}{4}$ where M is the total mass of the car. In the following analysis, the motion of the spring-dashpot system on an uneven road is modeled as shown in **Figure 2.27**.

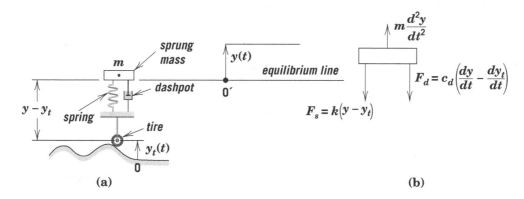

Figure 2.27 Suspension system model on an uneven road: (a) displacement from an absolute reference frame; and (b) external forces on sprung mass.

The tire is modeled as a point by ignoring its elasticity. The tire instantaneous displacement $y_t(t)$ is measured from a fixed reference point O. The sprung mass instantaneous displacement, $y(t)$, is measured from the static equilibrium line, at point O', as shown in **Figure 2.27a**.

Here, relative to the tire displacement $y_t(t)$, the spring force is given by

$$F_s = k\left(y - y_t\right) \tag{2.93}$$

The same reasoning applies to the dashpot; that is, the magnitude of the damping force is given by

$$F_d = c_d\left(\frac{dy}{dt} - \frac{dy_t}{dt}\right) \tag{2.94}$$

As in **Figure 2.25**, the weight is eliminated from the ODE. For that reason, the weight of the sprung mass, as an external force, is omitted from the free body diagram in **Figure 2.27b**, so that Newton's second law of motion is written as $-F_s - F_d = ma$.

Substituting Eqs. (2.93) and (2.94) together with $a = \dfrac{d^2 y}{dt^2}$ into Newton's law yields

$$-k\left(y - y_t\right) - c_d\left(\frac{dy}{dt} - \frac{dy_t}{dt}\right) = m\frac{d^2 y}{dt^2}$$

or

$$\frac{d^2 y}{dt^2} + \frac{c_d}{m}\frac{dy}{dt} + \frac{ky}{m} = \frac{k}{m}y_t + \frac{c_d}{m}\frac{dy_t}{dt} \tag{2.95}$$

Equation (2.95) is a second-order linear nonhomogeneous ODE that governs the displacement of the sprung mass of a car suspension system. A comparison of this ODE with the engineering form, Eq. (2.6) yields the same natural frequency ω_n and damping factor ζ that were derived already in Section 2.5 for the spring-dashpot model. Furthermore, the forcing function is

$$f\left(t\right) = \frac{k}{m}y_t + \frac{c_d}{m}\frac{dy_t}{dt} \tag{2.96}$$

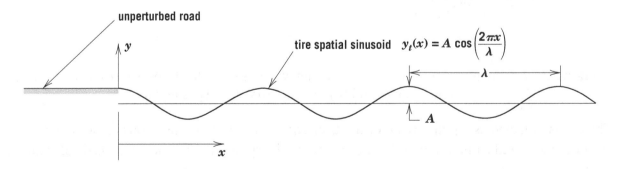

Figure 2.28 A sinusoid function with a wavelength λ and amplitude A.

As you may know, uneven roads are too complex to model by a simple function. Notwithstanding such complexity, a road geometry is a function of space. Since the tire is modeled as a point on the road, its displacement will have the same spatial function as the road. Thus, the question that you may ask is given a spatial function of the tire $y_t(x)$, as shown in **Figure 2.28**, how do we find the tire displacement in time, $y_t(t)$, as required in Eq. (2.96)?

The forcing function that is of interest to us for the study of resonance and beat is the sinusoid function. For this model we will write the tire spatial sinusoid as

$$y_t(x) = A\cos\left(\frac{2\pi x}{\lambda}\right) \tag{2.97}$$

where wavelength λ. If a car travels on a road with velocity v, then the time required to traverse one wavelength λ at velocity v is the period p. Similarly, the time required to traverse a distance x at velocity v is t. Therefore,

$$\frac{x}{t} = \frac{\lambda}{p} = v \tag{2.98}$$

Substituting x from Eq. (2.98) in Eq. (2.97) yields

$$y_t(t) = A\cos\left(\frac{2\pi vt}{\lambda}\right) \tag{2.99}$$

which is the desired result for the tire displacement in the time domain. As you can see from Eq. (2.99), the tire displacement depends on how fast the car travels and the road shape. A comparison of Eq. (2.99) with the standard sinusoid in the time domain $y_t(t) = A\cos(\omega t)$ shows that the angular frequency of the forcing function is

$$\omega = \frac{2\pi v}{\lambda} \tag{2.100}$$

Equation (2.100) demonstrates that when you drive at a velocity v, the smaller the wavelength of a road, λ, the grater the oscillation frequency, ω, and vice versa. If one of the shock absorbers fails, it is possible under certain conditions for the natural frequency of the mass-spring system to match the forcing frequency given by Eq. (2.100). When this happens, the suspension system is said to be resonating. In that case, the sprung mass displacement will become unbounded and may pose a hazardous condition for the car and passengers. Thus, under resonance $\omega_n = \omega$ or $\sqrt{\dfrac{k}{m}} = \dfrac{2\pi v}{\lambda}$, the car velocity is

$$v = \frac{\lambda}{2\pi}\sqrt{\frac{k}{m}} \tag{2.101}$$

Notice that under resonance the car velocity is independent of the road amplitude. The pertinent equations are summarized in **Table 2.5**.

ODE model	$\dfrac{d^2y}{dt^2} + \dfrac{c_d}{m}\dfrac{dy}{dt} + \dfrac{ky}{m} = f(t)$; Eq. (2.95)	$y(t)$, car displacement; c_d, damping coefficient; k, spring constant; $m = \dfrac{M}{4}$ where M is the total mass of the car.
Forcing function	$f(t) = \dfrac{k}{m}y_t + \dfrac{c_d}{m}\dfrac{dy_t}{dt}$; Eq. (2.96)	$y_t(t)$, tire displacement.
Tire model	$y_t(t) = A\cos(\omega t)$	A, road amplitude; λ, road wavelength; v, car velocity. $\omega = \dfrac{2\pi v}{\lambda}$, Eq. (2.100).

Table 2.5 Summary of second-order model.

Example 2.9 The Resonating Suspension System

Consider a car whose shock absorber on one side failed, which reduces the system to a mass-spring system. The car travels on an uneven road modeled by a sinusoid, as shown in **Figure 2.29**.

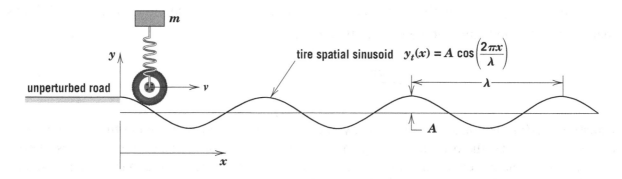

Figure 2.29 Spring-mass system moving at velocity v on a sinusoidal road.

1. Find the car velocity that causes resonance. The road wavelength and amplitude are $5\,m$ and $0.01\,m$, respectively, the sprung mass is $225\,kg$, and the spring stiffness is $87\left(10^3\right)\dfrac{N}{m}$.
2. What is the governing ODE under these circumstances?

Solution:

1. From Eq. (2.101), $v = \dfrac{5}{2\pi}\sqrt{\dfrac{87\left(10^3\right)}{225}} = 15.65\,\dfrac{m}{s}$, or $v = 35\ mph$.

2. For a failed shock absorber, the mass-spring system is derived from Eqs. (2.95) and (2.99)

    with $c_d = 0$, $\dfrac{d^2y}{dt^2} + \dfrac{k}{m}y = \dfrac{kA}{m}\cos\left(\dfrac{2\pi vt}{\lambda}\right)$. At resonance, the ODE becomes

$$\frac{d^2y}{dt^2} + 386.67y = 3.867\cos\left(19.66t\right).$$

In this example, the failed suspension system of a car is modeled at resonance by a second-order nonhomogeneous ODE without damping. The dependent variable is the sprung mass displacement $y(t)$, and the independent variable is time t.

■

2.7 Undamped Second-Order Systems with a Sinusoid Forcing Function

The solution of a nonhomogeneous second-order ODE with no damping and a sinusoid forcing function has very interesting characteristics. For a system whose natural frequency is ω_n and which is forced by a sinusoid function with a frequency ω, you will discover the following:

1. When $\omega = \omega_n$ the amplitude of the response becomes unbounded. This phenomenon is known as resonance.
2. When $\left|\omega - \omega_n\right|$ is finite but small in comparison to either ω or ω_n, the amplitude of the response varies such that the oscillation is characterized by a bursting- or pulsating-like behavior. This phenomenon is known as beats.

In order to explore these interesting phenomena we will let mathematics guide you by solving Eq. (2.6) with a damping factor $\zeta = 0$ and a forcing function

$$f(t) = F\cos(\omega t) \tag{2.102}$$

where F and ω are the amplitude and the angular frequency of the forcing function, respectively. You may use $f(t) = F\sin(\omega t)$ as the forcing function. However, the cosine function yields simpler results than the sine function. The results that you will explore will be the same in either case.

For $\zeta = 0$, Eqs. (2.6) and (2.102) yield

$$\frac{d^2y}{dt^2} + \omega_n^2 y = F\cos(\omega t) \tag{2.103}$$

Resonating systems: $\omega = \omega_n$

Resonance is a phenomenon that occurs when the system is forced with a sinusoid at a frequency ω equal to the system natural frequency ω_n. When $\omega = \omega_n$, Eq.(2.103) becomes

$$\frac{d^2 y}{dt^2} + \omega_n^2 y = F\cos(\omega_n t) \qquad (2.104)$$

For simplicity, the solution of Eq. (2.104) is found with zero ICs, that is, $y(0) = 0$ and $\frac{dy(0)}{dt} = 0$. For nonzero ICs, the behavior of the system is qualitatively the same. The homogeneous solution is given by row 1 in **Table 2.2** as

$$y_h(t) = a\cos(\omega_n t) + b\sin(\omega_n t) \qquad (2.105)$$

For the nonhomogeneous part, the particular solution is given by row 2 in **Table 2.3**. Since this particular solution is linearly dependent on the homogeneous solution, you will multiply the suggested particular solution by t; thus

$$y_p(t) = t\left[B_1 \cos(\omega_n t) + B_2 \sin(\omega_n t) \right] \qquad (2.106)$$

Substituting Eq. (2.106) in Eq. (2.104) yields $B_1 = 0$ and $B_2 = \dfrac{F}{2\omega_n}$. Thus, the particular solution is given by

$$y_p(t) = \left(\frac{F}{2\omega_n} \right) t\sin(\omega_n t) \qquad (2.107)$$

You can immediately see that $y_p(t)$ is unbounded as t approaches infinity. The complete solution $y(t)$ is the sum of Eqs. (2.105) and (2.107),

$$y(t) = a\cos(\omega_n t) + b\sin(\omega_n t) + \frac{F}{2\omega_n} t\sin(\omega_n t) \qquad (2.108)$$

Using the ICs $y(0) = 0$ and $\frac{dy(0)}{dt} = 0$ with Eq. (2.108) yields $a = 0$ and $b = 0$. Thus,

$$y(t) = \frac{F}{2\omega_n} t\sin(\omega_n t) \qquad (2.109)$$

In order to study the influence of the forcing function, Eq. (2.102), on the response, $y(t)$, you should express Eq. (2.109) as a cosine function with amplitude A_o as

$$y(t) = A_o \cos\left(\omega_n t - \frac{\pi}{2}\right) \qquad (2.110)$$

where

$$A_o = \left(\frac{1}{2}\frac{F}{\omega_n}\right)t \qquad (2.111)$$

From Eq. (2.110) you can see that the response lags behind $f(t)$ by a phase angle $\phi = \frac{\pi}{2}$, with a

time lag given by $t_{lag} = \dfrac{|\phi|}{\omega_n} = \dfrac{\pi}{2\omega_n}$. From Eq. (2.111) observe that the amplitude of the response

grows linearly with time. So when $t \to \infty$, $A_o \to \infty$, that is, the amplitude becomes unbounded. Physically, when a system resonates with the forcing function, it absorbs the maximum amount of energy. In mechanical systems, resonance, leading to unbounded amplitudes, is catastrophic. However, in electrical systems, resonance can be useful, as in the design of a receiver that requires the charges moving up and down its antenna to resonate with the transmitter electric field. In the foregoing analysis, damping was ignored. In reality, there is always damping that will limit the amplitude of oscillation. However, when damping is very small, that is $\zeta \to 0$ the amplitude remains bounded but can be very large.

The response $y(t)$ is compared with the forcing function $f(t)$, as shown in **Figure 2.30**. The lag between the two functions can be clearly seen graphically, as well as the linear increase of the response amplitude with time.

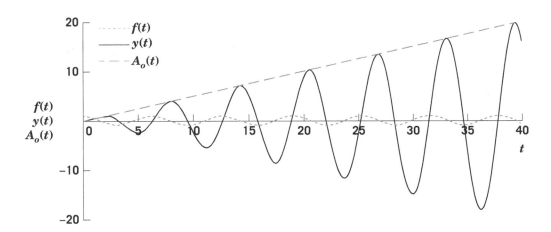

Figure 2.30 The response $y(t)$ for a resonating system, $\omega = \omega_n$.

The pertinent equations are summarized in **Table 2.6**.

ODE model	$\dfrac{d^2y}{dt^2} + \omega_n^2 y = f(t)$; Eq. (2.103)	$y(t)$, dependent variable; ω_n, natural frequency.
Forcing function	$f(t) = F\cos(\omega t)$; Eq. (2.102)	$\omega = \omega_n$, resonance.
ICs	$y(0) = 0$; $\dfrac{dy(0)}{dt} = 0$	
Solution	$y(t) = A_o \cos\!\left(\omega_n t - \dfrac{\pi}{2}\right)$; Eq. (2.110)	$A_o = \left(\dfrac{1}{2}\dfrac{F}{\omega_n}\right) t$; Eq. (2.111)

Table 2.6 Summary of second-order resonance.

Example 2.10 The Resonating LC Circuit

Solve for the current $i(t)$ based on the ODE you obtained in Example 2.8. The ICs are $i(0) = 0$ and $\dfrac{di(0)}{dt} = 0$. Plot the forcing function and the response in the range $0 \le t < \infty$.

Solution:

From Example 2.8 the governing ODE for $i(t)$ for an undamped resonating circuit was found to be

$$\frac{d^2i}{dt^2} + \omega_n^2 i(t) = F\cos(\omega t) \tag{2.112}$$

where $\omega_n = \omega = 6.16(10^6)\dfrac{rad}{s}$ and $F = 1.742(10^{10})\dfrac{amp}{s^2}$. Since Eq. (2.112) has the same form as Eq.(2.104), then for the same ICs $i(t) = 0$ and $\dfrac{di(0)}{dt} = 0$ the solution for $i(t)$ is given by Eqs.(2.110) and (2.111),

$$i(t) = \frac{1}{2}\frac{F}{\omega_n} t \cos\!\left(\omega_n t - \frac{\pi}{2}\right) \tag{2.113}$$

Substituting the values of ω_n and F in Eq. (2.113) yields

$$i(t) = 1414t \cos\!\left(6.16(10^6)t - \frac{\pi}{2}\right) \tag{2.114}$$

The forcing function is given by $f(t) = F\cos(\omega t)$. From Eq. (2.112) you may recognize that the units of F, the amplitude of $f(t)$, must be the same as $\omega_n^2 i(t)$. Therefore, for consistency, plot

$f(t)$ and $\omega_n^2 i(t)$, both in $\dfrac{amp}{s^2}$ versus t, as shown in **Figure 2.31**. Clearly, the current grows unboundedly in time, which is not realistic. In reality, the current will be large, but bounded, as soon as you introduce some damping into the system, that is, by taking into consideration the resistance of the inductor.

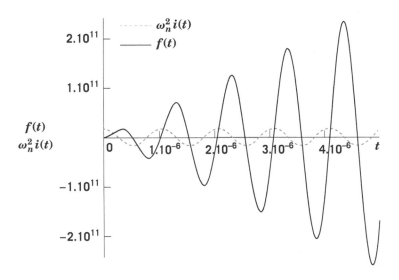

Figure 2.31 The forcing function and response of an undamped circuit.

Nonresonating systems: The phenomenon of beats

For a nonresonating undamped system, the homogeneous solution is still given Eq. (2.105), and for $\omega \neq \omega_n$ the particular solution is given by

$$y_p(t) = B_1 \cos(\omega t) + B_2 \sin(\omega t) \qquad (2.115)$$

Substituting $y_p(t)$ and its second derivative in Eq. (2.103) yields $B_1 = \dfrac{F}{\omega_n^2 - \omega^2}$ and $B_2 = 0$.

Thus, the general solution is

$$y(t) = a \cos(\omega_n t) + b \sin(\omega_n t) + \frac{F}{\omega_n^2 - \omega^2} \cos(\omega t) \qquad (2.116)$$

Applying the ICs $y(0) = 0$ and $\dfrac{dy(0)}{dt} = 0$ to Eq. (2.116) yields $a = -\dfrac{F}{\omega_n^2 - \omega^2}$ and $b = 0$. Thus, the solution $y(t)$ becomes

$$y(t) = \frac{F}{\omega_n^2 - \omega^2} \left[\cos(\omega t) - \cos(\omega_n t) \right] \qquad (2.117)$$

Equation (2.117) displays an interesting physical behavior when $|\omega - \omega_n|$ is finite, but small in comparison to either ω or ω_n. This behavior can be seen by plotting $\cos(\omega t)$ and $\cos(\omega_n t)$ for close values of ω and ω_n, and then plotting $y(t)$ proportional to the linear superposition of $\cos(\omega t)$ and $\cos(\omega_n t)$, as required by Eq. (2.117), where the proportionality factor is $\dfrac{F}{\omega_n^2 - \omega^2}$.
The result is shown in **Figure 2.32**.

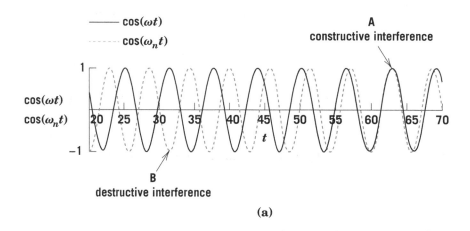

(a)

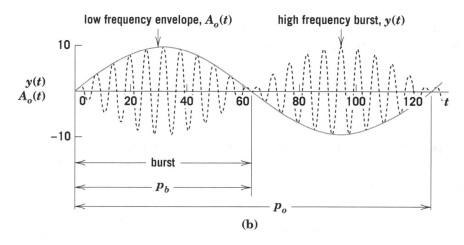

(b)

Figure 2.32 Interfering sinusoids: (a) two sinusoids of equal amplitude and close values; and (b) $y(t)$ proportional to two sinusoids superposed linearly.

Notice in **Figure 2.32a** how $\cos(\omega t)$ and $\cos(\omega_n t)$ interact as a result of their close frequencies. For example, at point A, the two sinusoids are in-phase and add constructively, also known as constructive interference. In other instances, as at point B for example, the two sinusoids are out-of-phase and add destructively, also known as destructive interference. Thus, at point A the linear superposition of two sinusoids, as required in Eq. (2.117), causes a maximum amplitude for $y(t)$. At point B, the amplitude of $y(t)$ is zero. Between a constructive interference and a destructive interference the amplitude of $y(t)$ varies with time.

Notice also in **Figure 2.32b** how $y(t)$ displays between two destructive interferences a bursting- or pulsating-like waves enclosed by an outer envelope. The outer envelope represents the amplitude variation of $y(t)$. You can visualize $y(t)$ as a train of bursting-like high-frequency waves inside a low-frequency envelope. In fact, you can show that the bursting-like waves and the low-frequency envelope can be described using $\sin\left(\dfrac{\omega + \omega_n}{2}t\right)$ and $\sin\left(\dfrac{\omega - \omega_n}{2}t\right)$, respectively, as follows. Using the trigonometric relationship

$$\cos(m) - \cos(n) = -2\sin\left(\frac{m-n}{2}\right)t\sin\left(\frac{m+n}{2}\right)t$$

you can write Eq. (2.117) as

$$y(t) = \frac{2F}{\omega^2 - \omega_n^2}\sin\left(\frac{\omega - \omega_n}{2}\right)t\sin\left(\frac{\omega + \omega_n}{2}\right)t \tag{2.118}$$

If you inspect Eq. (2.118), you notice that one sinusoid contains the low-frequency $\dfrac{\omega - \omega_n}{2}$ and a second sinusoid contains the high-frequency $\dfrac{\omega + \omega_n}{2}$. Thus, if you let $\sin\left(\dfrac{\omega + \omega_n}{2}\right)t$ describe the high-frequency bursting-like waves and $\sin\left(\dfrac{\omega - \omega_n}{2}\right)t$ describe the low-frequency outer envelope, you may write Eq. (2.118) as

$$y(t) = A_o(t)\sin\omega_i t \tag{2.119}$$

where

$$A_o(t) = \frac{2F}{\omega^2 - \omega_n^2}\sin\omega_o t \tag{2.120}$$

is the time-varying amplitude, and

$$\omega_i = \frac{\omega + \omega_n}{2} \qquad (2.121)$$

and

$$\omega_o = \frac{\omega - \omega_n}{2} \qquad (2.122)$$

are defined as the high frequency of the inner bursts, and the low frequency of the outer envelope, respectively. The family of bursts between two destructive interferences will be defined as beat. Let p_b be the period of a beat, as shown in **Figure 2.32b**. You can easily see that

$$p_b = \frac{p_o}{2} \qquad (2.123)$$

where p_o is the period of the outer envelope. Equation (2.123) can be written in terms of frequency f as $\dfrac{1}{f_b} = \dfrac{1}{2f_o}$ or

$$f_b = 2f_o \qquad (2.124)$$

Equation (2.124) can be written in terms of the angular frequency as

$$\omega_b = 2\omega_o \qquad (2.125)$$

when ω_b is defined as the angular frequency of a beat. Substituting Eq. (2.122) in Eq. (2.125) yields

$$\omega_b = \omega - \omega_n \qquad (2.126)$$

Thus, given a forcing frequency ω that is close to the natural frequency of a system ω_n the response $y(t)$ will display a pattern of repeated beats at a frequency ω_b given by Eq. (2.126). The pertinent equations are summarized in **Table 2.7**.

ODE model	$$\frac{d^2 y}{dt^2} + \omega_n^2 y = f(t)$$	$y(t)$, independent variable; ω_n, natural frequency.
Forcing function	$f(t) = F\cos(\omega t)$; Eq. (2.102)	$\omega \neq \omega_n$; $\lvert \omega - \omega_n \rvert$ finite & small.
ICs	$$y(0) = 0\ ;\ \frac{dy(0)}{dt} = 0$$	
Solution (form 1)	$$y(t) = \frac{F}{\omega_n^2 - \omega^2}\left[\cos(\omega t) - \cos(\omega_n t)\right];$$ Eq. (2.118)	
Solution (form 2)	$y(t) = A_o(t)\sin\omega_i t$; Eq. (2.119) $$A_o(t) = \frac{2F}{\omega^2 - \omega_n^2}\sin\omega_o t\ ;\ \text{Eq. (2.120)}$$	$\omega_i = \dfrac{\omega + \omega_n}{2}$, Eq. (2.121) $\omega_o = \dfrac{\omega - \omega_n}{2}$; Eq. (2.122)

Table 2.7 Summary of second-order beat.

Example 2.11 The Suspension System: Beat Phenomenon

Consider a car traveling on an uneven road modeled by a sinusoid as shown in **Figure 2.33**.

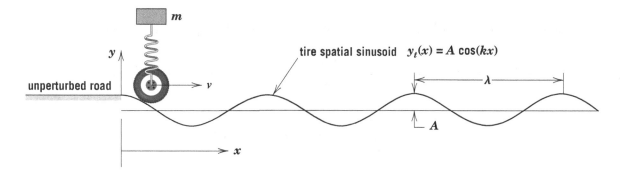

Figure 2.33 Spring-mass system moving at velocity v on a sinusoidal road.

Due to poor car maintenance, the front shock absorbers failed. Thus, the sprung mass is only supported by the springs of the suspension system.

The road wavelength and amplitude are $5m$ and $0.01m$, respectively. The sprung mass is 225 kg, and the spring stiffness is $87(10^3)\dfrac{N}{m}$. The car velocity is 40 mph.

1. Find the displacement of the sprung mass $y(t)$.

2. Plot $y(t)$ and indicate whether or not the driver experiences beats.

3. Determine the maximum amplitude the driver experiences at 40 mph.

Solution:

1. Find the displacement of the sprung mass $y(t)$.

With failed shock absorbers, the damping coefficient $c_d = 0$, and the sprung mass displacement given from Eq. (2.95) reduces to $\dfrac{d^2 y}{dt^2} + \dfrac{k}{m} y = \dfrac{k}{m} y_t$, where y_t is given by $y_t = A \cos(\omega t)$. Thus, $\dfrac{d^2 y}{dt^2} + \dfrac{k}{m} y = \dfrac{Ak}{m} \cos(\omega t)$, which can be put in the engineering form as

$$\frac{d^2 y}{dt^2} + \omega_n^2 y = F \cos(\omega t) \tag{2.127}$$

where $\omega_n^2 = \dfrac{k}{m}$ and $F = \dfrac{Ak}{m} = A\omega_n^2$.

The solution of Eq. (2.127) requires that you determine whether or not ω and ω_n differ from each other. The natural frequency of the system is $\omega_n = \sqrt{\dfrac{k}{m}} = \sqrt{\dfrac{87(10^3)}{225}} = 19.66 \dfrac{rad}{s}$. In the metric system (SI), the car velocity of 40 mph is $v = 17.88 \dfrac{m}{s}$. Using v and $\lambda = 5m$ in Eq. (2.100) yields a forcing angular frequency $\omega = 22.47 \dfrac{rad}{s}$. For $\omega \neq \omega_n$ the solution of Eq. (2.127) is given by

$$y(t) = \frac{A\omega_n^2}{\omega_n^2 - \omega^2} \left[\cos(\omega t) - \cos(\omega_n t) \right] \tag{2.128}$$

assuming zero ICs, that is, $y(0) = 0$ and $\dfrac{dy(0)}{dt} = 0$. Equation (2.128) reduces to

$$y(t) = -0.033 \left[\cos(22.47t) - \cos(19.66t) \right]$$

2. Plot $y(t)$ and indicate whether or not the driver experiences beats.

Since $\left|\omega - \omega_n\right| = \left|22.47 - 19.66\right| = 2.81$ is finite but small in comparison to either ω or ω_n, we expect the beat phenomenon to occur. If you plot $y(t)$ versus t, as shown in **Figure 2.34**, you can clearly see the beat phenomenon at an angular frequency $\omega_b = 2.81 \dfrac{rad}{s}$.

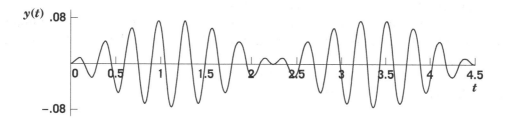

Figure 2.34 Beat phenomenon experienced by driver.

3. Determine the maximum amplitude the drive experiences at $40 \; mph$.

If you express $y(t)$ using Eqs. (2.119) and (2.120), you can see that the maximum amplitude is given by $A_{o,\max} = \dfrac{2F}{\omega^2 - \omega_n^2} = \dfrac{2A\omega_n^2}{\omega^2 - \omega_n^2}$. Numerically, $A_{o,\max} = 0.0653m$ or $6.53 \; cm$. This amplitude is quite excessive even though the road amplitude is only $1 \; cm$.

This problem illustrates how to detect the beat phenomenon in an undamped oscillating system with a natural frequency ω_n close to the forcing frequency ω. This can be done by comparing $\left|\omega - \omega_n\right|$ to either ω or ω_n and by plotting the dependent variable $y(t)$ versus the independent variable t.

HOMEWORK PROBLEMS FOR CHAPTER TWO

Problems for Sections 2.2–2.4

For Problems 1–6,

a. Cast the governing ODE in the engineering form, Eq. (2.6).
b. Classify the response as undamped, underdamped, critically damped, or overdamped.
c. Find the undamped natural frequency, the damping factor, and where applicable the damped natural frequency.
d. Find the solution that satisfies the given ICs.
e. Plot the solution in the interval $0 \le t < \infty$.

1. $3\dfrac{d^2 y}{dt^2} + y = 0$; $y(0) = 0$ and $\dfrac{dy(0)}{dt} = 1$.

2. $\dfrac{d^2 y}{dt^2} + 2y = 0$; $y(0) = 1$ and $\dfrac{dy(0)}{dt} = 0$.

3. $9\dfrac{d^2 y}{dt^2} + 12\dfrac{dy}{dt} + 16y = 0$; $y(0) = 0$ and $\dfrac{dy(0)}{dt} = -1$.

4. $4\dfrac{d^2 y}{dt^2} + 20\dfrac{dy}{dt} + 25y = 0$; $y(0) = 1$ and $\dfrac{dy(0)}{dt} = 0$.

5. $4\dfrac{d^2 y}{dt^2} + 6\dfrac{dy}{dt} + y = 0$; $y(0) = 0$ and $\dfrac{dy(0)}{dt} = -1$.

6. $4\dfrac{d^2 y}{dt^2} + 4\dfrac{dy}{dt} + y = 0$; $y(0) = 1$ and $\dfrac{dy(0)}{dt} = 0$.

For Problems 7–11, consider a mass attached to the spring, which causes it to stretch to a static equilibrium position, y_0, from its free length, as shown in **Figure 2.25**. $g = 9.81 \dfrac{m}{s^2}$.

a. Find the spring stiffness, k in *N/m*.
b. Cast the equation of motion in the engineering form, Eq. (2.6) for a mass-spring-damper in free vibration.
c. Find the undamped natural frequency, the damping factor, and where applicable the damped natural frequency and period.
d. Classify the response as undamped, underdamped, critically damped, or overdamped.
e. Find the solution that satisfies the ICs $y(0) = 0$ and $\dfrac{dy(0)}{dt} = 1$ using one of the forms in **Table 2.2**.
f. Plot the solution in the interval $0 \le t < \infty$.

7. $m = 20 \ kg$; $y_0 = 0.2m$; $c_d = 100 \ N.s/m$.

8. $m = 2 \ kg$; $y_0 = 0.2m$; $c_d = 100 \ N.s/m$.

9. $m = 1 \ kg$; $y_0 = 0.5m$; $c_d = 0$.

10. $m = 0.102 \ kg$; $y_0 = 0.102m$; $c_d = 2.001 \ N.s/m$.

11. $m = 2 \ kg$; $y_0 = 0.8m$; $c_d = 10 \ N.s/m$.

12. Consider the cylinder of radius r rolling on a cylindrical surface of radius R, as shown in **Figure 2.35**. This device is part of a high-speed printer. The cylinder equation of motion is given by $\dfrac{3}{2g}(R-r)^2 \dfrac{d\theta^2}{dt^2} = (r-R)\theta$, where g is the acceleration of gravity. The angle $\theta(t)$ describes the angular motion of the cylinder as a function of time t.

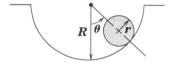

Figure 2.35 A rolling cylinder of radius r on a cylindrical surface of radius R.

a. Cast the governing equation of the cylinder motion in the engineering form, Eq. (2.6).
b. Is the ODE homogeneous or nonhomogeneous?
c. Identify the damping factor of the system.
d. Identify the undamped natural frequency of the system.
e. Identify the forcing function and response.

13. The instantaneous angular displacement, $\theta(t)$ of a control arm of a tire suspension system, as shown on the left in **Figure 2.36**, is modeled as an arm in 2D connected to a spring with stiffness k, and a viscous damper with damping coefficient c_d, as shown in the schematic on the right.

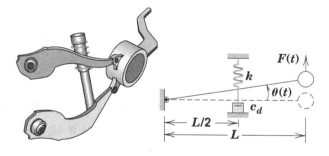

Figure 2.36 Control arm of a tire suspension system.

Newton's law of motion states that the moments about the pivoting point of the forces acting on the arm are balanced by the inertial moment, $I\alpha$, where I is the mass moment of inertia of the arm and α is the angular acceleration. From physics, $\alpha = \dfrac{d^2\theta}{dt^2}$. Thus, Newton's law yields

$$-\frac{L}{2}ky - \frac{L}{2}c_d\frac{dy}{dt} + LF(t) = I\frac{d^2\theta}{dt^2}$$

where $F(t)$ is an external force as a function of time t, L is the length of the rod, y is the spring vertical displacement, and $\dfrac{dy}{dt}$ is the vertical velocity required in modeling the viscous damping force, $c_d\dfrac{dy}{dt}$. The first term of the left-hand side of the equation consists of ky, which is the spring force given by Hooke's law, times the moment arm, $\dfrac{L}{2}$. The minus sign is due to the force of the spring acting downward against the arm. The second and third term are interpreted likewise.

From geometry, $y = \dfrac{L}{2}\tan(\theta)$. For small angular displacement, by design, $\tan(\theta) \simeq \theta$ in radians.

Thus, $\dfrac{dy}{dt} = \dfrac{L}{2}\dfrac{d\theta}{dt}$. Using this information in the equation of motion yields an ODE for $\theta(t)$ as

$$-\frac{L}{2}k\left(\frac{L}{2}\theta\right) - \frac{L}{2}c_d\left(\frac{L}{2}\frac{d\theta}{dt}\right) + LF(t) = I\frac{d^2\theta}{dt^2}$$

a. Cast the equation of motion as an ODE in the engineering form, Eq. (2.6).
b. Is the ODE homogeneous or nonhomogeneous?
c. Identify the damping factor of the system and show that it is dimensionless.
d. Identify the undamped natural frequency of the system and show that it has the dimensions of 1/time.
e. Identify the forcing function and response.

14. The instantaneous angular displacement, $\theta(t)$ of a foot pedal, as shown on the left in **Figure 2.37**, is modeled as an arm in 2D connected to a spring with stiffness k, and a viscous damper with damping coefficient c_d, as shown in the schematic on the right.

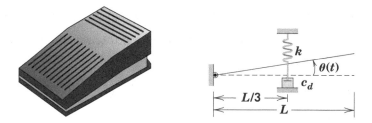

Figure 2.37 Foot pedal.

Following a theoretical approach similar to that in Problem 13, Newton's law of motion leads to

$$-\frac{L}{3}k\left(\frac{L}{3}\theta\right)-\frac{L}{3}c_d\left(\frac{L}{3}\frac{d\theta}{dt}\right)=I\frac{d^2\theta}{dt^2}$$

after the pedal is pressed and released from rest.

a. Cast the equation of motion as an ODE in the engineering form, Eq. (2.6).
b. Is the ODE homogeneous or nonhomogeneous?
c. Identify the undamped natural frequency and damping factor of the system.
d. Provide a relationship between c_d, I, k, and L for an overdamped system using the constrain $\zeta>2$.

15. A DC motor is modeled as shown in **Figure 2.38**.

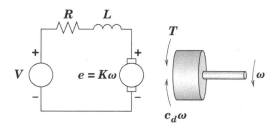

Figure 2.38 Model of a DC motor.

Mechanically, the torque applied to the motor armature, $T(t)$ as a function of time t, and the dissipative force, $c_d\omega$, where ω is the angular velocity, are balanced by the inertial force $J\alpha$, where J is the mass moment of inertia of the armature and α is the angular acceleration. From physics, $\alpha=\frac{d\omega}{dt}$. Thus, Newton's law yields $T(t)-c_d\omega=J\frac{d\omega}{dt}$. In turn, the torque is shown empirically to be proportional to current, expressed as $T(t)=Ki$, where K is a constant, so that Newton's law of motion equation becomes

$$Ki-c_d\omega=J\frac{d\omega}{dt} \tag{2.129}$$

Electrically, the motor armature, $V(t)$, is modeled as a resistor and an inductor in series with the applied voltage. In rotation, the motor induces a voltage, known as back electromotive force (emf), which is shown as voltage e in the schematic, so that Ohm's law yields $V(t) = iR + L\dfrac{di}{dt} + e(t)$. Empirically, it is also shown that $e(t)$ is proportional to the angular velocity, expressed as $e(t) = K\omega$, so that

$$V(t) = iR + L\frac{di}{dt} + K\omega \tag{2.130}$$

a. Eliminate the current, i, from Eqs. (2.129) and (2.130) in favor of an ODE that governs the angular velocity, $\omega(t)$.
b. Cast the ODE in the engineering form, Eq. (2.6).
c. Is the ODE homogeneous or nonhomogeneous?
d. Identify the damping factor of the system.
e. Identify the undamped natural frequency of the system.
f. Identify the forcing function and response.
g. Report the undamped natural frequency in rad/s, the damping factor and the damped natural frequency in rad/s for $c_d = 0.08 N.m.s$, $J = 0.475(10^{-3}) kgm^2$, $K = 0.0415\dfrac{N.m}{amp} = 0.0415 volt$, $R = 0.55 ohm$ and $L = 10^{-3} henry$.
h. For what value of R in ohm is $\omega(t)$ critically damped? The rest of the parameters are the same.

16. A helicopter company that specializes in retrieving cylindrical buoys from the ocean tried to attach a cable to the buoy, as shown on the left-hand side in **Figure 2.39**. In the first attempt, it knocks it in the calm sea with an initial displacement of $0.5\ m$ relative to the equilibrium position, with zero initial velocity. The initial buoyancy force, which is the weight of the displaced water of volume Ad, where A is the cross-sectional area of the buoy and d is the buoy submerged, is given by $F_{bo} = (\rho Ad)g$, where ρ is the density of the salt water, and g is the acceleration of gravity. The buoy weight is $W = mg$. In the initial configuration, the buoyancy force balances the weight, which leads to

$$W = \rho Adg \tag{2.131}$$

Equation (2.131) is consistent with buoy mass, $m = \rho Ad$.

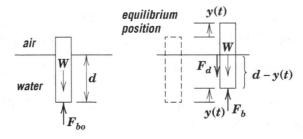

Figure 2.39 Model of a cylindrical buoy.

After the buoy is knocked and set in motion, it oscillates vertically with an instantaneous displacement $y(t)$ relative to the initial position, as shown in the schematic on the right. In this case, the instantaneous upward buoyancy force is given by

$$F_b = \rho A (d-y) g \qquad (2.132)$$

As the buoy oscillates, it experiences a viscous or damping force proportional to velocity, expressed as

$$F_d = c_d \frac{dy}{dt} \qquad (2.133)$$

where c_d is a damping coefficient. Newton's second law of motion requires that

$$F_b - W - F_d = ma = m \frac{d^2 y}{dt^2} \qquad (2.134)$$

a. Substitute Eqs. (2.131)–(2.133) in Eq. (2.134), and derive the governing ODE for the instantaneous displacement of the buoy response, $y(t)$.
b. Is the ODE homogeneous or nonhomogeneous?
c. State the ICs.
d. Define the natural frequency of oscillation, the damping factor, and the damped natural frequency.
e. Find the response, $y(t)$, which satisfies the ICs.
f. Plot the response over the domain $t \geq 0$.
g. Calculate the period of oscillation in s.
h. How long should the helicopter wait before the maximum amplitude reaches 10% of the initial displacement?

The buoy equilibrium submerged depth is $d = 1.5m$. The density of salt water is $\rho = 1025 \frac{kg}{m^3}$.

The buoy cross-sectional area is $A = 1m^2$. The damping coefficient is $c_d = 100 \frac{kg}{s}$ and $g = 10 \frac{m}{s^2}$.

17. In Example 2.6, a hinge is modeled as an underdamped system in which the damped natural frequency and damping factor were found to be $\omega_d = 3 \frac{rad}{s}$ and $\zeta = 0.287$. The response of the system, $\theta(t)$ was found with initial displacement $\theta(t) = 10^o$ and zero initial angular speed, $\frac{d\theta(0)}{dt} = 0$. Using the same damped natural frequency and damping factor, the engineer tested the response of the system, $\theta(t)$, with different ICs. In the new test, the pendulum is initially vertical, that is, $\theta(0) = 0$. However, an initial angular velocity of 1 rad/s is imparted on the pendulum, that is, $\frac{d\theta(0)}{dt} = 1 \frac{rad}{s}$.

a. Find the response of the pendulum, $\theta(t)$ using the new ICs.

b. Plot $\theta(t)$ in the domain $t \geq 0$.

c. Determine the maximum amplitude in radians and degrees, and discuss whether or not the small-angle approximation used in modeling the pendulum motion is violated. For definiteness, assume that an error of less than 1% between $\sin(\theta)$ and θ in radian as an acceptable criterion for small-angle approximation.

18. Repeat Problem 17 with $\theta(0) = 0$ and $\dfrac{d\theta(0)}{dt} = 2 \dfrac{rad}{s}$.

Problems for Section 2.5

19. In a sheet metal manufacturing company, a strip of metal, as shown **Figure 2.40**, is used to make aluminum cans. The sheet metal drum is driven by a motor. The motor voltage, $V_a(t)$, is regulated in order to accelerate the sheet metal uniformly, then move it at uniform velocity, and finally decelerate it uniformly to rest. This control strategy is modeled by the applied voltage to the motor as a forcing function, as shown on the right side. At $t = 3s$, the sheet metal is cut by a hydraulic ram. The system dwells for $0.5s$; that is, no motion takes place during that time before the cycle is repeated.

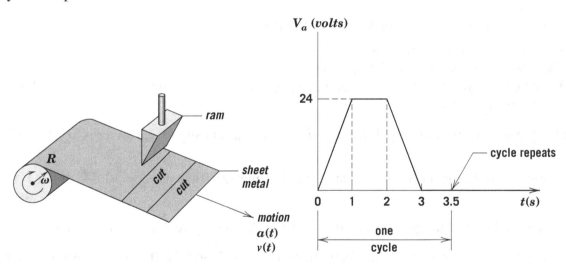

Figure 2.40 Voltage motor model for a sheet metal drum.

The motor angular velocity is modeled by the ODE

$1.145\left(10^{-5}\right)\dfrac{d^2\omega}{dt^2} + 7\left(10^{-3}\right)\dfrac{d\omega}{dt} + 0.9\omega = V_a(t)$, where $V_a(t)$ is the applied voltage in *volts*. It is given as a piecewise function in the above schematic.

a. Express $V_a(t)$ as a piecewise function over the interval $0 \leq t \leq 3.5s$.

b. Find the angular velocity of the motor by solving the ODE over the interval $0 \le t \le 3.5s$ subject to the following ICs: the initial angular velocity and acceleration are $\omega(0) = 0 \ rad/s$, and $\dfrac{d\omega(0)}{dt} = 3000 \dfrac{rad}{s^2}$, respectively. You should have four solutions over the four time subintervals [0,1], [1,2], [2,3], and [3,3.5].

c. Plot $\omega(t)$ over the time interval $0 \le t \le 3.5s$.

d. The sheet metal linear velocity, v, is given by $v = \omega R$, where $R = 0.0254m$ is the motor radius. Plot $v(t)$ over the time interval $0 \le t \le 3.5s$, and report the sheet metal maximum velocity in m/s.

Problems for Section 2.6

20. Revisit the RLC circuit in Section 2.6.

a. State the relationships between R, L, and C for an underdamped, critically damped, and overdamped circuit.

b. Find the minimum value of R in ohm using $L = 10^{-4} \ henry$ and $C = 10^{-12} \ farad$ for the overdamped case.

21. Consider a car whose shock absorber on one side failed while traveling at 50 mph. The failed side is reduced to a mass-spring system. The car travels on a road modeled by a sinusoid, as shown in **Figure 2.29**.

a. Find the road wavelength in m that causes the car to experience resonance.

b. Find the frequency at resonance in rad/s and Hz.

c. What is the governing ODE of the tire instantaneous displacement, $y(t)$, under these circumstances? The road amplitude is $0.01m$. The sprung mass is 250 kg, and the spring stiffness is $75000 \dfrac{N}{m}$.

Problems for Section 2.7

For Problems 22–24,

a. Cast the ODE in the engineering form, Eq. (2.6).

b. Report ω_n and identify the forcing function $f(t)$.

c. Find the response $y(t)$ for the case $\omega = \omega_n$.

d. Plot the response $y(t)$ and the forcing function $f(t)$ over the domain $t \ge 0$.

e. Determine the lag between the response and the forcing function.

22. $3\dfrac{d^2y}{dt^2} + y = 2\sin(\omega t)$; zero ICs.

23. $5\dfrac{d^2y}{dt^2}+y=5\cos(\omega t)$; zero ICs.

24. $\dfrac{d^2y}{dt^2}+y=\cos(\omega t)+\sin(\omega t)$; zero ICs.

25. Consider the ODE $2\dfrac{d^2y}{dt^2}+y=2\sin(\omega t)$ with zero ICs.

a. Cast the ODE in the engineering form, Eq. (2.6).
b. Report ω_n and identify the forcing function $f(t)$.
c. Find the response $y(t)$ for the case $\omega=0.95\omega_n$.
d. Plot the response $y(t)$ over the domain $t\geq0$ and adjust the t-axis appropriately in order to show the beat phenomenon.
e. Report the beat angular frequency.

26. Consider the ODE $5\dfrac{d^2y}{dt^2}+y=5\cos(\omega t)$ with zero ICs.

a. Cast the ODE in the engineering form, Eq. (2.6).
b. Report ω_n and identify the forcing function $f(t)$.
c. Find the response $y(t)$ for the case $\omega=1.05\omega_n$.
d. Plot the response $y(t)$ over the domain $t\geq0$ and adjust the t-axis appropriately in order to show the beat phenomenon.
e. Report the beat angular frequency.

27. Revisit the RLC circuit in Section 2.7 as part of a radio receiver and tuner circuit. Consider the case with zero resistance, $L=0.01 henry$ and $C=10^{-6}\,farad$. The receiver signal voltage in volts is modeled as a sinusoid given by $V_S(t)=\sqrt{2}\sin(\omega t)$, where t is in s and $\omega=9500\dfrac{rad}{s}$.

a. Find the solution of the instantaneous current, $i(t)$ with zero ICs.
b. Plot $i(t)$ over the domain $t\geq0$ and indicate whether the phenomenon is resonance or beat.

28. A 500 kg spring-mounted trailer, supported by two wheels, travels over a road at velocity v. The spring stiffness of the combined wheels is $250(10^3)\dfrac{N}{m}$. The road is modeled by a sinusoid whose amplitude and wavelength are $0.10m$ and $1.5m$, respectively, as shown **Figure 2.41**.

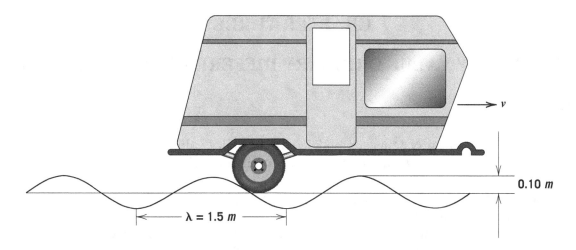

Figure 2.41 A spring-mounted trailer moving on a road modeled as a sinusoid.

a. At what velocity in *mph* will the trailer experience the beat phenomenon at a frequency equal to 5% of the natural frequency?

b. Find the equation of the trailer instantaneous displacement $y(t)$ with zero ICs.

c. Plot the response $y(t)$ over the domain $t \geq 0$.

CHAPTER THREE

BOUNDARY VALUE ORDINARY DIFFERENTIAL EQUATIONS

3.1 Introduction

In this chapter, engineering systems are modeled by boundary value ordinary differential equations, which are also known as boundary value problems (BVPs).

BVPs appear in many engineering applications. An example is the catenary cable, as shown in **Figure 3.1**, for bridge support; other applications include transmission power lines, telephone lines, and tether lines for hot-air balloons. As a design engineer, do you know how much tension is required in the cable in order to safely support a bridge? Have you ever wondered why the cable sags? These questions are answered in this chapter as part of BVPs.

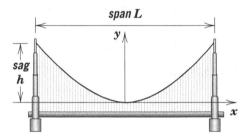

Figure 3.1 Catenary cable as part of a bridge design.

You will learn in Section 3.3 that when the cable weight is much smaller than the structure weight, the cable local deflection, or displacement, $y(x)$ is governed by the following second-order linear ODE,

$$\frac{d^2 y}{dx^2} = f(x) \tag{3.1}$$

over the domain $-\frac{L}{2} < x < \frac{L}{2}$, where $f(x) = \frac{w}{T_o}$, w is the weight per unit length of the cable, and T_o is the horizontal component of the tension force at the reference frame origin, as shown in **Figure 3.2**.

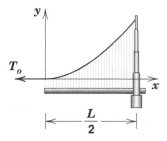

Figure 3.2 Cable section showing the horizontal tension T_o at the cable lowest point.

To have a sense for w, consider a bridge section that weighs $12,000\ N$ over a span of $300\ m$. If the weight is assumed to be distributed uniformly, then $w = \dfrac{12,000N}{300m} = 40\dfrac{N}{m}$.

With respect to the choice of the reference frame in **Figure 3.1**, the boundary conditions (BCs) for Eq. (3.1) are

$$y\left(\frac{L}{2}\right) = h \tag{3.2}$$

and

$$y\left(-\frac{L}{2}\right) = h \tag{3.3}$$

Equations (3.1)–(3.3) constitute a BVP. However, there are other engineering applications where the cable does not support a structure. For example, in the design of electrical power transmission towers, cables may have a span in the order of kilometers, and their weights are not negligible. The weight per length can be as much as $w = 50N\ /\ m$. For these cables, the local displacement, $y(x)$, is governed by the following second-order nonlinear ODE

$$\left(\frac{d^2y}{dx^2}\right)^2 - f(x)^2\left(\frac{dy}{dx}\right)^2 = f(x)^2 \tag{3.4}$$

over the domain $-\dfrac{L}{2} < x < \dfrac{L}{2}$. The ODE, Eq. (3.4), and the BCs given by Eqs. (3.2)–(3.3) constitute a nonlinear BVP. The nonlinearity is due to the square of either the first derivative or the second derivative. You may ask at this point why a cable with negligible weight in comparison to a heavy structure is modeled by a linear ODE, whereas a cable that supports its own weight leads to a nonlinear ODE? This question is explored in Section 3.3, where you will learn that the answer lies in how the weight per unit length $w(x)$ is modeled.

In a homework problem, you may show that $y(x) = \frac{1}{f}\cosh(xf) - \frac{1}{f}\cosh\left(\frac{fL}{2}\right) + h$ is a solution

that satisfies the ODE, Eq. (3.4), and the BCs, Eqs. (3.2) and (3.3), where $f = \frac{w}{T_o}$. As you can

see from this solution, the shape of a catenary cable is a hyperbolic cosine function, which is often mistaken for a parabola. The parabolic shape is only an approximation for a sag h, when it is approximately 10% less than the span L. The catenary cable tension is usually underestimated using a parabolic solution. In a homework problem, you may explore the error magnitude between the hyperbolic cosine and parabolic models.

BVPs are encountered in other areas of engineering. An example is a heat sink to transfer heat from a microprocessor as required by the operating temperature of the device as analyzed in Section 3.4. Have you ever wondered how to predict the amount of heat removal from a microprocessor and at what rate?

In electrical engineering, the electric potential in a coaxial cable, as shown in **Figure 3.3a**, is modeled as a BVP. Coaxial cables are widely used in data transmission over the Internet, radio, and video signals, to name a few applications. The signal quality depends in part on the capacitance of the cable, C. In this configuration, the central copper wire and the copper mesh are separated by insulation, also known as a dielectric material. The solution of the BVP allows you to figure out the necessary geometry as you may explore in a homework problem. Whenever there is an electric current, the two wires form a capacitor, known as a coaxial capacitor, as shown in **Figure 3.3b**.

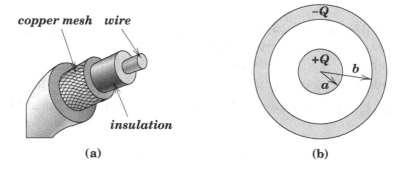

(a)   (b)

Figure 3.3 Coaxial cable: (a) inner copper wire and copper mesh separated by dielectric material (insulation); and (b) cross section with the copper wire of radius a with charge Q, and the dielectric material with radius b, surrounded by the copper mesh with charge $-Q$.

As a result, there is a positive charge Q in Coulomb at the inner wire with voltage V_a and a negative charge $-Q$ at the outer copper mesh with voltage $V_b = 0$, which is set at zero as a reference voltage, that is, by assuming that the copper mesh is grounded. By definition, the capacitance is $C = \frac{Q}{V_a - V_b} = \frac{Q}{V_a}$ in *farad*. In turn, the charge Q is given by Gauss's law

$Q = -(2\pi\varepsilon L)r\dfrac{dV}{dr}$, where ε is a dielectric property, known as permittivity, and L is the length of

the cable. Thus, $C = -\left(\dfrac{2\pi\varepsilon L}{V_a}\right)r\dfrac{dV}{dr}$. In the foregoing equation, the only unknown is the voltage

V, which is found by solving the second-order linear BVP $r\dfrac{d^2V}{dr^2} + \dfrac{dV}{dr} = 0$ over the domain

$a < r < b$, where r is the radial position. The BCs are $V = V_a$ at $r = a$ and $V = V_b = 0$ at $r = b$.

BVPs appear in problems dealing with fluid flow, for example, air flow in ducts, water flow in tubes or pipes, and blood flow in arteries that can be modeled as a network of tubes. Have you ever wondered what the percent reduction in blood flow is when the artery radius is reduced due to plaque, say, by 10%? For an average speed $\bar{v}$ of blood flow through an artery of cross-sectional area A, as shown in **Figure 3.4a**, the blood flow is quantified by the volumetric flow

rate, defined as $\dot{V} = A\bar{v}$ in $\dfrac{m^3}{s}$, which requires knowledge of $\bar{v}$. The average speed $\bar{v}$ is obtained

from knowledge of the radial velocity distribution, $v(r)$, as shown in **Figure 3.4b**, which is given

by the average value theorem as $\bar{v} = \dfrac{1}{A}\int v(r)\,dA = \dfrac{1}{\pi R^2}\int_0^R v(r)2\pi r\,dr = \dfrac{2}{R^2}\int_0^R v(r)r\,dr$, where R

is the tube radius. In the first integral the element area is $dA = 2\pi r\,dr$.

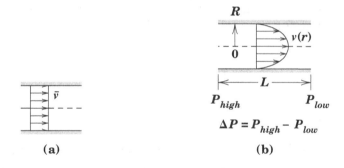

Figure 3.4 Flow in a tube: (a) average speed $\bar{v}$; and (b) local velocity $v(r)$.

For certain fluid properties and flow speeds, the velocity distribution, $v(r)$, is parabolic, as shown in

Figure 3.4b, which is obtained by solving the second-order BVP, $\dfrac{d^2v}{dr^2} + \dfrac{1}{r}\dfrac{dv}{dr} = -\dfrac{1}{\mu}\dfrac{\Delta P}{L}$, over the

domain $0 < r < R$, with BCs $\dfrac{dv(0)}{dr} = 0$ and $v(R) = 0$, where μ is a property of the fluid, known

as viscosity, L is the length of the artery, and ΔP is the pressure drop between the entrance and exit of the artery. Notice that the BC at $r = 0$ requires the first derivative of velocity to vanish. Notice also that the speed at $r = R$ is zero, which is based on a hypothesis in fluid mechanics, known as the no-slip condition. This condition requires that fluid at a surface have the same speed as that of the surface. In this model, the blood is assumed stationary relative to a stationary artery.

In this introduction, you can see that BVPs appear in many engineering problems with BCs that may require specifying the function or its derivative at either end of the domain. The emphasis in this chapter is on second-order BVP for its simplicity in comparison to higher-order BVPs. BVPs with orders higher than two are very important in engineering. For example, the deflection of the rear axle of cars is described by a fourth-order ODE. However, these high-order BVPs are covered in specialized areas, for example, solid mechanics, which necessitate a background in various types of end supports in order to pose the mathematical boundary conditions properly.

3.2 Basic Concepts

In general, a second-order BVP is described by an ODE in the dependent variable y as a function of the independent variable x as

$$\frac{d^2y}{dx^2} + a_1\frac{dy}{dx} + a_0 y = f(x)$$

over a domain $a < x < b$ with two BCs specified at the left and right ends, $x = a$ and $x = b$, respectively. When the coefficients a_0 and a_1 are at most function of x, the ODE is linear; otherwise, the ODE is nonlinear. We assume that a_0 and a_1 are continuous and bounded over the interval of interest, $a < x < b$. We also assume that $f(x)$ is twice integrable over the same interval. In order to completely classify a problem as a BVP, you must specify the BCs. At least three types of BCs arise in engineering for a second-order ODE.

Boundary conditions

BC of the first kind or Dirichlet type

One type of BCs is classified as the first kind or Dirichlet type by specifying the function at either end of the interval of interest, for example, $y(a) = y_a$ or $y(b) = y_b$. In reference to **Figure 3.1**, for instance, it is natural to specify Dirichlet BCs at the fixed ends, $y\left(-\frac{L}{2}\right) = h$ and $y\left(\frac{L}{2}\right) = h$.

BC of the second kind or Neumann type

Another type of BCs is classified as the second kind or Neumann type by specifying the derivative of y at a or b, that is, $\frac{dy(a)}{dx} = y'_a$ or $\frac{dy(b)}{dx} = y'_b$.

If you revisit **Figure 3.1**, notice that the cable is symmetric with respect to the y axis. As a result, you may solve the ODE over one-half of the domain, $0 < x < \frac{L}{2}$, as shown in **Figure 3.2**. A minimum displacement at $x = 0$ requires $\dfrac{dy(0)}{dx} = 0$, which is one BC. The other BC is

$y\left(\dfrac{L}{2}\right) = h$. In this case, the BCs are mixed, Neumann and Dirichlet types.

BC of the third kind or Robin type

A third type of BCs is classified as the third kind or Robin type by specifying a linear combination of the function and its derivative at a or b; for example, specifying $k_1 y(a) + k_2\, y'(a)$ or $l_1 y(b) + l_2\, y'(b)$, where k_1, k_2, l_1, and l_2 are real constants fixed by the physics of the problem. You will learn about the Robin BC in Section 3.4 by modeling a heat transfer problem.

In general, a second-order linear BVP is described by the following ODE over the interval $a < x < b$,

$$\frac{d^2 y}{dx^2} + a_1 \frac{dy}{dx} + a_0 y = f(x) \tag{3.5}$$

and the two generalized BCs

$$k_1 y(a) + k_2\, y'(a) = r_a \tag{3.6}$$

and

$$l_1 y(b) + l_2\, y'(b) = r_b \tag{3.7}$$

where r_a and r_b are specified values. Note that when k_1 or l_1 is equal to zero, the corresponding BC reduces to a Neumann type. Similarly, when k_2 or l_2 is equal to zero, the corresponding BC reduces to a Dirichlet type. However, k_1 and k_2, or l_1 and l_2 cannot vanish simultaneously. **Table 3.1** is a summary of the boundary conditions for boundary value ODEs in various forms.

$$\frac{d^2 y}{dx^2} + a_1 \frac{dy}{dx} + a_0 y = f(x) \text{ over the interval } a < x < b.$$

Type of boundary condition	Specified boundary conditions	Engineering examples
Dirichlet	$y(a) = y_a$; $y(b) = y_b$	Figure 3.1
Neumann	$\dfrac{dy(a)}{dx} = y'_a$; $\dfrac{dy(b)}{dx} = y'_b$	Figure 3.2
Robin	$k_1 y(a) + k_2\, y'(a) = r_a$; $l_1 y(b) + l_2\, y'(b) = r_b$	Figure 3.10

Table 3.1 Boundary conditions for the boundary value ODE.

In the following sections, the modeling of engineering problems as BVPs will be limited to the spatial domain. However, you can analyze certain IVPs in the time domain as BVPs from a mathematical point of view. Conversely, some BVPs can be solved as IVPs in space.

The solution of the general linear BVP Eq. (3.5) with specified BCs will be found by the standard method used in Chapter 2, that is, by finding the homogeneous solution, $y_h(x)$, and the particular solution, $y_p(x)$. However, for the cases where the coefficients a_0 and a_1 vanish, the solution of the ODE can be found by direct integration, as you will learn next. Finally, in Section 3.5 you will learn how to solve a special class of homogeneous BVPs known as the eigenvalue problem.

3.3 Solution of Linear BVPs: Direct Integration

For the cases where the coefficients a_0 and a_1 vanish in Eq. (3.5), the solution of the general ODE can be found by direct integration. That is, given

$$\frac{d^2y}{dx^2} = f(x) \tag{3.8}$$

integrate twice to obtain

$$y(x) = \int\left(\int f(x)\,dx\right)dx + C_1 x + C_2 \tag{3.9}$$

where C_1 and C_2 are constants of integration. Notice that $f(x)$ must be twice integrable as stated earlier.

Modeling of a second-order BVP: The displacement of a catenary cable

An important engineering problem that is modeled by a second-order ODE and classified as a BVP is the displacement of a catenary cable with a constant cross section, as shown in **Figure 3.5** and **Figure 3.6**, where $y(x)$ is the transverse displacement.

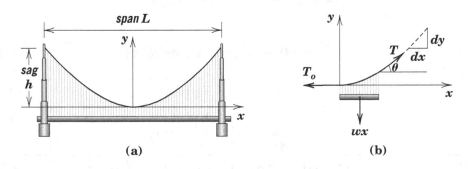

(a)   **(b)**

Figure 3.5 Cable supporting a structure: (a) structure of span L and cable with sag h; and (b) free body diagram of a cable section showing the structure weight per unit length x in the horizontal direction.

In **Figure 3.5a**, the cable supports a structure such as the section of a bridge. In that configuration, the structure weight is much larger than the cable weight. In **Figure 3.6a**, the cable does not support any external structure. Such a cable is used in many engineering designs as in electrical power transmission lines. Here, the cable supports its own weight.

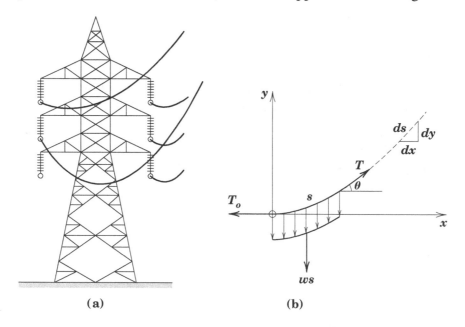

(a)   (b)

Figure 3.6 Catenary cable: (a) electrical power transmission cable with significant weight; and (b) free body diagram of a cable section showing the cable weight per unit cable length s.

In order to predict the transverse displacement or simply the deflection $y(x)$ under different loading conditions, the cable is analyzed in a state of static equilibrium. The governing ODE will be derived by applying the following three concepts: (1) modeling of weight distribution; (2) free body diagram; and (3) Newton's first law.

Modeling of weight distribution

For a cable supporting a structure of span L and weight W_S, the weight per unit length is defined as $w = \dfrac{W_S}{L}$. For a local span x, the weight per unit length is defined as $w = \dfrac{W_S}{x}$ so that the local structure weight is $W_S(x) = wx$.

Free body diagram

A free body diagram (FBD) is a technique widely used in mechanics to identify important forces and moments in a system. In this chapter, the FBD is created by: (1) isolating a suitable differential element from its surroundings, and (2) identifying and applying all forces or moments that act directly on the differential element, such as weight, reactions from supports, and external loadings. In **Figure 3.5b**, the isolated FBD is chosen between the lowest point of cable at $x = 0$ and a point located at distance x to the right. By definition, the tension acts tangentially to the cable. Thus, at $x = 0$, the tension is T_o and acting horizontally, as shown in **Figure 3.5b**. At distance x, the tension is T and acting along the tangent at slope $\dfrac{dy}{dx}$, and the cable section has a weight $W_S(x) = wx$.

Newton's first law

Newton's first law is required to balance the forces acting on the differential element. The first law states that if a body is at rest, that is, in static equilibrium, the sum of the external forces applied to the body is equal to zero, $\sum \vec{F} = 0$, where the vector $\vec{F}$ represents all external forces. Newton's first law can be resolved into horizontal and vertical components as follows. In the horizontal direction, that is, the x-direction in **Figure 3.5b**, $-T_o + T\cos\theta = 0$ or

$$T\cos\theta = T_o \tag{3.10}$$

In the vertical direction, $T\sin\theta - wx = 0$ or

$$T\sin\theta = wx \tag{3.11}$$

The ratio of Eqs. (3.10)–(3.11) yields

$$\tan\theta = \frac{w}{T_o}x \tag{3.12}$$

The local slope at x can be expressed as

$$\tan\theta = \frac{dy}{dx} \tag{3.13}$$

Combining Eqs. (3.12) and (3.13) yields

$$\frac{dy}{dx} = \frac{w}{T_o}x \tag{3.14}$$

Equation (3.14) is converted to a second-order ODE by taking its derivative with respect to x,

$$\frac{d^2y}{dx^2} = \frac{w}{T_o} \tag{3.15}$$

Equation (3.15) is a linear second-order nonhomogeneous ODE that has the form of Eq. (3.8) with $f(x) = \dfrac{w}{T_o}$. The boundary conditions corresponding to fixed ends are of Dirichlet type,

$$y\left(-\frac{L}{2}\right) = h \text{ and } y\left(\frac{L}{2}\right) = h.$$

Notice upon inspecting Eq. (3.14) that a horizontally taut cable will have zero slope, which requires T_o to approach infinity. This is not possible due to material strength limitation. For that reason, a finite slope is required, which leads to the necessary sag.

Example 3.1 Cable Supporting a Structure

Consider a cable fastened between two towers, $250m$ apart. The endpoints are at the same height, as shown in **Figure 3.5a**. The cable supports a structure that has a weight per unit length equal to $4500\dfrac{N}{m}$. As an engineer, you are charged with determining the tension force in the cable in order to limit the sag to 10% of the span.

1. Determine the tension at the bottom of the cable, T_o.
2. Determine the tension force T at the fastened points.
3. Determine the length of the cable, L_{cable}.

Solution:

1. Determine the tension at the bottom of the cable, T_o.

The cable displacement $y(x)$ is given by Eq. (3.15), $\dfrac{d^2 y}{dx^2} = \dfrac{w}{T_o}$ over the domain $-\dfrac{L}{2} < x < \dfrac{L}{2}$, where $L = 250m$. The solution of this ODE is given by Eq. (3.9),

$$y(x) = \frac{w}{2T_o}x^2 + C_1 x + C_2 \tag{3.16}$$

The BCs $y\left(-\dfrac{L}{2}\right) = y\left(\dfrac{L}{2}\right) = h$ yield $h = \dfrac{wL^2}{8T_o} - \dfrac{L}{2}C_1 + C_2$ and $h = \dfrac{wL^2}{8T_o} + \dfrac{L}{2}C_1 + C_2$.

Adding the two foregoing equations yields a solution for $C_2 = h - \dfrac{wL^2}{8T_o}$, and substituting C_2 in one of the equations yields $C_1 = 0$. Thus, Eq. (3.16) becomes

$$y(x) = \frac{w}{2T_o}\left(x^2 - \frac{L^2}{4}\right) + h \tag{3.17}$$

While Eq. (3.17) satisfies the BCs, it does not automatically satisfy $y(0) = 0$ at the reference frame origin for arbitrary h, L, w, and T_o. At $x = 0$, Eq. (3.17) yields $y(0) = -\dfrac{wL^2}{8T_o} + h$. Thus, out of the set h, L, w, and T_o, three elements may be arbitrary, and the fourth element is fixed by satisfying $-\dfrac{wL^2}{8T_o} + h = 0$. For definiteness, fix h, L, and w and determine T_o from the foregoing equation. Note that the data is given with that in mind, that is, h, L, and w, which yields

$T_o = \dfrac{wL^2}{8h} = 4500\dfrac{N}{m}(250m)^2\dfrac{1}{8(0.1)250m} = 1406.25kN$. This value for tension is necessary at the origin for a well-posed problem that satisfies the BCs as well as $y(0) = 0$. Substituting

$T_o = \dfrac{wL^2}{8h}$ in Eq. (3.17) yields the shape of the cable $y(x) = \dfrac{4h}{L^2}x^2$.

2. Determine the tension force T at the fastened points.

Applying Eq. (3.11) at $x = \dfrac{L}{2}$ yields the tension at the fastened end,

$$T = \frac{w}{\sin\theta}\frac{L}{2} \tag{3.18}$$

At $x = \dfrac{L}{2}$, θ is found from Eq. (3.12), $\theta = \tan^{-1}\left(\dfrac{wL}{2T_o}\right)$. Thus, $\theta = 21.8^o$.

From Eq. (3.18), $T = \dfrac{4500}{\sin(21.8^o)}\dfrac{250}{2} = 1515kN$. Notice that this tension is approximately 8% greater than T_o.

3. Determine the length of the cable, L_{cable}.

The cable length is given by $L_{cable} = \int dL_c$ where dL_c is a differential cable element. From calculus, the differential length of a curve is given by $dL_c = \sqrt{1 + y'^2}\, dx$. Thus,

$L_{cable} = \int_{-\frac{L}{2}}^{\frac{L}{2}} \sqrt{1 + y'^2}\, dx$, where $y' = \dfrac{dy}{dx}$ is the local slope. From $y(x) = \dfrac{4h}{L^2} x^2$, $\dfrac{dy}{dx} = \dfrac{8h}{L^2} x$. Thus

$$L_{cable} = \int_{-\frac{L}{2}}^{\frac{L}{2}} \sqrt{1 + \left(\frac{8h}{L^2} x\right)^2}\, dx = 2\int_{0}^{\frac{L}{2}} \sqrt{1 + \left(\frac{8h}{L^2} x\right)^2}\, dx \qquad (3.19)$$

Using CAS to evaluate the integral in Eq. (3.19) yields

$$L_{cable} = \frac{L}{2}\sqrt{1 + \left(\frac{4h}{L}\right)^2} + \frac{L^2}{8h}\ln\left(\frac{4h}{L} + \sqrt{1 + \left(\frac{4h}{L}\right)^2}\right)$$

Using $\dfrac{h}{L} = 0.1$ yields $L_{cable} = 256.52m$.

In this engineering application, the cable deflection is modeled by a second-order linear ODE, where the dependent variable is the cable displacement $y(x)$ and the independent variable is x in the span direction. The general solution of the differential equation was found by direct integration, and the constants of integration were calculated by applying Dirichlet BCs at the fixed ends. You also learned that while the solution satisfies the BCs $y\left(-\dfrac{L}{2}\right) = y\left(\dfrac{L}{2}\right) = h$, it was necessary to impose further conditions, for example, the value of T_o in order to satisfy $y(0) = 0$.

Interestingly, $\dfrac{d^2 y}{dx^2} = \dfrac{w}{T_o}$ can also be solved as an initial value problem, IVP, using the ICs $y(0) = 0$ and $\dfrac{dy(0)}{dx} = 0$, as required at the origin in **Figure 3.5**. In this case, T_o is fixed by satisfying $y\left(\dfrac{L}{2}\right) = h$.

The analysis of a cable that supports its own weight, as in **Figure 3.6a**, differs from the previous analysis in how the distributed load is modeled. From **Figure 3.6b**, the cable weight over a section of length s is $W_S(s) = ws$. In the horizontal direction, the balance of force components is the same as Eq. (3.10). However, in the vertical direction, the equation is similar to Eq. (3.11) except for the right-hand-side term, which is ws in this case instead of wx. That is,

$$T \sin \theta = ws \tag{3.20}$$

Dividing Eq. (3.20) by Eq. (3.10) yields $\tan \theta = \dfrac{w}{T_o} s$, and combining with the slope expression, Eq. (3.13) yields

$$\frac{dy}{dx} = \frac{w}{T_o} s \tag{3.21}$$

Differentiating Eq. (3.21) with respect to x yields

$$\frac{d^2 y}{dx^2} = \frac{w}{T_o} \frac{ds}{dx} \tag{3.22}$$

From **Figure 3.6b**, observe that $ds^2 = dx^2 + dy^2$. This relationship yields $\left(\dfrac{ds}{dx} \right)^2 = 1 + \left(\dfrac{dy}{dx} \right)^2$. Combining the foregoing equation with the square of Eq. (3.22) yields $\left(\dfrac{d^2 y}{dx^2} \right)^2 = \left(\dfrac{w}{T_o} \right)^2 \left[1 + \left(\dfrac{dy}{dx} \right)^2 \right]$, or

$$\left(\frac{d^2 y}{dx^2} \right)^2 - \left(\frac{w}{T_o} \right)^2 \left(\frac{dy}{dx} \right)^2 = \left(\frac{w}{T_o} \right)^2 \tag{3.23}$$

Equation (3.23) is a second-order nonlinear ODE due to the square of either the first or the second derivative. This nonlinear ODE possesses an analytical solution that is addressed in the homework problems. Its solution consists of a hyperbolic cosine function, hence the hyperbolic shape of the cable rather than the mistakenly parabolic shape. This nonlinear ODE requires a trial and error solution to find the necessary tension at the fastened ends. For small deflections, the cable displacement is approximated parabolically in order to find the tension directly as you learned in the previous example. However, it is possible to find the tension using the hyperbolic cosine exact solution with the aid of CAS.

Next, you will explore the solution of BVPs using homogeneous and particular solutions, as you learned in Chapter 2.

3.4 General Solution of Second-Order Linear BVPs: Homogeneous and Particular Solutions

The BVP defined in Eqs. (3.5)–(3.7) is classified as a nonhomogeneous BVP when either $f(x)$, r_a, or r_b is not zero. The solution of the general second-order linear boundary value ODE, Eq. (3.5), is limited to constant coefficients a_0 and a_1. The method is similar to what you learned in Chapter 2, second-order IVPs. That is, you construct the solution as the sum of a homogeneous solution $y_h(x)$ corresponding to $f(x) = 0$, and a particular solution $y_p(x)$ corresponding to $f(x) \neq 0$.

For constant coefficients, the homogeneous solution $y_h(x)$ is found by proposing an exponential solution, $y_h(x) = e^{\lambda x}$ where λ is a constant to be determined. The particular solution $y_p(x)$ is found using the method of undetermined coefficients as outlined in Chapter 2. The general solution $y(x)$ is formed by summing $y_h(x)$ and $y_p(x)$, that is, $y(x) = y_h(x) + y_p(x)$. A summary of the nonhomogeneous BVP is shown in **Table 3.2**.

Nonhomogeneous ODE	Boundary conditions	Solution
$\dfrac{d^2y}{dx^2} = f(x)$	$k_1 y(a) + k_2\, y'(a) = r_a$ $l_1 y(b) + l_2\, y'(b) = r_b$	$y(x) = \int\left(\int f(x)\,dx\right)dx + C_1 x + C_2$
$\dfrac{d^2y}{dx^2} + a_1\dfrac{dy}{dx} + a_0 y = f(x)$		$y(x) = y_h(x) + y_p(x)$, see Chapter 2

Table 3.2 The nonhomogeneous BVP.

Modeling a second-order BVP: Cooling of a computer fin

The applied engineering problem that you will model as a BVP with nonzero coefficients a_0 and a_1 is the cooling fin of a microprocessor, as shown in **Figure 3.7**.

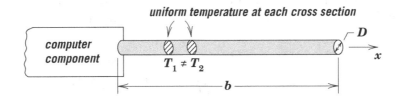

Figure 3.7 Fin temperature modeling as one-dimensional problem.

During operation, high-speed microprocessors generate a great amount of heat that may cause the microprocessor temperature to exceed the design limit and compromises the reliability of the computer as a result of overheating. One method to circumvent overheating problems is to design the microprocessor with fins mounted on its surface. The fins may have various cross sections; however, the fin studied in this section has a circular cross section. Thus, for a given fin geometry, the amount of heat transfer is found indirectly by describing the spatial distribution of temperature $T(x)$ along the fin. To arrive at $T(x)$, you will apply the following three laws:

(1) conservation of energy, given by the first law of thermodynamics; (2) Fourier's law of conduction; and (3) Newton's law of cooling.

Conservation of energy

Consider a fin as a rod whose diameter D is much smaller than its length b, as shown in **Figure 3.7**. The constraint $D << b$ allows you to model the temperature, $T(x)$ as one-dimensional by treating the temperature as uniform at a given cross section, but which varies along the rod in the x-direction. Furthermore, consider a steady-state case; that is, the computer is turned on for a long time, and its temperature is at a steady value. The analysis proceeds by considering a differential element of thickness Δx, as shown in **Figure 3.8a**. At the surfaces of this element there are two different modes of heat transfer. At the left surface at x, as shown in **Figure 3.8b**, heat is by virtue of molecular interaction, known as conduction heat transfer. The rate of this heat transfer is labeled $\dot{Q}_x$. Similarly, heat transfer is in the form of conduction at the right surface, $x + \Delta x$, labeled $\dot{Q}_{x+\Delta x}$. On the lateral surface of the differential element, the mode of heat transfer is by virtue of forcing air around the fin, which is defined as convection heat transfer, labeled $\dot{Q}_c$.

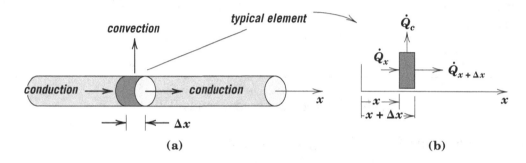

Figure 3.8 Fin heat transfer model: (a) differential element in the fin; and (b) energy balance across a differential element.

The first law of thermodynamics requires conservation of energy, so that the rate of energy entering the differential element is equal to the rate of energy leaving the element under steady-state conditions, that is,

$$\underbrace{\dot{Q}_x}_{\substack{\text{rate of energy} \\ \text{at the left face}}} = \underbrace{\dot{Q}_{x+\Delta x}}_{\substack{\text{rate of energy} \\ \text{at the right face}}} + \underbrace{\dot{Q}_c}_{\substack{\text{rate of energy at} \\ \text{the lateral surface}}} \tag{3.24}$$

Fourier's law of conduction

In order to introduce temperature into Eq. (3.24), Fourier's law of conduction is invoked, which states that the rate of heat conduction, $\dot{Q}_x$, is proportional to the product of the temperature gradient and the cross-sectional area of the rod, A, as shown in **Figure 3.9**. The temperature gradient at a local point x is the slope $\dfrac{dT}{dx}$. In other words, Fourier's law of conduction is expressed as

$$\dot{Q}_x = -kA\frac{dT}{dx} \tag{3.25}$$

where k is defined as the thermal conductivity coefficient, which depends on the rod material.

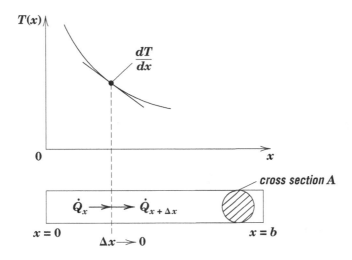

Figure 3.9 Temperature gradient or slope at a local point x.

In **Figure 3.9**, the temperature gradient as shown is negative. For that reason, a minus sign is introduced in Eq. (3.25) in order to define the heat transfer as a positive quantity.

Newton's law of cooling

Newton's law of cooling states that the rate of energy convected from the lateral surface of the element at temperature $T(x)$ to the surrounding air at temperature T_f, such that $T(x) > T_f$, is proportional to the difference $T(x) - T_f$; that is,

$$\dot{Q}_c = hA_l\left(T - T_f\right) \tag{3.26}$$

where h is defined as the average heat transfer coefficient. $A_l = P_e \Delta x$ is the lateral surface area exposed to the fluid, where P_e is the perimeter of the element, as shown in **Figure 3.8a**.

Substituting Eq. (3.26) in (3.24) yields $\dot{Q}_x = \dot{Q}_{x+\Delta x} + hP_e\Delta x(T - T_f)$ or

$-\dfrac{\left(\dot{Q}_{x+\Delta x} - \dot{Q}_x\right)}{\Delta x} = hP_e\left(T - T_f\right)$. Noting that $\displaystyle\lim_{\Delta x \to 0} \dfrac{\dot{Q}_{x+\Delta x} - \dot{Q}_x}{\Delta x} = \dfrac{d\dot{Q}_x}{dx}$, then

$$-\frac{d\dot{Q}_x}{dx} = hP_e\left(T - T_f\right) \tag{3.27}$$

Substitute Fourier's law of conduction Eq. (3.25) into Eq. (3.27),

$$-\frac{d}{dx}\left(-kA\frac{dT}{dx}\right) = hP_e\left(T - T_f\right) \tag{3.28}$$

For a constant cross-sectional area, divide Eq. (3.28) by A, and rearrange the terms,

$$\frac{d}{dx}\left(k\frac{dT}{dx}\right) - \left(\frac{hP_e}{A}\right)T = -\frac{hP_e}{A}T_f \tag{3.29}$$

Expand the left term in Eq. (3.29) using the product rule of derivatives,

$$k\frac{d^2T}{dx^2} + \frac{dk}{dx}\frac{dT}{dx} - \frac{hP_e}{A}T = -\frac{hP_e}{A}T_f$$

which can be put in the general form of Eq. (3.5) as

$$\frac{d^2T}{dx^2} + a_1(x)\frac{dT}{dx} + a_0(x)T = f(x) \tag{3.30}$$

where $a_1(x) = \frac{1}{k}\frac{dk}{dx}$, $a_0(x) = -\frac{hP_e}{kA}$, and $f(x) = -\frac{hP_e}{kA}T_f$.

In certain heat transfer applications, the medium is fabricated by varying the properties in space, for example by varying the conductivity k in the x-direction. In this case, the second term in Eq. (3.30) requires knowledge of $a_1(x)$ from $a_1(x) = \frac{1}{k}\frac{dk}{dx}$. However, for general-purpose heat transfer applications, the conductivity k is constant throughout the domain. In this case $a_1(x)$ varnishes because $\frac{dk}{dx} = 0$, and Eq. (3.30) reduces to

$$\frac{d^2T}{dx^2} + a_0T = f(x) \tag{3.31}$$

over the domain $0 < x < b$. Equation (3.30) or (3.31) is incomplete without the BCs.

Boundary conditions

In this particular application, the aim is to maintain the microprocessor at a design temperature, which is the same value at the fin base, $x = 0$. Thus, the BC at $x = 0$ is of Dirichlet type, that is, $T(0) = T_o$, where T_o is a specified design temperature.

Specified temperature BC at $x = b$: Dirichlet type

For a very long fin one may assume that the fin end at $x = b$ is in thermal equilibrium with the surrounding air, that is, $T(b) = T_f$, where T_f is the air temperature. This BC is of Dirichlet type.

Insulated BC at $x=b$: Neumann type

If heat transfer at the fin end, $x=b$, is negligible in comparison to the heat transfer from the lateral surface of the fin, then you can state that $\dot{Q}_{x=b} \approx 0$. Combining this approximation with Fourier's law of conduction, Eq. (3.25) yields $\dfrac{dT(b)}{dx}=0$. The BC also corresponds to an insulated surface known as an adiabatic surface. This BC is of Neumann type.

Convective BC at $x=b$: Robin type

If heat transfer is allowed to be transferred from the fin end at $x=b$ to the surroundings, as shown in **Figure 3.10**, then you can state that heat transfer by conduction at $x=b^{-}$ is equal to heat transfer by convection at $x=b^{+}$,

$$\dot{Q}_{conduction}\Big|_{x=b^{-}} = \dot{Q}_{convection}\Big|_{x=b^{+}} \tag{3.32}$$

where b^{-} is the limit $x \rightarrow b$ from the left, and b^{+} is the limit $x \rightarrow b$ from the right.

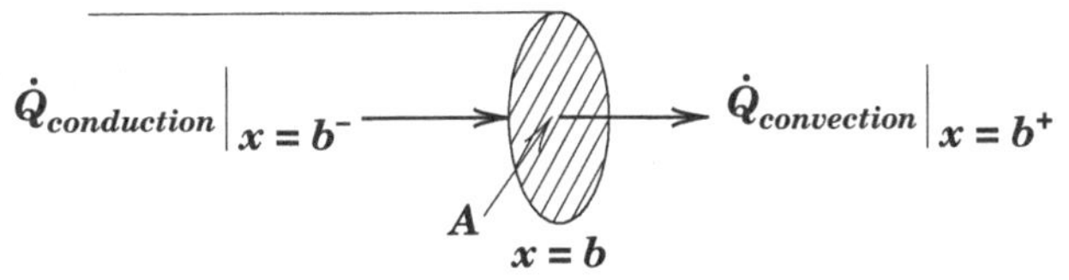

Figure 3.10 Balanced conduction and convection heat transfers at the end of the rod, $x=b$.

Writing $x=b$ for definiteness, and substituting Fourier's law of conduction and Newton's cooling law, Eqs. (3.25) and (3.26), respectively, into Eq. (3.32) yield

$$-kA\frac{dT}{dx}\Big|_{x=b} = hA(T-T_f)\Big|_{x=b} \tag{3.33}$$

Rearrange Eq. (3.33) to read

$$T(b)+\frac{k}{h}\frac{dT(b)}{dx}=T_f \tag{3.34}$$

which is a BC of the Robin type in the form of Eq. (3.7), $l_1 T(b)+l_2\dfrac{dT(b)}{dx}=r_b$, where $l_1=1$,

$l_2=\dfrac{k}{h}$ and $r_b=T_f$.

In summary, the steady-state, one-dimensional temperature distribution in a homogeneous-material fin is governed by Eq. (3.31), $\dfrac{d^2T}{dx^2} + a_0 T = f(x)$ over the interval $a < x < b$, with possible boundary conditions, as shown in **Table 3.3**. For case 1, the temperature is specified at both ends of the fin; for case 2, the temperature is specified on the left end, whereas the right end is insulated; and for case 3, the temperature is specified on the left end, whereas heat is allowed by convection on the right end.

Boundary conditions	Dirichlet		Neumann		Robin	
	$x = a$	$x = b$	$x = a$	$x = b$	$x = a$	$x = b$
Case 1	●	●				
Case 2	●			●		
Case 3	●					●

Table 3.3 Boundary conditions for the ODE $\dfrac{d^2T}{dx^2} + a_0 T = f(x)$ over the interval $a < x < b$.

Heat removal from a fin

Once you solve the fin BVP, Eq. (3.31) with the specified BCs, you can quantify the rate of heat transfer from the microprocessor as $\dot{Q}_{x=0} = -kA \dfrac{dT(0)}{dx}$, since all of the heat removed must be through the base of the fin, which is naturally by conduction.

Example 3.2 Computer Cooling Fin Device

Consider a microprocessor that operates at a steady design temperature of $80\,^oC$, which is maintained by an array of cylindrical fins mounted directly on the surface, as shown in **Figure 3.11**. Each fin has a length $b = 2.54cm$ and a diameter $D = 2mm$. The thermal conductivity of the fin is $k = 200\,\dfrac{W}{m\,^oC}$. The fins are cooled by forcing air through them. The forced air has an average heat transfer coefficient $h = 200\,\dfrac{W}{m^2\,^oC}$ and a temperature $T_f = 25\,^oC$.

1. Describe the temperature distribution along the fin. Assume the fin to be very long.
2. Determine the rate of heat transfer per fin removed from the microprocessor in *Watt* (*W*).

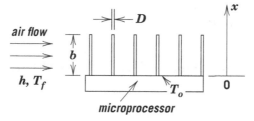

Figure 3.11 Array of cylindrical fins mounted on the surface of a microprocessor.

Solution:

1. For a fin with uniform conductivity, k, and cross-sectional area, A, the temperature distribution, $T(x)$, over the domain $0 < x < b$ is given by Eq. (3.31),

$$\frac{d^2T}{dx^2} + a_0 T = f(x) \tag{3.35}$$

where $a_0 = -\dfrac{hP_e}{kA}$ and $f(x) = -\dfrac{hP_e}{kA}T_f$. The BC at $x = 0$ is $T(0) = T_o = 80\,^{\circ}C$. Given that the fin is very long, then the BC at $x = b$ is $T(b) = T_f = 25\,^{\circ}C$. The general solution of Eq. (3.35) is

$$T(x) = T_h(x) + T_p(x) \tag{3.36}$$

where the homogeneous solution satisfies

$$\frac{d^2T_h}{dx^2} + a_0 T_h = 0 \tag{3.37}$$

and the particular solution satisfies

$$\frac{d^2T_p}{dx^2} + a_0 T_p = f(x) \tag{3.38}$$

Using $P_e = \pi D$ and $A = \dfrac{\pi D^2}{4}$, then $a_o = -4\dfrac{h}{kD}$ and $f = -4\dfrac{hT_f}{kD}$. Using the given data yields $a_0 = -2000$ and $f(x) = -5(10^4)$.

The homogeneous solution is generally given as $y_h(x) = e^{\lambda x}$. Substituting this solution in Eq. (3.37) yields $\lambda^2 - 2000 = 0$ or $\lambda_{1,2} = \pm 20\sqrt{5}$. Thus,

$$T_h(x) = C_1 e^{20\sqrt{5}x} + C_2 e^{-20\sqrt{5}x} \tag{3.39}$$

For constant $f(x)$, the particular solution is given as $y_p(x) = A$, based on the undetermined coefficients method. Substituting this solution in Eq. (3.38) yields

$$T_p(x) = 25 \tag{3.40}$$

3-20

Thus, the general solution is constructed by substituting Eqs. (3.39) and (3.40) into Eq. (3.36),

$$T(x) = C_1 e^{20\sqrt{5}x} + C_2 e^{-20\sqrt{5}x} + 25 \tag{3.41}$$

Next, apply the BCs. Substituting $T(0) = 80$ and $T(0.0254) = 25$ into Eq. (3.41) yields

$$55 = C_1 + C_2$$
$$0 = 3.1140 C_1 + 0.32113 C_2$$

Solving for C_1 and C_2 yields $C_1 = -6.2338$ and $C_2 = 61.324$. Thus, the solution that satisfies the ODE and the BCs is

$$T(x) = -6.2338 e^{20\sqrt{5}x} + 61.324 e^{-20\sqrt{5}x} + 25 \tag{3.42}$$

over the interval $0 \le x \le 0.0254 m$. The temperature distribution is shown in **Figure 3.12**.

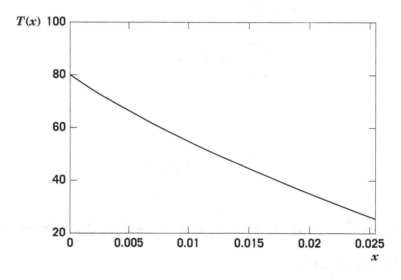

Figure 3.12 The temperature distribution along the fin.

2. The rate of heat transfer from the microprocessor per fin is given by $\dot{Q}_{x=0} = -kA \dfrac{dT(0)}{dx}$.

From Eq. (3.41), $\dfrac{dT}{dx} = 20\sqrt{5}\left(C_1 e^{20\sqrt{5}x} - C_2 e^{-20\sqrt{5}x}\right)$ so that $\dfrac{dT(0)}{dx} = 20\sqrt{5}\left(C_1 - C_2\right) = -3021.27$.
Thus,

$$\dot{Q}_{x=0} = -\left(200 \frac{W}{m\,^oC}\right)\frac{\pi}{4}(0.002m)^2\left(-3021.27\frac{^oC}{m}\right) = 1.9W$$

which is the rate of heat transfer from a single fin.

In this engineering application, a cooling fin is modeled as a second-order nonhomogeneous BVP with Dirichlet BCs. The dependent variable is temperature T, and the independent variable is x.

In a homework problem you may revisit this problem by treating the fin tip at $x = b$ as adiabatic and applying the Neumann BC in order to compare the rate of heat transfer from the fin with the value obtained in this example. In another homework problem you may apply a more realistic BC, namely the Robin BC at the fin tip.

Next you will learn how to find the solution to a homogeneous BVP.

3.5 Homogeneous BVP: The Eigenvalue Problem

The BVP is classified as homogeneous when $f(x) = 0$, $r_a = 0$ and $r_b = 0$ in Eqs. (3.5)–(3.7), that is,

$$\frac{d^2 y}{dx^2} + a_1 \frac{dy}{dx} + a_0 y = 0 \tag{3.43}$$

over $a < x < b$, and the BCs

$$k_1 y(a) + k_2\, y'(a) = 0 \tag{3.44}$$

and

$$l_1 y(b) + l_2\, y'(b) = 0 \tag{3.45}$$

The homogeneous ODE, Eq. (3.43), will have a general solution consisting of two linearly independent solutions, $y_{h_1}(x)$ and $y_{h_2}(x)$; that is,

$$y_h(x) = C_1 y_{h_1}(x) + C_2 y_{h_2}(x) \tag{3.46}$$

where the general constants C_1 and C_2 are fixed by the homogeneous BCs, Eqs. (3.44)–(3.45). A summary of the homogeneous BVP is shown in **Table 3.4**.

Homogeneous ODE	Homogeneous boundary conditions	Solution
$\dfrac{d^2 y}{dx^2} + a_1 \dfrac{dy}{dx} + a_0 y = 0$	$k_1 y(a) + k_2\, y'(a) = 0$ $l_1 y(b) + l_2\, y'(b) = 0$	$y_h(x) = C_1 y_{h_1}(x) + C_2 y_{h_2}(x)$

Table 3.4 The homogeneous BVP.

The solution $y_h(x)$ will either degenerate to the trivial solution $y(x) = 0$ or will have a nonunique solution by possessing an infinite number of solutions. The nonunique solution is of interest because it yields important physical information about the system under consideration. The class of homogeneous BVPs with infinitely nontrivial solutions is known as an eigenvalue problem, where the coefficient a_0 is a free parameter, as you will learn next.

Column buckling

A practical problem that is classified as an eigenvalue problem is related to the design of columns that are important structural elements to support bridges, and very large ceilings, just to name two applications. If you consider a column of length L, as shown in **Figure 3.13a**, it is possible that the column bends, as in **Figure 3.13b**, or breaks, as in **Figure 3.13c**, under a load, F_L, depending on its material and geometry. The result is catastrophic for bridges or large-scale structures. Thus, for various designs and conditions, it is important for engineers to predict the critical load, F_L, that leads to buckling.

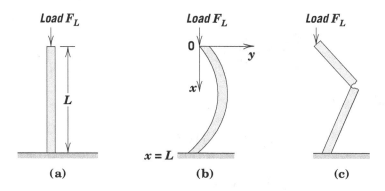

Figure 3.13 Column design: (a) desirable design; (b) failure by bending; and (c) failure by breaking.

At the onset of buckling, as in **Figure 3.13b**, the local deflection y is a function of column depth x, and governed by the second-order linear ODE

$$\frac{d^2 y}{dx^2} + a_0 y = 0 \tag{3.47}$$

where $a_0 = \dfrac{F_L}{E_m I}$; E_m is a material property, known as the modulus of elasticity, which has units of $\dfrac{N}{m^2}$; and I is a geometric factor in m^4, known as the cross-sectional moment of inertia. Note that $a_0 > 0$, so that the general solution of Eq. (3.47) consists of sine and cosine functions. The BCs for this ODE are of Dirichlet type at $x = 0$ and $x = L$; that is,

$$y(0) = 0 \tag{3.48}$$

and

$$y(L) = 0 \tag{3.49}$$

which correspond to a column pinned at $x = 0$ and $x = L$. As you can see from Eqs. (3.47)–(3.49), the ODE and BCs constitute a homogeneous BVP. Note that the load F_L necessary to cause bucking is unknown, which is equivalent to treating the coefficient a_0 in Eq. (3.47) as a parameter. Using the general homogeneous solution $y_h = e^{\lambda x}$, you can show that for $a_0 > 0$,

$$y_h(x) = C_1 \sin\left(\sqrt{a_0} x\right) + C_2 \cos\left(\sqrt{a_0} x\right) \tag{3.50}$$

Applying the BC Eq. (3.48) to Eq. (3.50) yields $C_2 = 0$. Thus,

$$y_h(x) = C_1 \sin\left(\sqrt{a_0} x\right) \tag{3.51}$$

Applying the BC Eq. (3.49) to Eq. (3.51) yields

$$C_1 \sin\left(\sqrt{a_0} L\right) = 0 \tag{3.52}$$

One of the choices to satisfy Eq. (3.52) is $C_1 = 0$. However, this leads to the trivial solution $y_h = 0$, which is not the case when the column buckles. Hence, Eq. (3.52) is satisfied by setting

$$\sin\left(\sqrt{a_0} L\right) = 0 \tag{3.53}$$

which will lead to specific or proper values of a_0. The argument $\sqrt{a_0} L$ that satisfies Eq. (3.53) must be an integer multiple of π, that is, $\sqrt{a_0} L = n\pi$, or

$$a_0 = \left(\frac{n\pi}{L}\right)^2 \qquad n = 1, 2, 3, \cdots \tag{3.54}$$

$n = 0$ is excluded because it leads to the trivial solution $y(x) = 0$. For each n, there corresponds a value of a_0 and a homogeneous solution y_h with a general constant. For example, for $n = 1$, $y_{h_1}(x) = C_{1,1} \sin\left(\frac{\pi}{L} x\right)$, which describes the beam shape, as shown in **Figure 3.14a**. For $n = 2$

and $n = 3$ the beam shapes are given by $y_{h_2}(x) = C_{1,2} \sin\left(\dfrac{2\pi}{L} x\right)$ and $y_{h_3}(x) = C_{1,3} \sin\left(\dfrac{3\pi}{L} x\right)$, respectively, and so on and so forth, as shown in **Figure 3.14b** and **Figure 3.14c**.

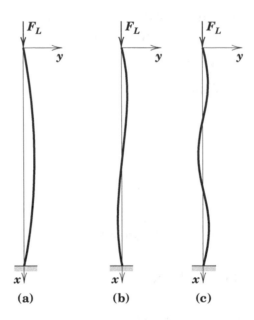

Figure 3.14 Buckling shapes for: (a) $n = 1$; (b) $n = 2$; and (c) $n = 3$.

Resetting the general constant as C_n for simplicity, the homogeneous BVP possesses an infinite set of nonunique solutions, whose nth element is given by

$$y_{h_n} = C_n \sin\left(\frac{n\pi x}{L}\right); \qquad n = 1, 2, 3, \cdots$$

Even though you could not calculate C_n, the information about a_0 is valuable by setting the critical load $a_0 = \dfrac{F_L}{E_m I}$ equal to $a_0 = \left(\dfrac{n\pi}{L}\right)^2$ so that $F_L = \dfrac{E_m I \pi^2 n^2}{L^2}$. The smallest critical load that will cause a column to buckle corresponds to $n = 1$; thus, $F_L = \dfrac{E_m I \pi^2}{L^2}$, which is known as the Euler load.

The column buckling model is not limited to bridges and structures only, but can extend to other applications such as screw jacks and cell phone towers, as demonstrated in the homework problems, and femur leg bones, as explained in the next example, to name just a few cases.

Example 3.3 Buckling of Femur Leg Bone

A bioengineer wants to determine the maximum buckling load that a femur leg bone can sustain, as shown in **Figure 3.15a**. As a first step in the analysis, the engineer models the bone as a slender cylinder pinned at the ends, that is, pinned at the hip joint at the top end and the knee cap at the bottom end as shown in **Figure 3.15b**.

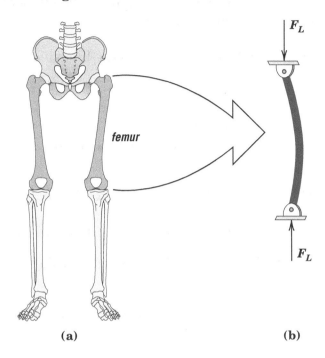

(a)   (b)

Figure 3.15 Femur bone model: (a) leg joint assembly; and (b) cylindrical column model.

The slender cylinder is $0.45\ m$ in length and has a moment of inertia $I = 3\left(10^{-8}\right)m^4$. The femur bone modulus of elasticity is $E_m = 15\left(10^9\right)Pa$, where Pa is N/m^2.

1. Use the governing ODE and boundary conditions for the model given by Eqs. (3.47)–(3.49), and determine the maximum buckling load that the femur can sustain.

Solution:

The set of Eqs. (3.47)–(3.49) requires $a_0 = \dfrac{F_L}{E_m I} = \left(\dfrac{n\pi}{L}\right)^2$ for nontrivial solutions. Therefore, the smallest critical load that will cause a femur to buckle corresponds to $n = 1$, so that $F_L = \dfrac{E_m I \pi^2}{L^2}$.

Substituting the given values yields the maximum buckling load that the femur can sustain, $F_L = 21932N$.

The condition given by Eq. (3.53) is known as the eigencondition, and the values of a_0 given by Eq. (3.54) are known as the eigenvalues. The word "eigen" is German for "own" or "proper." However, in the mathematical context, the eigenvalues are the proper values that a_0 can take in order to satisfy a homogeneous BVP with nontrivial solutions. The foregoing eigencondition and eigenvalues are derived from the given BCs, in this case Dirichlet type. Other types of homogeneous BCs, for example, Neumann or Robin type, yield different eigenconditions and eigenvalues, as demonstrated in the homework problems.

The eigenvalue BVP is generalized by the so-called Sturm–Liouville problem, which is formulated as

$$\frac{d\left(p(x)\dfrac{dy}{dx}\right)}{dx}+q(x)y+\lambda r(x)y=0 \tag{3.55}$$

along with the generalized BCs, Eqs. (3.44) and (3.45), over the interval $a<x<b$, where $p(x)$, $\dfrac{dp}{dx}$, $q(x)$, and $r(x)$ are continuous over $a\le x\le b$, and $p(x)>0$ and $r(x)>0$ over $a\le x\le b$. Here, the eigenvalues are set by the parameter λ. Expand the leading term in Eq. (3.55),

$$p(x)\frac{d^2 y}{dx^2}+\frac{dp}{dx}\frac{dy}{dx}+q(x)y+\lambda r(x)y=0$$

and divide by $p(x)$ so that the Sturm–Liouville ODE is cast as

$$\frac{d^2 y}{dx^2}+\frac{1}{p(x)}\frac{dp}{dx}\frac{dy}{dx}+\frac{q(x)}{p(x)}y+\lambda\frac{r(x)}{p(x)}y=0 \tag{3.56}$$

A comparison of Eq. (3.56) with Eq. (3.43), such that $q(x)$ and $r(x)$ are linearly independent, yields $a_1=\dfrac{1}{p(x)}\dfrac{dp}{dx}$, $\dfrac{q(x)}{p(x)}=0$, $\lambda=a_0$, and $\dfrac{r(x)}{p(x)}=1$.

Example 3.4

Consider the column buckling problem given by Eqs. (3.47) and BC Eqs. (3.48) and (3.49) as a Sturm–Liouville problem.

1. Find $p(x)$, $q(x)$, $r(x)$ and λ.

Solution:

Comparing Eq. (3.47) with Eq. (3.56) yields $a_1 = \dfrac{1}{p(x)}\dfrac{dp}{dx} = 0$, or $\dfrac{dp}{dx} = 0$, so that $p = C$, and

$\dfrac{q(x)}{p(x)} = 0$, $\lambda = a_0 = \dfrac{F_L}{E_m I}$, and $\dfrac{r(x)}{p(x)} = 1$, so that $q(x) = 0$ and $r(x) = C$. **Table 3.5** is a summary

of the Sturm–Liouville coefficients.

Sturm–Liouville formulation	Column buckling ODE
$\dfrac{d^2 y}{dx^2} + \dfrac{1}{p(x)}\dfrac{dp}{dx}\dfrac{dy}{dx} + \dfrac{q(x)}{p(x)}y + \lambda \dfrac{r(x)}{p(x)}y = 0$	$\dfrac{d^2 y}{dx^2} + a_0 y = 0$; $a_0 = \dfrac{F_L}{E_m I}$
$p(x) > 0$; $r(x) > 0$	$p(x) = r(x) = C$; $q(x) = 0$; $\lambda = a_0 = \dfrac{F_L}{E_m I}$

Table 3.5 Comparison of the column buckling ODE, Eq. (3.47), with the Sturm–Liouville formulation, Eq. (3.56).

HOMEWORK PROBLEMS FOR CHAPTER THREE

Problems for Sections 3.2–3.3

For Problems 1–3, classify the BCs as Dirichlet, Neumann, or Robin and find the $y(x)$ over $0 < x < 1$.

1. ODE: $\dfrac{d^2y}{dx^2} = 1$; $y(0) = 0$ and $y(1) = 4$.

2. ODE: $\dfrac{d^2y}{dx^2} = e^x$; $y(0) = 5$ and $y'(1) = 2$.

3. ODE: $5\dfrac{d^2y}{dx^2} = 15x^3 + 10x^2 + 5x$; $y'(0) = 0$ and $y(0) - y'(1) = 0$.

4. In the plastic industry, tapes are coated with chemicals by pulling the tape of width w, and length L horizontally through a lubricant bath, as shown in **Figure 3.16a**. As the tape is pulled with force, F_{pull} at speed v_{tape} through the lubricant with viscosity μ, it is resisted by two forces on top and bottom of its surface. In addition, the fluid in the gap of height h moves with different speeds, as shown in **Figure 3.16b**. At $y = 0$, the speed is zero, $v(0) = 0$, and at $y = h$, the speed is $v(h) = v_{tape}$.

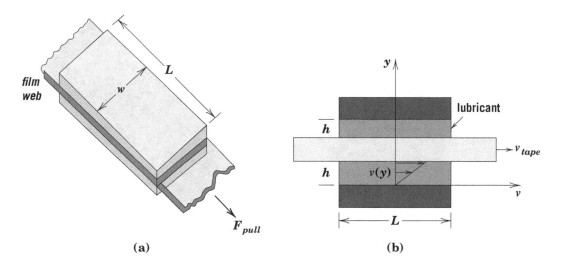

(a)　　　　　　　　　　　　(b)

Figure 3.16 Coated tape speed modeled horizontally as a BVP: (a) actual system; and (b) linear velocity model in each lubricant gap of height h.

The force exerted by the lubricant on the tape is given by Newton's law of viscosity as

$\mu A \dfrac{dv(h)}{dy}$, where $A = wL$, and $\dfrac{dv(h)}{dy}$ is the slope at $y = h$. Thus, $F_{pull} = 2\mu A \dfrac{dv(h)}{dy}$.

The fluid motion in the gap is governed by $\dfrac{d^2 v}{dy^2} = 0$, derived from the equation of motion in fluid mechanics.

a. Find $v(y)$ by solving the ODE $\dfrac{d^2 v}{dy^2} = 0$ over the domain $0 < y < h$, which satisfies the BCs

$v(0) = 0$ and $v(h) = v_{tape}$.

b. From your solution find $F_{pull} = 2\mu A \dfrac{dv(h)}{dy}$ using $h = 0.01m$, $w = 1m$, $L = 10m$ $v_{tape} = 0.1\dfrac{m}{s}$,

and $\mu = 0.2 \dfrac{kg}{m.s}$.

5. What if the coating process in Problem 4 is carried out vertically, as shown in **Figure 3.17**. In this case, the fluid motion in the gap is governed by $\dfrac{d^2 v}{dx^2} = \dfrac{\rho g}{\mu}$, derived from the equation of motion in fluid mechanics, where ρ is the lubricant density, and g is the gravitational constant.

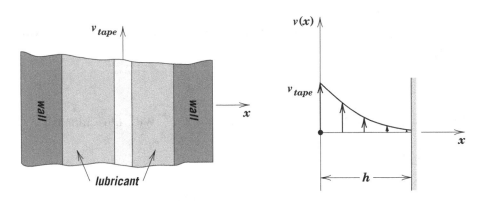

Figure 3.17 Coated tape speed modeled vertically as a BVP.

a. Find $v(x)$ by solving the ODE $\dfrac{d^2 v}{dx^2} = \dfrac{\rho g}{\mu}$ over the domain $0 < x < h$, which satisfies the BCs

$v(0) = v_{tape}$ and $v(h) = 0$.

b. From your solution find $F_{pull} = 2\mu A \dfrac{dv(h)}{dx}$ using $h = 0.01m$, $w = 1m$, $L = 10m$ $v_{tape} = 0.1\dfrac{m}{s}$, $\mu = 0.2\dfrac{kg}{m.s}$, $\rho = 800\dfrac{kg}{m^3}$, and $g = 9.81\dfrac{m}{s^2}$.

6. Consider a cable fastened between two towers, $200m$ apart. The endpoints are at the same height, as shown in **Figure 3.5**. The cable supports a structure that has a uniform weight per unit length w in $\dfrac{N}{m}$. As an engineer, you are charged with determining the maximum w so that the tension force at the bottom of the cable does not exceed $2(10^6)N$ while limiting the sag to 5% of the span.

a. Determine the uniform weight per unit length in $\dfrac{N}{m}$.

b. Determine the tension force T at the fastened points.

c. Determine the length of the cable.

7. Revisit Example 3.1.

a. Solve the ODE using symmetry over the domain $0 < x < \dfrac{L}{2}$ with Neumann BC, $\dfrac{dy(0)}{dx} = 0$ and Dirichlet BC $y\left(\dfrac{L}{2}\right) = h$.

b. Find the necessary expression for T_o by satisfying $y(0) = 0$. Is the expression different from that found in Example 3.1?

8. Revisit Example 3.1.

a. Solve the ODE using symmetry over the domain $0 < x < \dfrac{L}{2}$ with Dirichlet BC, $y(0) = 0$ and Dirichlet BC $y\left(\dfrac{L}{2}\right) = h$.

b. Find the necessary expression for T_o by satisfying $\dfrac{dy(0)}{dx} = 0$. Is the expression different from that found in Example 3.1?

9. Revisit Example 3.1.

a. Solve the ODE as an initial value problem, IVP, using the ICs $y(0) = 0$ and $\dfrac{dy(0)}{dx} = 0$, as required at the origin in **Figure 3.5**.

b. Find the necessary expression for T_o by satisfying $y\left(\dfrac{L}{2}\right)=h$. Is the expression different from that found in Example 3.1?

10. Consider a high-voltage cable that spans between two towers, $50m$ apart. The endpoints are at the same height, as shown in **Figure 3.5**. However, the wire is covered with snow and ice. Based on observations, the load per unit length is given empirically as $w(x)=ax^2$ in N/m, where x is in m, and $a=0.1$ with appropriate units. The sag h is limited to 1% of the span.

a. Use the parabolic model to find the general solution of the cable displacement by direct integration of Eq. (3.15), $\dfrac{d^2y}{dx^2}=\dfrac{w(x)}{T_o}$ over the domain $-\dfrac{L}{2}<x<\dfrac{L}{2}$.

b. Fix the constants of integration by applying the BCs $y\left(\dfrac{L}{2}\right)=y\left(-\dfrac{L}{2}\right)=h$.

c. Determine the tension at the bottom of the cable, T_o.
d. Use Eq. (3.13) to determine the angle θ at the fastened ends.
e. Use Eq. (3.10) to determine the tension force T at the fastened points.
f. Determine the length of the cable.

11. Consider a high-voltage wire that spans $30m$ between two buildings, The endpoints are at the same height h, as shown in **Figure 3.18**. However, the wire is covered with snow and ice. Due to wind, the snow and ice load causes asymmetry in the cable shape, that is, $x_r\neq\dfrac{L}{2}$ where x_r is the right end distance from the origin. The left end is at x_r-L from the origin

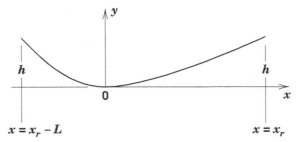

Figure 3.18 High-voltage wire modeled as a BVP.

Although there is a tendency to apply the BCs $y(x_r)=h$ and $y(x_r-L)=h$, x_r is now unknown. What remains true, however, is that $y(0)=0$ and $\dfrac{dy(0)}{dx}=0$ at the origin. Based on observations, the load per unit length is given empirically as $w(x)=ax^2+bx+c$ in N/m, where x is in m and $a=10^{-3}$, $b=-0.05$, and $c=0.525$ with appropriate units. The sag h is limited to 5% of the span.

a. Use the parabolic model to find the general solution of the cable displacement by direct integration of Eq. (3.15), $\dfrac{d^2 y}{dx^2} = \dfrac{w(x)}{T_o}$ over the domain $x_r - L < x < x_r$.

b. Fix the constants of integration using $y(0) = 0$ and $\dfrac{dy(0)}{dx} = 0$, that is, treat the problem as an IVP.

c. Apply the BCs at x_r and $x_r - L$, which result in two nonlinear equations in x_r and T_o as unknown. Use CAS to find x_r and T_o.

d. Use Eq. (3.13) to determine the angle θ at the fastened ends.

e. Use Eq. (3.10) to determine the tension force T at the fastened points.

f. Determine the length of the cable using $L_{cable} = \int_{x_r - L}^{x_r} \sqrt{1 + y'^2}\, dx$.

12. Consider the second-order linear BVP $r\dfrac{d^2 V}{dr^2} + \dfrac{dV}{dr} = 0$ over the domain $a < r < b$, with the following BCs: $V = V_a$ at $r = a$ and $V = V_b = 0$ at $r = b$, which governs the coaxial cable voltage $V(r)$, as shown in **Figure 3.3**.

a. Let $z = \dfrac{dV}{dr}$ and rewrite the foregoing ODE as $r\dfrac{dz}{dr} + z = 0$, which is a first-order ODE. Solve the ODE and set $z(r) = F(r)$ where $F(r)$ is the solution of the first-order ODE.

b. Write $z(r) = F(r)$ as $\dfrac{dV}{dr} = F(r)$ and integrate this ODE to find $V(r)$, which should have two constants of integration.

c. Fix the constants of integration by applying the Dirichlet BCs given above.

d. From your last result find an expression for $r\dfrac{dV}{dr}$ and substitute it into the capacitance expression $C = -\left(\dfrac{2\pi\varepsilon L}{V_a}\right) r\dfrac{dV}{dr}$, which should be a function of ε, L, radii a and b.

e. Find the ratio of the radii, $\dfrac{b}{a}$, necessary for the coaxial geometry for $L = 1m$, $\varepsilon = 20\left(10^{-12}\right)\dfrac{farad}{m}$ and $C = 75\left(10^{-12}\right) farad$.

13. Consider the flow of fluid in a pipe, as shown in **Figure 3.4b**, which has a speed distribution, $v(r)$ obtained by solving the second-order BVP $\dfrac{d^2 v}{dr^2} + \dfrac{1}{r}\dfrac{dv}{dr} = -\dfrac{1}{\mu}\dfrac{\Delta P}{L}$, over the domain $0 < r < R$, with BCs $\dfrac{dv(0)}{dr} = 0$ and $v(R) = 0$.

a. Let $z = \dfrac{dv}{dr}$ and rewrite the foregoing ODE as $\dfrac{dz}{dr} + \dfrac{1}{r} z = a$, which is a first-order ODE,

   where $a = -\dfrac{1}{\mu}\dfrac{\Delta P}{L}$ is a constant. Solve the first-order ODE and set $z(r) = F(r)$ where $F(r)$ is the solution.

b. Write $z(r) = F(r)$ as $\dfrac{dv}{dr} = F(r)$ and integrate the ODE to find $v(r)$, which should have two constants of integration.

c. Fix the constants of integration by applying the Dirichlet and Neumann BCs given above.

d. From the solution, find expressions for the average speed in the pipe using

   $$\bar{v} = \dfrac{2}{R^2} \int_0^R v(r)\, r\, dr \text{ and the volumetric flow rate } \dot{V} = A\bar{v}.$$

e. If this case is blood flow in an artery, what is the percent reduction in the volumetric flow rate if the artery radius is reduced by 10% due to plaque?

14. Consider the catenary cable as shown in **Figure 3.6b**, where w is the weight per unit cable length and T_o is the tension at $x = 0$ and $y = 0$.

a. Show that the hyperbolic cosine equation, $y(x) = \dfrac{1}{f}\cosh(xf) - \dfrac{1}{f}\cosh\left(\dfrac{fL}{2}\right) + h$ of a

   catenary cable satisfies the ODE, Eq. (3.4) and the BCs, Eqs. (3.2) and (3.3), over the domain

   $-\dfrac{L}{2} \le x \le \dfrac{L}{2}$, where $f = \dfrac{w}{T_o}$ is a constant.

b. Find an expression for $y(0)$. Use CAS to find the necessary value of f and T_o so that

   $y(0) = 0$ for $L = 250m$, $w = 100\dfrac{N}{m}$ and $\dfrac{h}{L} = 25\%$.

c. What is T_o for $w = 100\dfrac{N}{m}$ and $\dfrac{h}{L} = 25\%$ if you were to model the catenary cable using a

   parabolic model, for example, $y(x) = \dfrac{w}{2T_o}\left(x^2 - \dfrac{L^2}{4}\right) + h$? Report the percent difference from

   that found using the hyperbolic cosine equation.

d. Explore in CAS h/L in percent, so that the difference in tension between the parabolic and the hyperbolic cosine models is no more than 1%.

15. Consider the nonlinear ODE of the catenary cable given by Eq. (3.23),

$$\left(\dfrac{d^2 y}{dx^2}\right)^2 - f^2\left(\dfrac{dy}{dx}\right)^2 = f^2$$

over the domain $0 < x < \dfrac{L}{2}$ with BCs $\dfrac{dy(0)}{dx} = 0$ and $y\left(\dfrac{L}{2}\right) = h$, where $f = \dfrac{w}{T_o}$.

Let $z = \dfrac{dy}{dx}$ and rewrite the foregoing ODE as $\left(\dfrac{dz}{dx}\right)^2 - f^2 z^2 = f^2$, or

$\left(\dfrac{dz}{dx}\right)^2 = f^2\left(1 + z^2\right)$, which yields $\dfrac{dz}{dx} = \pm f\sqrt{\left(1 + z^2\right)}$. Choose the positive root, $\dfrac{dz}{dx} = f\sqrt{\left(1 + z^2\right)}$,

because $\dfrac{dz}{dx} = \dfrac{d^2 y}{dx^2}$ is concave up by virtue of cable geometry.

a. Use CAS to solve the foregoing equation by separation of variables, that is, $\dfrac{dz}{\sqrt{1 + z^2}} = f dx$,

which yields $z(x) = F(x)$ where $F(x)$ is the solution of the ODE. Your answer should be a hyperbolic sine function.

b. Apply the BC $\dfrac{dy(0)}{dx} = 0$ to $z(x) = F(x)$ in order to fix the constant of integration. Note that

$\dfrac{dy(0)}{dx}$ is equivalent to $z(0)$.

c. Next, write $z(x) = F(x)$ as $\dfrac{dy}{dx} = F(x)$ and integrate this first-order ODE to find the solution

$y(x)$. Apply the BC $y\left(\dfrac{L}{2}\right) = h$ in order to fix the constant of integration.

d. State the nonlinear relationship that governs f in order to satisfy $y(0) = 0$.

16. Revisit the ODE $\left(\dfrac{dz}{dx}\right)^2 - f^2 z^2 = f^2$ in Problem 15. Solve the ODE as an initial value

problem (IVP) using the ICs $y(0) = 0$ and $\dfrac{dy(0)}{dx} = 0$ required at the bottom of the cable at $x = 0$,

as follows.

a. Use CAS to solve $\dfrac{dz}{\sqrt{1 + z^2}} = f dx$, which yields $z(x) = F(x)$ where $F(x)$ is the solution of

the ODE. Apply the IC $z(0) = \dfrac{dy(0)}{dx} = 0$ to $z(x) = F(x)$ in order to fix the constant of

integration.

b. Next, write $z(x) = F(x)$ as $\dfrac{dy}{dx} = F(x)$ and integrate this first-order ODE to find the solution

$y(x)$. Apply the IC $y(0) = 0$ in order to fix the constant of integration.

c. State the nonlinear relationship that governs f in order to satisfy $y\left(\dfrac{L}{2}\right) = h$.

Problems for Section 3.4

17. Revisit Example 3.2. Assume that the heat transfer at the lateral surface is more dominant than that at the tip so that the heat transfer at $x = b$ is modeled by the Neumann BC $\dfrac{dT(b)}{dx} = 0$.

a. Describe the temperature distribution along the fin by solving the ODE $\dfrac{d^2T}{dx^2} + a_0 T = f(x)$

over the domain $0 < x < b$ subject to the BCs $T(0) = 80\,°C$ and $\dfrac{dT(b)}{dx} = 0$.

b. Determine the rate of heat transfer per fin removed from the microprocessor in *Watt (W)*.

18. Revisit Example 3.2. Assume that the heat transfers at the lateral surface and tip are equally important so that the heat transfer at $x = b$ is modeled by the Robin BC $T(b) + \dfrac{k}{h}\dfrac{dT(b)}{dx} = T_f$.

a. Describe the temperature distribution along the fin by solving the ODE $\dfrac{d^2T}{dx^2} + a_0 T = f(x)$

over the domain $0 < x < b$ subject to the BCs $T(0) = 80\,°C$ and $T(b) + \dfrac{k}{h}\dfrac{dT(b)}{dx} = T_f$.

b. Determine the rate of heat transfer per fin removed from the microprocessor in *Watt (W)*.

19. Consider a microprocessor that operates at a steady design temperature of $80\,°C$, which is maintained by an array of cylindrical fins mounted directly on the surface. Each fin has a length $b = 0.0254m$ and a diameter $D = 2\left(10^{-3}\right)m$. The fin thermal conductivity is

$k = 200\dfrac{W}{m\,°C}$. The fins are cooled by forcing air through them with an average heat transfer

coefficient $h = 200\dfrac{W}{m^2\,°C}$. However, the fluid temperature varies in the x-direction as

$T_f = 4\left(10^4\right)x^2 - 2000x + 50$ in $°C$ over $0 < x < b$.

a. Plot T_f over $0 < x < b$ and report $T_f(0)$ and $T_f(b)$ in $°C$.

b. Find an expression for $f(x)$ using its definition $f(x) = -\dfrac{4h}{kD}T_f(x)$.

c. Describe the temperature distribution along the fin by solving the ODE $\dfrac{d^2T}{dx^2} + a_0 T = f(x)$

over the domain $0 < x < b$ subject to the BCs $T(0) = 80\ ^\circ C$ and $\dfrac{dT(b)}{dx} = 0$.

d. Determine the rate of heat transfer per fin removed from the microprocessor in *Watt (W)*.

Problems for Section 3.5

20. An engineer is charged with the design of an aluminum screw jack, as shown in **Figure 3.19a**. The jack is modeled as a slender column, as shown in **Figure 3.19b**, so that the tip deflection is δ at the onset of buckling under a minimum critical load. Thus, as part of the design, the engineer needs to determine the screw length, L, in m so that the jack does not buckle under the action of a critical load $F_L = 2(10^4)N$ when fully extended. The jack moment of inertia and modulus of elasticity are $I = 3(10^{-8})m^4$ and $E_m = 70(10^9)Pa$, respectively.

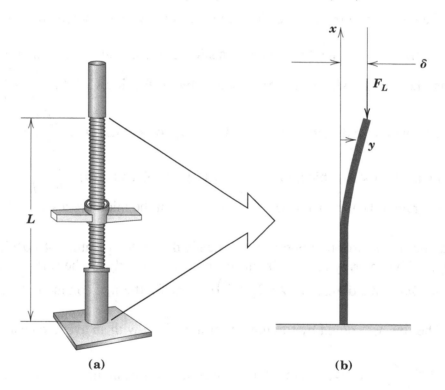

(a)   (b)

Figure 3.19 Screw jack: (a) basic geometry; and (b) column model with Neumann BC at the bottom, $\dfrac{dy(0)}{dx} = 0,$ and Dirichlet BC at the top, $y(L) = \delta$.

At the onset of buckling, the local deflection, y, as a function of depth, x, is governed by the nonhomogeneous ODE $\dfrac{d^2y}{dx^2} + a_0 y = \dfrac{F_L \delta}{E_m I}$, where $a_o = \dfrac{F_L}{E_m I}$. The BCs for this ODE are of the

Neumann type at $x = 0$ and Dirichlet type at $x = L$, that is, $\dfrac{dy(0)}{dx} = 0$ and $y(L) = \delta$, respectively.

a.   Define the transformation variable $z(x) = y(x) - \delta$ and transform the nonhomogeneous BVP in $y(x)$ to a homogeneous BVP in $z(x)$. State the necessary BCs for $z(x)$.

b.   Find the solution of the ODE in $z(x)$, eigencondition and eigenvalues for nontrivial solutions.

c.   Use the eigenvalues to determine the screw length that satisfies the design data.

21.   An engineer needs to determine if a 60-m aluminum cell phone tower will buckle when loaded with heavy equipment at the top, as shown in **Figure 3.20a**. The tower is fixed to the ground at the bottom end and free to move at the top end with a deflection δ, and is modeled as a column, as shown in **Figure 3.20b**. The tower column moment of inertia and modulus of elasticity are $I = 832.5\left(10^{-6}\right) m^4$ and $E_m = 70\left(10^9\right) Pa$, respectively.

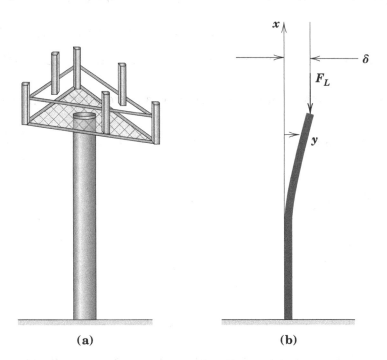

(a)   (b)

Figure 3.20 Cell phone tower: (a) basic geometry; and (b) column model with Neumann BC at the bottom, $\dfrac{dy(0)}{dx} = 0$, and Dirichlet BC at the top, $y(L) = \delta$.

At the onset of buckling, the local deflection, y, as a function of depth, x, is governed by the nonhomogeneous ODE $\dfrac{d^2 y}{dx^2} + a_0 y = \dfrac{F_L \delta}{E_m I}$, where $a_o = \dfrac{F_L}{E_m I}$. The BCs for this ODE are of the Neumann type at $x = 0$ and Dirichlet type at $x = L$, that is, $\dfrac{dy(0)}{dx} = 0$ and $y(L) = \delta$, respectively.

a. Define the transformation variable $z(x) = y(x) - \delta$ and transform the nonhomogeneous BVP in $y(x)$ to a homogeneous BVP in $z(x)$. State the necessary BCs for $z(x)$.

b. Find the solution of the ODE in $z(x)$, eigencondition and eigenvalues for nontrivial solutions.

c. Use the eigenvalues to determine the maximum buckling load or weight the tower can carry.

22. In Chapter 2 you analyzed IVPs in the time domain. One IVP case is the free vibration of a mass-spring system without damping, governed by the ODE $\dfrac{d^2 y}{dt^2} + \omega_n^2 y = 0$, where $y(t)$ is the instantaneous displacement and $\omega_n = \sqrt{\dfrac{k}{m}}$. For certain ICs, for example, $y(0) = 1$ and $\dfrac{dy(0)}{dt} = 0$, this IVP problem is nonhomogeneous by virtue of $y(0) = 1$ and possesses a unique solution $y(t) = \cos(\omega_n t)$. That is, if you know the initial displacement of the mass, $y(0) = 1$, and its initial speed, $\dfrac{dy(0)}{dt} = 0$, at rest in this case, then you can determine exactly where the mass is in the domain $t \geq 0$, as shown in **Figure 3.21**, over one time period for $\omega_n = \dfrac{1}{2} \dfrac{rad}{s}$.

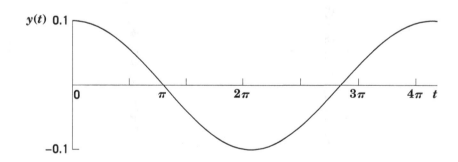

Figure 3.21 Mass-spring system initial value problem modeled as a BVP.

This time, you want to design a mass-spring system by choosing k and m so that regardless of where the system starts, it must have zero displacement at $t = \pi$; that is, $y(\pi) = 0$, which is consistent with the graph in **Figure 3.21**. However, the initial speed is still zero, $\frac{dy(0)}{dt}=0$. For that purpose, solve the homogeneous BVP consisting of

$$\frac{d^2y}{dt^2}+\omega_n^2 y=0$$

and the BCs $\frac{dy(0)}{dt}=0$ and $y(\pi)=0$. This BVP is now an eigenvalue problem that no longer has a unique solution; that is, an infinite number of natural frequencies ω_n satisfy the BVP.

a. Find the solution of the eigenvalue problem over the domain $0<t<\pi$, where t is in s.
b. Find the eigencondition for nontrivial solutions.
c. Find the eigenvalues.
d. For a mass of 1 kg, find the spring constant in N/m for the first eigenvalue.
e. For a mass of 1 kg, find the spring constant in N/m for the second eigenvalue.

CHAPTER FOUR

SYSTEMS OF ORDINARY DIFFERENTIAL EQUATIONS

4.1 Introduction

In Chapter 1, you studied first-order ODEs of the form $\dfrac{dy}{dt} + a_0 y = r(t)$. This form can be extended to model engineering problems that have more than one dependent variable, for example, $y_1(t)$ and $y_2(t)$, as a system of nonhomogeneous first-order ODEs,

$$\frac{dy_1}{dt} + ay_1 + by_2 = g_1(t)$$
$$\frac{dy_2}{dt} + cy_1 + dy_2 = g_2(t)$$

where a, b, c, and d are coefficients, $y_1(t)$ and $y_2(t)$ are the dependent variables, and $g_1(t)$ and $g_2(t)$ are forcing functions. This system is designated coupled first-order ODEs because $y_1(t)$ and $y_2(t)$ appear in the equations simultaneously, and the first derivative is the highest order.

Similarly, you studied in Chapter 2, second-order ODEs of the form $\dfrac{d^2 y}{dt^2} + a_1 \dfrac{dy}{dt} + a_0 y = r(t)$ that can also be extended to engineering problems with more than one dependent variable, for example, $y_1(t)$ and $y_2(t)$, as a second-order and coupled system of nonhomogeneous ODEs,

$$\frac{d^2 y_1}{dt^2} + a\frac{dy_1}{dt} + b\frac{dy_2}{dt} + cy_1 + dy_2 = g_1(t)$$
$$\frac{d^2 y_2}{dt^2} + e\frac{dy_1}{dt} + f\frac{dy_2}{dt} + gy_1 + hy_2 = g_2(t)$$

where a, b, c, d, e, f, g, and h are constant coefficients, $y_1(t)$ and $y_2(t)$ are the dependent variable, and $g_1(t)$ and $g_2(t)$ are the forcing functions.

As an example, consider a two-story building in a seismic area, which may displace laterally as a result of ground motion $x_g(t)$, as shown in **Figure 4.1a**. It is known that an earthquake may shake structures of any type with oscillation frequencies in the range of $0.05\dfrac{rad}{s}$ to $1\dfrac{rad}{s}$. In order to develop the governing ODEs for this structure, each floor is modeled as a single point mass, m, as shown in **Figure 4.1b**. Like any other material that resists motion, think of the walls and ceiling of the first floor, and the walls of the second floor and roof as springs with effective stiffness that is a function of the building materials. Thus, the simplified model consists of masses, m_1 and m_2, which are separated by the stiffnesses of the first and second floors, k_1, and k_2, respectively, as shown in **Figure 4.1c**. The lateral displacements of the floors caused by ground motion, x_1 and x_2, measured relative to the static equilibrium position, as shown in **Figure 4.1b**, are the dependent variables. x_1 and x_2 constitute a two-degree-of-freedom vibration system. Furthermore, $x_g(t)$ models the lateral support motion. In the absence of an earthquake, $x_g(t) = 0$.

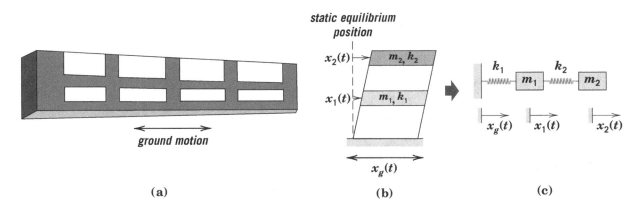

Figure 4.1 Two-story building in a seismic area: (a) building subject to ground motion; (b) building lateral displacement model; and (c) its equivalent mass-spring system.

This two-degree-of-freedom system is modeled in detail in Section 4.6, which leads to two coupled ODEs of the general form

$$\frac{d^2 x_1}{dt^2} + a x_1 + b x_2 = g_1(t)$$

$$\frac{d^2 x_2}{dt^2} + c x_1 + d x_2 = g_2(t)$$

where a, b, c, and d are parameters that depend on m_1, m_2, k_1, and k_2, and $g_1(t)$ and $g_2(t)$ are forcing functions that depend on m_1, m_2, k_1, k_2, and $x_g(t)$. From the foregoing coupled ODEs, you can obtain the natural frequencies of the building and the shape the oscillating building will take at these frequencies.

Nonhomogeneous and coupled ODEs are solved in Chapter 5 by the Laplace transform method. The focus in this chapter is on the solution of coupled and homogeneous ODEs. Specifically, the emphasis is on the transformation of second-order and higher-order ODEs to a system of coupled first-order ODEs; solution of coupled and homogeneous first-order ODEs as an eigenvalue problem; stability of coupled first-order ODEs in the phase plane; and solution of coupled and homogeneous second-order ODEs as an eigenvalue problem.

4.2 Basic Concepts

First-order systems: State-space variables

Any nth-order ODE can be transformed to a system of n coupled and nonhomogeneous first-order ODEs in n dependent variables, y_1, y_2, ..., y_n, given by

$$\frac{dy_1}{dt} = a_{11}y_1 + a_{12}y_2 + \cdots + a_{1n}y_n + g_1(t)$$

$$\frac{dy_2}{dt} = a_{21}y_1 + a_{22}y_2 + \cdots + a_{2n}y_n + g_2(t)$$

$$\vdots \qquad\qquad \vdots \qquad\qquad\qquad\qquad (4.1)$$

$$\frac{dy_n}{dt} = a_{n1}y_1 + a_{n2}y_2 + \cdots + a_{nn}y_n + g_n(t)$$

or in matrix form

$$\frac{dY}{dt} = AY + G \qquad\qquad (4.2)$$

where $Y = \begin{bmatrix} y_1 \\ y_2 \\ \cdot \\ \cdot \\ \cdot \\ y_n \end{bmatrix}$ is the dependent variable column matrix, $A = \begin{bmatrix} a_{11} & a_{12} & \cdot & \cdot & \cdot & a_{1n} \\ a_{21} & a_{22} & \cdot & \cdot & \cdot & a_{2n} \\ \cdot & & & \cdot & \\ \cdot & & & & \cdot \\ \cdot & & & & & \cdot \\ a_{n1} & a_{n2} & \cdot & \cdot & \cdot & a_{nn} \end{bmatrix}$ is the

coefficient matrix, and $G = \begin{bmatrix} g_1(t) \\ g_2(t) \\ \cdot \\ \cdot \\ \cdot \\ g_n(t) \end{bmatrix}$ is the forcing function column matrix. The initial conditions

(ICs) for the system of equations are given by the column matrix

$$Y(t_o) = \begin{bmatrix} y_1(t_o) \\ y_2(t_o) \\ \cdot \\ \cdot \\ \cdot \\ y_n(t_o) \end{bmatrix}$$

The system of ODEs in Eq. (4.2) is homogeneous when $G = Z$, and nonhomogeneous when $G \neq Z$, where Z is the zero matrix; that is, all the elements in the Z matrix are zero. Furthermore, matrix Y constitutes what is coined in control theory as n state-space variables in y_1, y_2, ...,y_n.

Example 4.1

Given the ODE
$$\frac{d^2 y_1}{dt^2} + \frac{dy_1}{dt} + y_1 = \cos(t) \qquad (4.3)$$

and the ICs
$$y_1(0) = 0 \text{ and } \frac{dy_1(0)}{dt} = 1 \qquad (4.4)$$

1. Rewrite Eq. (4.3) as a system of first-order ODEs by defining

$$y_2 = \frac{dy_1}{dt} \qquad (4.5)$$

2. Cast the coupled ODEs in the form of Eq. (4.2).
3. Identify Y, A, and G.
4. Specify the IC column matrix, $Y(t_o)$.

Solution:

1. Rewrite Eq. (4.3) as a system of first-order ODEs.

Substituting Eq. (4.5) into Eq. (4.3) yields

$$\frac{dy_2}{dt} + y_2 + y_1 = \cos(t) \qquad (4.6)$$

2. Cast the coupled ODEs in the form of Eq. (4.2).

Rewrite Eq. (4.5) as

$$\frac{dy_1}{dt} = y_2 \tag{4.7}$$

and Eq. (4.6) as

$$\frac{dy_2}{dt} = -y_1 - y_2 + \cos(t) \tag{4.8}$$

Equations (4.7) and (4.8) in matrix form are

$$\frac{d}{dt}\begin{bmatrix} y_1 \\ y_2 \end{bmatrix} = \begin{bmatrix} 0 & 1 \\ -1 & -1 \end{bmatrix}\begin{bmatrix} y_1 \\ y_2 \end{bmatrix} + \begin{bmatrix} 0 \\ \cos(t) \end{bmatrix} \tag{4.9}$$

3. Identify Y, A, and G.

Equation (4.9) is cast in the form of Eq. (4.2), where $Y = \begin{bmatrix} y_1 \\ y_2 \end{bmatrix}$, $A = \begin{bmatrix} 0 & 1 \\ -1 & -1 \end{bmatrix}$, and $G = \begin{bmatrix} 0 \\ \cos(t) \end{bmatrix}$.

4. Specify the IC column matrix, $Y(t_o)$.

The ICs for Eq. (4.9) are found by using Eqs. (4.4) and (4.5); that is, $y_1(0) = 0$ and $y_2(0) = \dfrac{dy_1(0)}{dt} = 1$, or in matrix form, $Y(t_o) = \begin{bmatrix} y_1(0) \\ y_2(0) \end{bmatrix} = \begin{bmatrix} 0 \\ 1 \end{bmatrix}$.

■

Example 4.2

Given the system of coupled ODEs

$$3\frac{d^2 y_1}{dt^2} + \frac{1}{2}\left(\frac{dy_1}{dt} - \frac{dy_2}{dt}\right) + y_1 - y_2 = \sin(3t) \tag{4.10}$$

$$\frac{d^2 y_2}{dt^2} + \frac{dy_2}{dt} - \frac{dy_1}{dt} + 4(y_2 - y_1) = 0 \tag{4.11}$$

and the ICs $y_1(0) = 0, \dfrac{dy_1(0)}{dt} = 1, y_2(0) = 0, \dfrac{dy_2(0)}{dt} = 0.$ $\tag{4.12}$

1. Rewrite Eqs. (4.10) and (4.11) as a system of first-order ODEs by defining

$$y_3 = \frac{dy_1}{dt} \tag{4.13}$$

and
$$y_4 = \frac{dy_2}{dt} \tag{4.14}$$

2. Cast the system of coupled ODEs in the form of Eq. (4.2).
3. Identify Y, A, and G.
4. Specify the IC column matrix, $Y(t_o)$.

Solution:

1. Rewrite Eqs. (4.10) and (4.11) as a system of first-order ODEs.

Substituting Eqs. (4.13) and (4.14) into Eqs. (4.10) and (4.11) yields

$$\frac{dy_3}{dt} = -\frac{1}{3}y_1 + \frac{1}{3}y_2 - \frac{1}{6}y_3 + \frac{1}{6}y_4 + \frac{1}{3}\sin(3t) \tag{4.15}$$

$$\frac{dy_4}{dt} = 4y_1 - 4y_2 + y_3 - y_4 \tag{4.16}$$

2. Cast the system of coupled ODEs in the form of Eq. (4.2).

Equations(4.13), (4.14), (4.15), and (4.16) in matrix form are

$$\frac{d}{dt}\begin{bmatrix} y_1 \\ y_2 \\ y_3 \\ y_4 \end{bmatrix} = \begin{bmatrix} 0 & 0 & 1 & 0 \\ 0 & 0 & 0 & 1 \\ -\frac{1}{3} & \frac{1}{3} & -\frac{1}{6} & \frac{1}{6} \\ 4 & -4 & 1 & -1 \end{bmatrix}\begin{bmatrix} y_1 \\ y_2 \\ y_3 \\ y_4 \end{bmatrix} + \begin{bmatrix} 0 \\ 0 \\ \frac{1}{3}\sin(3t) \\ 0 \end{bmatrix} \tag{4.17}$$

3. Identify Y, A, and G from Eq. (4.2).

Equation (4.17) is cast in the form of Eq. (4.2), where $Y = \begin{bmatrix} y_1 \\ y_2 \\ y_3 \\ y_4 \end{bmatrix}$, $A = \begin{bmatrix} 0 & 0 & 1 & 0 \\ 0 & 0 & 0 & 1 \\ -\dfrac{1}{3} & \dfrac{1}{3} & -\dfrac{1}{6} & \dfrac{1}{6} \\ 4 & -4 & 1 & -1 \end{bmatrix}$,

and $G = \begin{bmatrix} 0 \\ 0 \\ \dfrac{1}{3}\sin(3t) \\ 0 \end{bmatrix}$

4. Specify the IC column matrix, $Y(t_o)$.

The ICs for Eq. (4.17) are found by using Eqs. (4.12), (4.13), and (4.14): $y_1(0) = 0$, $y_2(0) = 0$,

$y_3(0) = \dfrac{dy_1(0)}{dt} = 1$, $y_4(0) = \dfrac{dy_2(0)}{dt} = 0$, or in matrix form $\begin{bmatrix} y_1(0) \\ y_2(0) \\ y_3(0) \\ y_4(0) \end{bmatrix} = \begin{bmatrix} 0 \\ 0 \\ 1 \\ 0 \end{bmatrix}$.

4.3 Eigenvalues and Stability of Homogeneous and Linear System of First-Order ODEs

Eigenvalues

The homogeneous counterpart of the system of first-order ODEs in Eq. (4.2) is

$$\frac{dY}{dt} = AY \tag{4.18}$$

Equation (4.18) constitutes an eigenvalue problem whose solution is given by

$$Y = Ve^{\lambda t} \tag{4.19}$$

where V is the set of n eigenvectors v_1, v_2, ..., v_n that corresponds to the eigenvalues λ_1, λ_2, ..., λ_n of the $n \times n$ matrix A. This particular system of equations constitutes an eigenvalue problem because substituting Eq. (4.19) into Eq. (4.18) yields $\lambda Ve^{\lambda t} = AVe^{\lambda t}$ or the classical eigenvalue formulation $AV = \lambda V$ after dividing both sides of the equation by $e^{\lambda t}$. The eigenvalues could be real or complex numbers. The eigenvalues of $AV = \lambda V$, or

$$(A - \lambda I)\, V = 0 \tag{4.20}$$

where I is the identity matrix, are found by setting the determinant of Eq. (4.20) equal to zero, that is, $|A - \lambda I| = 0$, as illustrated in the following example.

Example 4.3

Find the eigenvalues of matrix $A = \begin{bmatrix} 0 & 1 \\ -2 & 2 \end{bmatrix}$.

Solution:

The eigenvalues are given by $\begin{vmatrix} -\lambda & 1 \\ -2 & 2-\lambda \end{vmatrix} = 0$ that yields $\lambda^2 - 2\lambda + 2 = 0$ with roots $\lambda_{1,2} = 1 \pm i$.

■

Based on Example 4.3, the general form of an eigenvalue is $\lambda = a + ib$, where a is the real part of λ, or simply $a = \mathrm{Re}(\lambda)$, and b is the imaginary part of λ, or simply $b = \mathrm{Im}(\lambda)$, so that $e^{\lambda t}$ in Eq. (4.19) takes the form of $e^{\lambda t} = e^{(a+ib)t} = e^{at}e^{ibt}$, or $e^{\lambda t} = e^{at}\left[\cos(bt) + i\sin(bt)\right]$. The foregoing equation indicates that $b = \mathrm{Im}(\lambda)$ causes the solution to oscillate via the sine and cosine terms, whereas $a = \mathrm{Re}(\lambda)$ causes the oscillating solution to either decay if $a < 0$, oscillate with constant amplitude if $a = 0$, or grow unbounded if $a > 0$.

In second-order engineering systems, $a = \mathrm{Re}(\lambda)$ and $b = \mathrm{Im}(\lambda)$ may be related to the natural frequency, ω_n, damping frequency, ω_d, and damping factor, ζ, as demonstrated by the next example.

Example 4.4

Consider the free vibration of a single-degree-of-freedom, undamped system as described in Chapter 2, Example 2.2. The system consists of a mass m and spring of stiffness k, as shown in **Figure 4.2**. The instantaneous displacement $y(t)$ is governed by the second-order homogeneous ODE,

$$\frac{d^2y}{dt^2} + \omega_n^2 y = 0 \tag{4.21}$$

where $\omega_n = \sqrt{\dfrac{k}{m}}$ is the natural frequency.

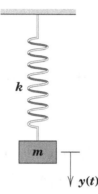

Figure 4.2 A single-degree-of-freedom undamped mass-spring system.

Rewrite Eq. (4.21) as a system of first-order ODEs using the velocity definition

$$v = \frac{dy}{dt} \tag{4.22}$$

Solution:

Substituting the velocity definition into Eq. (4.21) yields

$$\frac{dv}{dt} = -\omega_n^2 y \tag{4.23}$$

In matrix form, Eqs. (4.22) and (4.23) yield

$$\frac{d}{dt}\begin{bmatrix} y \\ v \end{bmatrix} = \begin{bmatrix} 0 & 1 \\ -\omega_n^2 & 0 \end{bmatrix}\begin{bmatrix} y \\ v \end{bmatrix} \tag{4.24}$$

where the state-space variables are the displacement, $y(t)$, and velocity, $v(t)$ and $A = \begin{bmatrix} 0 & 1 \\ -\omega_n^2 & 0 \end{bmatrix}$.

Equation (4.24) has the same form as Eq. (4.18), which leads to an eigenvalue formulation. The eigenvalues are found by setting $\begin{vmatrix} -\lambda & 1 \\ -\omega_n^2 & -\lambda \end{vmatrix} = 0$, which yields $\lambda^2 = -\omega_n^2$, or two complex roots $\lambda_{1,2} = \pm i\omega_n$ with magnitude $|\lambda_1| = |\lambda_2| = \omega_n$. Thus, for a single-degree-of-freedom undamped mass-spring system, the eigenvalue magnitudes match the system's natural frequency.

∎

Stability of a system of first-order ODEs using eigenvalues

For an autonomous system of n coupled first-order linear ODEs described by Eq. (4.18), with n eigenvalues, λ_i, for $i = 1, 2, \ldots, n$, the system is generally classified as:

- neutrally stable if the real part of every eigenvalue of matrix A is zero, that is, $\mathrm{Re}(\lambda_i) = 0$,

- asymptotically stable if the real part of every eigenvalue of matrix A is negative, that is, $\mathrm{Re}(\lambda_i) < 0$, and

- unstable if the real part of one or more eigenvalues of matrix A is positive, that is, $\mathrm{Re}(\lambda_i) > 0$.

A simple autonomous system of two first-order coupled system ODEs can be derived from a second-order system with no forcing function,

$$\frac{d^2 y}{dt^2} + a_1 \frac{dy}{dt} + a_0 y = 0 \tag{4.25}$$

In engineering systems, for example, mechanical or electrical, $a_1 = 2\zeta\omega_n$ and $a_0 = \omega_n^2$.

Let $y = y_1$ and $y_2 = \dfrac{dy_1}{dt}$ so that Eq. (4.25) is transformed to a system of first-order ODEs,

$$\frac{dy_1}{dt} = y_2 \tag{4.26}$$

$$\frac{dy_2}{dt} = -a_0 y_1 - a_1 y_2 \tag{4.27}$$

$y_1(t)$ and $y_2(t)$ represent displacement and velocity, respectively, in a mechanical system in linear motion. On the other hand, $y_1(t)$ and $y_2(t)$ represent angular displacement and angular velocity, respectively, in a mechanical system in circular motion. In an electrical system, for example, an RLC circuit, $y_1(t)$ and $y_2(t)$ represent charge, q, and current, $i = \dfrac{dq}{dt}$, respectively, if the ODE is formulated in terms of charge, $\dfrac{d^2q}{dt^2} + a_1\dfrac{dq}{dt} + a_0q = 0$. If, on the other hand, the electrical system is formulated as $\dfrac{d^2i}{dt^2} + a_1\dfrac{di}{dt} + a_0i = 0$, then $y_1(t)$ represents current, but $y_2(t) = \dfrac{di}{dt}$ generally has no name; however, it may be viewed as the voltage across an inductor per unit *henry*.

In matrix form,

$$\frac{d}{dt}\begin{bmatrix} y_1 \\ y_2 \end{bmatrix} = \begin{bmatrix} 0 & 1 \\ -a_0 & -a_1 \end{bmatrix}\begin{bmatrix} y_1 \\ y_2 \end{bmatrix} = A\begin{bmatrix} y_1 \\ y_2 \end{bmatrix}$$

where $A = \begin{bmatrix} 0 & 1 \\ -a_0 & -a_1 \end{bmatrix}$. The eigenvalues are found by setting the determinant of matrix A equal to zero, that is, $\begin{vmatrix} -\lambda & 1 \\ -a_0 & -a_1 - \lambda \end{vmatrix} = 0$ that yields $\lambda^2 + a_1\lambda + a_0 = 0$ with roots $\lambda_{1,2}$. Using $\lambda_{1,2}$, the system is classified as:

- undamped when $\mathrm{Re}(\lambda_1) = \mathrm{Re}(\lambda_2) = 0$,

- underdamped, critically damped, or overdamped, when $\mathrm{Re}(\lambda_1) < 0$ and $\mathrm{Re}(\lambda_2) < 0$, and

- negatively damped when $\mathrm{Re}(\lambda_1) > 0$ or $\mathrm{Re}(\lambda_2) > 0$. For negative damping, you may refer to Figure 2.4 and the explanation given for wing motion.

These cases are summarized in **Table 4.1**.

Eigenvalues	Stability state	Damping
$\mathrm{Re}(\lambda_1) = \mathrm{Re}(\lambda_2) = 0$	neutrally stable	undamped; Figure 2.11
$\mathrm{Re}(\lambda_1) < 0$ and $\mathrm{Re}(\lambda_2) < 0$	asymptotically stable	underdamped, critically damped, or overdamped; Figure 2.11
$\mathrm{Re}(\lambda_1) > 0$ or $\mathrm{Re}(\lambda_2) > 0$	unstable	negatively damped; Figure 2.4

Table 4.1 Stability of Eq. (4.25) using the eigenvalues of a system of first-order ODEs.

Example 4.5

Consider the oscillating tuner circuit described in Figure 2.3, Chapter 2, without the forcing function, $V_S(t)$, as shown in **Figure 4.3**, where $L = 5(10^{-4})\,henry$, $C = 52.75(10^{-12})\,farad$, and $R = 500\,ohm$.

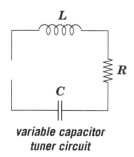

variable capacitor
tuner circuit

Figure 4.3 Oscillating tuner circuit without a forcing function for the purpose of stability analysis.

The instantaneous current, $i(t)$, in this circuit was derived in Chapter 2, Eq. (2.88), as

$$\frac{d^2 i}{dt^2} + \frac{R}{L}\frac{di}{dt} + \frac{1}{LC}i = 0 \qquad (4.28)$$

where t is in s.

1. Show that this circuit is stable in free oscillation.
2. Find the eigenvalues by transforming the ODE to a system of two coupled first-order ODEs, and classify the system as neutrally stable or asymptotically stable.

Solution:

1. Show that this system is stable in free oscillation.

Based on Chapter 2, the ODE yields, $\omega_n^2 = \frac{1}{LC}$ and $2\zeta\omega_n = \frac{R}{L}$. Using the given values of R, L,

and C yields $\omega = 6.16\left(10^6\right)\frac{rad}{s}$ and $\zeta = 0.081$. For $0 < \zeta < 1$, this circuit is underdamped so that

the instantaneous current, $i(t)$, is expected to decay asymptotically in time, that is, $i(t) \to 0$ as

$t \to \infty$. Naturally, such a system is stable.

2. Find the eigenvalues by transforming Eq. (4.28) to a system of two coupled first-order ODEs, and classify the system as neutrally stable or asymptotically stable.

Let $y_1 = i$ $y_2 = \frac{dy_1}{dt}$ so that the Eq. (4.28) is transformed to $\frac{dy_2}{dt} + \frac{R}{L}y_2 + \frac{1}{LC}y_1 = 0$, or

$\frac{dy_2}{dt} = -\frac{1}{LC}y_1 - \frac{R}{L}y_2$.

In matrix form, $\dfrac{d}{dt}\begin{bmatrix} y_1 \\ y_2 \end{bmatrix} = \begin{bmatrix} 0 & 1 \\ -\dfrac{1}{LC} & -\dfrac{R}{L} \end{bmatrix}\begin{bmatrix} y_1 \\ y_2 \end{bmatrix}$ so that $A = \begin{bmatrix} 0 & 1 \\ -\dfrac{1}{LC} & -\dfrac{R}{L} \end{bmatrix}$.

The eigenvalues of matrix A are found from $\begin{vmatrix} 0-\lambda & 1 \\ -\dfrac{1}{LC} & -\dfrac{R}{L}-\lambda \end{vmatrix} = 0$. Carrying out the operation

manually, or using Computer-Algebra System (CAS), yields the characteristic equation

$\lambda^2 + \dfrac{R}{L}\lambda + \dfrac{1}{LC} = 0$ with roots $\lambda_1 = -5\times10^5 + \left(6.14\times10^6\right)i$ and $\lambda_2 = -5\times10^5 - \left(6.14\times10^6\right)i$. Since

$\mathrm{Re}\left(\lambda_1\right) < 0$ and $\mathrm{Re}\left(\lambda_2\right) < 0$, then based on **Table 4.1** the system is classified as asymptotically stable.

$\blacksquare$

Example 4.6

Consider a mechanical system whose instantaneous displacement, $y(t)$, is described by the second-order ODE

$$\frac{d^2y}{dt^2} + y = 0 \tag{4.29}$$

where t is in s.

1. Show that this system is stable in free vibration.
2. Find the eigenvalues by transforming Eq. (4.29) to a system of two coupled first-order ODEs, and classify the system as neutrally stable or asymptotically stable.

Solution:

1. Show that this system is stable in free vibration.

Based on Chapter 2, the ODE yields $\omega_n = 1\dfrac{rad}{s}$ and $\zeta = 0$, which implies that this system is

undamped so that the instantaneous displacement, $y(t)$, is expected to have constant amplitude

in time; that is, $y(t)$ is bounded as $t \to \infty$. Naturally, such a system is stable.

2. Find the eigenvalues by transforming Eq. (4.29) to a system of two coupled first-order ODEs, and classify the system as neutrally stable or asymptotically stable.

Following a similar approach to that in Example 4.4, the ODE can be cast as $v = \dfrac{dy}{dt}$ and

$\dfrac{dv}{dt} = -y$, or in matrix form $\dfrac{d}{dt}\begin{bmatrix} y \\ v \end{bmatrix} = \begin{bmatrix} 0 & 1 \\ -1 & 0 \end{bmatrix}\begin{bmatrix} y \\ v \end{bmatrix}$ so that $A = \begin{bmatrix} 0 & 1 \\ -1 & 0 \end{bmatrix}$.

The eigenvalues of matrix A are found from $\begin{vmatrix} -\lambda & 1 \\ -1 & -\lambda \end{vmatrix} = 0$, which yields the characteristic

equation $\lambda^2 + 1 = 0$ with roots $\lambda_1 = i$ and $\lambda_2 = -i$. Since $\text{Re}(\lambda_1) = 0$ and $\text{Re}(\lambda_2) = 0$, then based on **Table 4.1**, the system is classified as neutrally stable.

4.4 Stability of a System of First-Order ODEs Using Phase Plane

The stability of the first-order system, Eqs. (4.26) and (4.27), can be described by plotting the state-space variable $y_2(t)$ as the ordinate versus the state-space variable $y_1(t)$ as the abscissa instead of plotting $y_1(t)$ and $y_2(t)$ versus t. The plane of the state-space variables is defined as the phase plane or phase portrait.

As a case in point, consider the single-degree-of-freedom mass and spring system in free vibration, as shown in **Figure 4.4**.

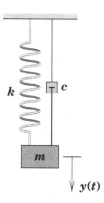

Figure 4.4 A single-degree-of-freedom mass and spring system in free vibration with possible damping.

If you were to solve for the displacement, $y(t)$, for some ICs $y(t_o) = y_o$ and $\dfrac{dy(t_o)}{dt} = v_o$, you

obtain one of the solutions as described in Table 2.2 and Figure 2.11. From $y(t)$, you can derive

an expression for the velocity using $v(t) = \dfrac{dy}{dt}$. If you plot $v(t)$ as the ordinate versus $y(t)$ as

the abscissa, you obtain a typical orbit about the equilibrium point (0,0), as shown in **Figure 4.5a** for an undamped system, classified as neutrally stable; **Figure 4.5b** for an underdamped system, classified as asymptotically stable; **Figure 4.5c** for an critically damped or overdamped system, also classified as asymptotically stable; and **Figure 4.5d** for a negatively damped system, classified as unstable.

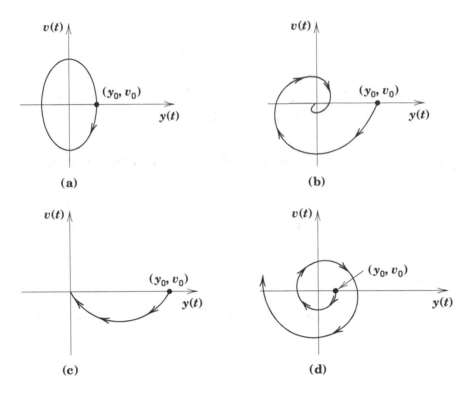

Figure 4.5 Phase planes and orbits of a second-order system: (a) neutrally stable; (b) and (c) asymptotically stable; and (d) unstable.

Each orbit is the locus of all instantaneous displacements and velocities as time increases from t_o to infinity. Furthermore, when $\zeta = 0$, the orbit of an undamped system is in neutral stability with regard to the equilibrium point; that is, it does not approach the (0,0) point, nor does it run away from it. For $0 < \zeta < 1$, the orbit of an underdamped system spirals inward toward the equilibrium point, as in **Figure 4.5b**, which is expected in free vibration as the displacement and velocity approach zero asymptotically as $t \to \infty$. For $\zeta > 1$, the orbit of an overdamped system plummets toward the equilibrium point without spiraling, as in **Figure 4.5c**. For $\zeta < 0$, the orbit of a negatively damped system spirals outward away from the equilibrium point, as in **Figure 4.5d**, so that the displacement grows unbounded as $t \to \infty$.

The graphical solutions in the phase plane must be consistent with the eigenvalues of the system, as summarized in **Table 4.2** as an expansion of **Table 4.1**.

Eigenvalues	Stability state	Damping	Typical orbit in the phase plane
$\mathrm{Re}(\lambda_1) = \mathrm{Re}(\lambda_2) = 0$	neutrally stable	undamped; Figure 2.11	Figure 4.5a
$\mathrm{Re}(\lambda_1) < 0$ and $\mathrm{Re}(\lambda_2) < 0$	asymptotically stable	underdamped, critically damped, or overdamped; Figure 2.11	Figure 4.5b Figure 4.5c
$\mathrm{Re}(\lambda_1) > 0$ or $\mathrm{Re}(\lambda_2) > 0$	unstable	negatively damped; Figure 2.4	Figure 4.5d

Table 4.2 Stability summary of a system governed by linear and homogeneous second-order ODE.

Example 4.7

Consider the free vibration of a single-degree-of-freedom, undamped system, as shown in **Figure 4.2**. The instantaneous displacement $y(t)$ is governed by the second-order homogeneous ODE,

$$\frac{d^2 y}{dt^2} + \omega_n^2 y = 0 \tag{4.30}$$

where $\omega_n = \sqrt{\dfrac{k}{m}}$ is the natural frequency, with ICs $y(0) = 0$ and $\dfrac{dy(0)}{dt} = 1$.

1. Find the orbit equation of this system from the displacement, $y(t)$ and velocity $v(t)$.
2. Plot the orbit in a phase plane for $\omega_n = 1$ and $\omega_n = 2$.

All units are assumed consistent.

Solution:

1. Find the orbit equation of this system from the displacement, $y(t)$ and velocity $v(t)$.

From Table 2.2 in Chapter 2, the general solution of Eq. (4.30) is given by $y(t) = a\cos(\omega_n t) + b\sin(\omega_n t)$ for an undamped system. Using the IC $y(0) = 0$ yields

$$y(t) = b\sin(\omega_n t) \tag{4.31}$$

From Eq. (4.31) the velocity is

$$v(t) = \frac{dy}{dt} = b\omega_n \cos(\omega_n t) \tag{4.32}$$

Using the IC $\dfrac{dy(0)}{dt} = 1$ with Eq. (4.32) yields $b = \dfrac{1}{\omega_n}$. Thus,

$$y(t) = \frac{1}{\omega_n}\sin(\omega_n t) \tag{4.33}$$

and

$$v(t) = \cos(\omega_n t) \tag{4.34}$$

The orbit equation is found by eliminating t from Eqs. (4.33) and (4.34). This can be done by squaring Eqs. (4.33) and (4.34) as $(y\omega_n)^2 = \sin^2(\omega_n t)$ and $v^2 = \cos^2(\omega_n t)$, and adding the squared equations, $(y\omega_n)^2 + v^2 = \sin^2(\omega_n t) + \cos^2(\omega_n t)$, or

$$y^2\omega_n^2 + v^2 = 1 \tag{4.35}$$

2. Plot the orbit in a phase plane for $\omega_n = 1$ and $\omega_n = 2$.

For $\omega_n = 1$, Eq. (4.35) yields $y^2 + v^2 = 1$, which is the equation of a circle of radius equal to 1, and centered at (0,0). For $\omega_n = 2$, Eq. (4.35) yields $4y^2 + v^2 = 1$, or $\dfrac{y^2}{0.25} + v^2 = 1$, which is the equation of an ellipse with major and minor axes equal to 1 and 0.5, respectively, and centered at (0,0), as shown in **Figure 4.6**. The orbits confirm the neutral stability of an undamped system with regard to the equilibrium point; that is, it does not approach the (0,0) point, nor does it run away from it.

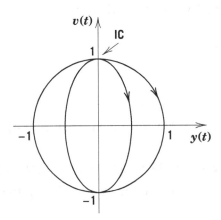

Figure 4.6 Circular and elliptic orbits of a single-degree-of-freedom undamped system in free vibration.

You should show from Eq. (4.34) that the velocity amplitude is equal to one whether $\omega_n = 1$ or $\omega_n = 2$, which is consistent with the circle radius matching the major axis of the ellipse. However, as the frequency increases from $\omega_n = 1$ to $\omega_n = 2$, Eq. (4.33) indicates that the displacement amplitude decreases from 1 to 0.5, which is consistent with the orbit radius of the circle decreasing to the minor axis.

█

The orbit equation for spiral orbits expected in an underdamped system in free vibration is not as straightforward as that obtained in Example 4.7. A case in point is illustrated in the following example.

Example 4.8

Consider the free vibration of a single-degree-of-freedom, underdamped system, as shown in **Figure 4.4**. The instantaneous displacement $y(t)$ is governed by the second-order homogeneous ODE,

$$\frac{d^2 y}{dt^2} + 2\zeta\omega_n \frac{dy}{dt} + \omega_n^2 y = 0 \qquad (4.36)$$

where $\omega_n = \sqrt{\dfrac{k}{m}}$ is the natural frequency and ζ is the damping factor, with ICs $y(0) = 0$ and $\dfrac{dy(0)}{dt} = 1$.

1. Find the orbit equation of this system from the displacement $y(t)$ and velocity $v(t)$.
2. Plot the orbit in a phase plane for $\omega_n = 1$ and $\zeta = 0.1$.

All units are assumed consistent.

Solution:

1. Find the orbit equation of this system from the displacement $y(t)$ and velocity $v(t)$.

From Table 2.2 in Chapter 2, the general solution of Eq. (4.36) is

$$y(t) = e^{-\zeta\omega_n t} \left[a\cos(\omega_d t) + b\sin(\omega_d t) \right]$$

for an underdamped system, where $\omega_d = \omega_n \sqrt{1-\zeta^2} = \sqrt{1-0.1^2} = 0.995$.

Using the IC $y(0) = 0$ yields

$$y(t) = e^{-\zeta\omega_n t} b\sin(\omega_d t) \qquad (4.37)$$

From Eq. (4.37) the velocity is

$$v(t) = \frac{dy}{dt} = be^{-\zeta\omega_n t} \left[-\zeta\omega_n \sin(\omega_d t) + \omega_d \cos(\omega_d t) \right] \qquad (4.38)$$

Using the IC $\dfrac{dy(0)}{dt}=1$ with Eq. (4.38) yields $b=\dfrac{1}{\omega_d}$. Thus,

$$y(t)=\frac{1}{\omega_d}e^{-\zeta\omega_n t}\sin(\omega_d t) \tag{4.39}$$

and

$$v(t)=e^{-\zeta\omega_n t}\left[-\zeta\frac{\omega_n}{\omega_d}\sin(\omega_d t)+\cos(\omega_d t)\right] \tag{4.40}$$

The orbit equation is found by eliminating t from Eqs. (4.39) and (4.40), which is obtained by squaring equations followed by several algebraic manipulations to eliminate t from the exponential argument. The result is an implicit equation between v and y,

$$\omega_d y+\sqrt{\left[\left(v+\zeta\omega_n y\right)^2+\left(\omega_d y\right)^2\right]}\sin\left(\frac{\omega_d}{2\zeta\omega_n}\ln\left[\left(v+\zeta\omega_n y\right)^2+\left(\omega_d y\right)^2\right]\right)=0 \tag{4.41}$$

2. Plot the orbit in a phase plane for $\omega_n=1$ and $\zeta=0.1$.

Some CAS may not have the capabilities to plot Eq. (4.41). However, it is possible in CAS to plot $v(t)$ versus $y(t)$ parametrically using Eqs. (4.39) and (4.40), where t is the parameter, as shown in **Figure 4.7a**, using $\omega_n=1$ and $\zeta=0.1$. That is, $y(t)=\dfrac{1}{0.995}e^{-0.1t}\sin(0.995t)$ and $v(t)=e^{-0.1t}\left[-0.1005\sin(0.995t)+\cos(0.995t)\right]$.

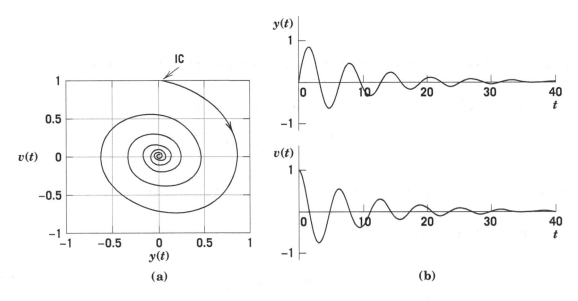

(a) (b)

Figure 4.7 Spiraling orbit of a single-degree-of-freedom underdamped system in free vibration: (a) velocity-displacement phase portrait generated from Eqs. (4.39) and (4.40); and (b) asymptotic decay of the displacement and velocity.

This inward spiraling starts at the IC $\left(y(0)=0, v(0)=1\right)$ and continues on clockwise until it approaches asymptotically the static equilibrium point, (0.0), as $t \to \infty$. You should also confirm graphically that $y(t) \to 0$ and $v(t) \to 0$ as $t \to \infty$, as shown in **Figure 4.7b**. Notice also that the number of spirals in **Figure 4.7a** is equal to the number of cycles in **Figure 4.7b**. For example, the spiral cuts $v(t)=0$ for $y(t)<0$ six times, which also is the number of cycles, or peak-to-peak, in **Figure 4.7b** over the time interval $0 \le t \le 40$.

For a critically damped or overdamped system, there is no oscillation. Thus, the orbit is expected to approach asymptotically the static equilibrium point, (0.0), as $t \to \infty$ without looping around (0,0), as illustrated in the following example.

Example 4.9

Consider the free vibration of a single-degree-of-freedom, critically damped system, $\zeta = 1$, as shown in **Figure 4.4**. The instantaneous displacement $y(t)$ is governed by the second-order homogeneous ODE,

$$\frac{d^2 y}{dt^2} + 2\omega_n \frac{dy}{dt} + \omega_n^2 y = 0 \tag{4.42}$$

where $\omega_n = \sqrt{\dfrac{k}{m}}$ is the natural frequency, with ICs $y(0)=0$ and $\dfrac{dy(0)}{dt}=1$.

1. Find the orbit equation of this system from the displacement $y(t)$ and velocity $v(t)$.
2. Plot the orbit in a phase plane for $\omega_n = 1$.

All units are assumed consistent.

Solution:

1. Find the orbit equation of this system from the displacement $y(t)$ and velocity $v(t)$.
From Table 2.2 in Chapter 2, the general solution of Eq. (4.42) is

$$y(t) = \left(a + bt\right)e^{-\omega_n t}$$

for a critically damped system. Using the IC $y(0)=0$ yields

$$y(t) = bt e^{-\omega_n t} \tag{4.43}$$

From Eq. (4.43) the velocity is

$$v(t) = \frac{dy}{dt} = be^{-\omega_n t} - b\omega_n te^{-\omega_n t} \qquad (4.44)$$

Using the IC $\frac{dy(0)}{dt} = 1$ with Eq. (4.44) yields $b = 1$. Thus,

$$y(t) = te^{-\omega_n t} \qquad (4.45)$$

and

$$v(t) = e^{-\omega_n t} - \omega_n te^{-\omega_n t} \qquad (4.46)$$

The orbit equation is found by eliminating t from Eqs. (4.45) and (4.46), which is an implicit equation between v and y,

$$\ln(v + \omega_n y) + \frac{\omega_n y}{v + \omega_n y} = 0 \qquad (4.47)$$

2. Plot the orbit in a phase plane for $\omega_n = 1$.

The plot of $v(t)$ versus $y(t)$ is generated in CAS parametrically using Eqs. (4.45) and (4.46), as shown in **Figure 4.8a**, with $\omega_n = 1$. That is, $y(t) = te^{-t}$ and $v(t) = e^{-t}(1-t)$.

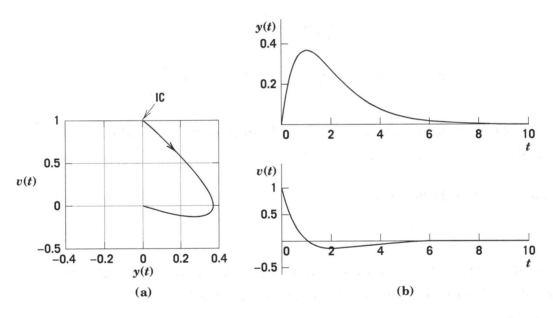

(a)   (b)

Figure 4.8 Orbit of a single-degree-of-freedom critically damped system in free vibration: (a) velocity-displacement phase portrait generated from Eqs. (4.45) and (4.46); and (b) asymptotic decay of the displacement and velocity.

This orbit starts at the IC $\left(y(0)=0, v(0)=1\right)$ and continues on clockwise until it approaches the static equilibrium point, (0,0), asymptotically as $t \to \infty$, but without looping around (0,0). This observation is consistent with $y(t) \to 0$ and $v(t) \to 0$ as $t \to \infty$, without oscillation, as shown in **Figure 4.8b**.

■

Differential equations of orbits

Another approach to find the orbit equation is by eliminating the differential dt from the system of first-order ODEs.

Orbit equation of an undamped system

For the case of an undamped system in free vibration, consider the displacement ODE $\dfrac{d^2 y}{dt^2} + \omega_n^2 y = 0$, which is transformed to two first-order ODEs,

$$\frac{dy}{dt} = v \tag{4.48}$$

and

$$\frac{dv}{dt} = -\omega_n^2 y \tag{4.49}$$

The differential time, dt, is eliminated from Eqs. (4.48) and (4.49) using the chain rule $\dfrac{dv}{dt} = \dfrac{dv}{dy}\dfrac{dy}{dt}$. Substituting Eqs. (4.48) and (4.49) in the foregoing equation yields $-\omega_n^2 y = \dfrac{dv}{dy}v$, or

$$\omega_n^2 y\, dy + v\, dv = 0 \tag{4.50}$$

Since the variables are now separate in Eq. (4.50), integrate to obtain

$$\frac{\omega_n^2 y^2}{2} + \frac{v^2}{2} = C \tag{4.51}$$

where C is a constant of integration. Equation (4.51) is the displacement–velocity orbit equation for an undamped system, which is similar to that found in Example 4.7. For $\omega_n = 1$, it is a family of circles, given by $y^2 + v^2 = 2C$, whereas for $\omega_n \neq 1$, Eq. (4.51) is a family of ellipses as shown in **Figure 4.9** for different values of C.

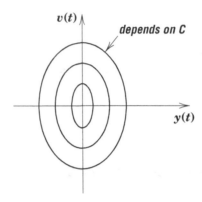

Figure 4.9 Family of elliptic orbits for a neutrally stable system given by Eq. (4.51).

Orbit equation of an underdamped system

For an underdamped system governed by the ODE $\dfrac{d^2y}{dt^2} + 2\zeta\omega_n\dfrac{dy}{dt} + \omega_n^2 y = 0$, the first-order ODEs are

$$\frac{dy}{dt} = v \tag{4.52}$$

and

$$\frac{dv}{dt} = -2\zeta\omega_n v - \omega_n^2 y \tag{4.53}$$

Using the chain rule as we did earlier, Eqs. (4.52) and (4.53) yield $-2\zeta\omega_n v - \omega_n^2 y = \dfrac{dv}{dy}v$, or

$$-2\zeta\omega_n v\,dy - \omega_n^2 y\,dy = v\,dv \tag{4.54}$$

Unfortunately, you cannot obtain the orbital equation directly from this equation because v and y are coupled as $v\,dy$ in the first term. However, it is possible to separate variables by introducing an intermediate variable, β, defined as $y = \beta v$ and eliminating y from Eq. (4.54), which yields

$$\frac{2\zeta\omega_n + \omega_n^2 \beta}{1 + 2\zeta\omega_n\beta + \omega_n^2\beta^2}\,d\beta = -\frac{dv}{v} \tag{4.55}$$

Equation (4.55) can be integrated using CAS, which leads to an implicit relationship between y and v, as that found in Example 4.8.

In the next section, you shall learn how to develop numerical solutions of orbits with relative ease using a system of first-order ODEs. This approach is useful not only for linear systems, but also for nonlinear ODEs when no analytical solutions exist.

4.5 Numerical Solution of Orbits

The numerical scheme for orbits is implemented in state-space variables governed by a system of first-order ODEs. Since you may not have a numerical background prior to this chapter, the simplest numerical scheme is used, namely the forward Euler method.

Numerical orbit of an undamped system

To start, consider a mass-spring undamped system, $\dfrac{d^2 y}{dt^2} + \omega_n^2 y = 0$, as shown in **Figure 4.2**, which is the simplest of all systems, where $y(t)$ is the displacement. The ICs are given by $y(0) = y_o$ and $\dfrac{dy(0)}{dt} = v_o$, where y_o and v_o are values for initial displacement and velocity, respectively.

Dynamically, the system begins to oscillate in time starting with y_o and v_o at $t = 0$, which are used to predict $y(t)$ and $v(t)$ numerically, for $t > 0$, from the two state-space ODEs, $\dfrac{dy}{dt} = v$ and $\dfrac{dv}{dt} = -\omega_n^2 y$. Thus, the numerical solution is no longer continuous, but rather consists of discrete points.

Near the ICs, y_o and v_o, you can approximate $\dfrac{dy}{dt} = v$ numerically as $\dfrac{\Delta y}{\Delta t} = v_o$, and $\dfrac{dv}{dt} = -\omega_n^2 y$ as $\dfrac{\Delta v}{\Delta t} = -\omega_n^2 y_o$, as shown in **Figure 4.10a** and **Figure 4.10b**, respectively where Δt is the run and Δy is the rise in the y–t plane, and Δv is the rise in the v–t plane. Δt is known numerically as the time step.

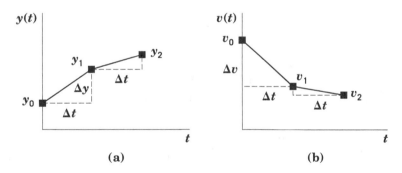

(a)   (b)

Figure 4.10 Numerical approach to state-space equations: (a) discrete point for the displacement; and (b) discrete points for velocity.

Near y_o, $\Delta y = y_1 - y_o$ so that $\dfrac{\Delta y}{\Delta t} = v_o$ is $\dfrac{y_1 - y_o}{\Delta t} = v_o$, which yields $y_1 = y_o + v_o \Delta t$, where y_1 is the predicted displacement after the first Δt, as shown in **Figure 4.10a**. Similarly, near v_o, $\Delta v = v_1 - v_o$ so that $\dfrac{\Delta v}{\Delta t} = -\omega_n^2 y_o$ is $\dfrac{v_1 - v_o}{\Delta t} = -\omega_n^2 y_o$, which yields $v_1 = v_o - \omega_n^2 y_o \Delta t$, where v_1 is the predicted velocity after the first Δt, as shown in **Figure 4.10b**.

Analogous to $y_1 = y_o + v_o \Delta t$ and $v_1 = v_o - \omega_n^2 y_o \Delta t$, write $y_2 = y_1 + v_1 \Delta t$ and $v_2 = v_1 - \omega_n^2 y_1 \Delta t$ in order to predict (y_2, v_2) from (y_1, v_1), as shown in **Figure 4.10**, after the second time interval Δt, and so on and so forth. The discrete equations for the state-space first-order ODEs described by Eqs. (4.48) and (4.49) are summarized in **Table 4.3** for $m = 1, 2, 3, ..., M$, where M is the total number of numerical steps.

ODE	Discrete equations	
$\dfrac{dy}{dt} = v$	$y_m = y_{m-1} + v_{m-1} \Delta t$	(4.56)
$\dfrac{dv}{dt} = -\omega_n^2 y$	$v_m = v_{m-1} - \omega_n^2 y_{m-1} \Delta t$	(4.57)

Table 4.3 Discrete equations for an undamped system represented by state-space ODEs.

The time step Δt should be chosen judiciously to insure that the numerical solution does not diverge. Otherwise, an unbounded numerical solution may lead you to conclude that the system in unstable. For example, you already know, based on analytical solutions, that an undamped

system governed by $\dfrac{d^2 y}{dt^2} + \omega_n^2 y = 0$ is neutrally stable; that is, its amplitude, $y(t)$ is bounded.

However, if the time step Δt is not chosen appropriately, the numerical solution of an undamped system, using Eqs. (4.56) and (4.57), may diverge in time, as shown in **Figure 4.11**, which is not representative of the physics of an undamped system.

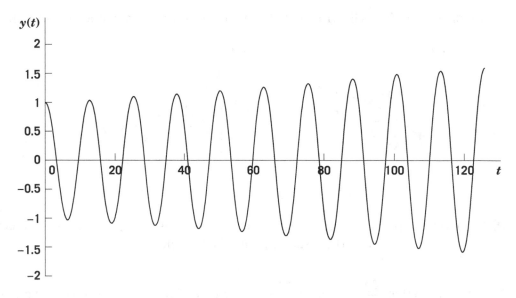

Figure 4.11 Illustration of time step poor choice leading to numerical divergence of a supposedly constant-amplitude undamped system.

A rule of thumb to avoid numerical divergence is to choose the time step such that $\Delta t \ll p$, where p is the damped or undamped oscillation period.

Example 4.10

Consider the discrete equations for the state-space variables for an undamped system, Eqs. (4.56) and (4.57). The natural frequency is 0.5 rad/s. The ICs are $y(0) = y_o = 1$ and $\dfrac{dy(0)}{dt} = v_o = 0$.

1. Choose the time interval such that $\dfrac{\Delta t}{p} = 0.25\left(10^{-3}\right)$ and determine M for one period.

2. Report the values of t, $y(t)$ and $v(t)$ for $m = 1$ and 2.

3. Plot the discrete points (y, v) in the phase plane.

Solution:

1. Choose the time interval such that $\dfrac{\Delta t}{p} = 0.25\left(10^{-3}\right)$ and determine M for one period.

Given that $\omega_n = 0.5\dfrac{rad}{s}$, the period is $p = \dfrac{2\pi}{0.5} = (4\pi)s$. Thus, $\Delta t = 0.25\left(10^{-3}\right)p = \pi\left(10^{-3}\right)$.

The number of points over one period is $M = \dfrac{p}{\Delta t} = \dfrac{1}{0.25\left(10^{-3}\right)} = 4000$ points. Thus, $m = 1, 2, ...,$ 4000.

2. Report the values of t, $y(t)$ and $v(t)$ for $m = 1$ and 2.

For $m = 1$, $t_1 = \Delta t = \pi\left(10^{-3}\right)s$.

From Eq. (4.56), $y_1 = y_o + v_o\Delta t = 1$, with $v_o = 0$.

From Eq. (4.57), $v_1 = v_o - \omega_n^2 y_o \Delta t = -(0.5)^2 \pi\left(10^{-3}\right) = -0.7854\left(10^{-3}\right)$.

For $m = 2$, $t_2 = 2\Delta t = 2\pi\left(10^{-3}\right)s$.

From Eq. (4.56), $y_2 = y_1 + v_1\Delta t = 1 - 0.7854\left(10^{-3}\right)\pi\left(10^{-3}\right) \simeq 1$.

From Eq. (4.57), $v_2 = v_1 - \omega_n^2 y_1 \Delta t = -0.7854\left(10^{-3}\right) - (0.5)^2 \pi\left(10^{-3}\right) = -1.571\left(10^{-3}\right)$, and so on and so forth.

You can easily reproduce Eqs. (4.56) and (4.57) in a spreadsheet, as shown by the first ten points, including the ICs, in **Table 4.4**.

m	Δt	$y(t)$	$v(t)$
IC	0	1	0
1	3.142E-03	1.000E+00	−7.854E-04
2	6.283E-03	1.000E+00	−1.571E-03
3	9.425E-03	1.000E+00	−2.356E-03
4	1.257E-02	1.000E+00	−3.142E-03
5	1.571E-02	1.000E+00	−3.927E-03
6	1.885E-02	1.000E+00	−4.712E-03
7	2.199E-02	9.999E-01	−5.498E-03
8	2.513E-02	9.999E-01	−6.283E-03
9	2.827E-02	9.999E-01	−7.068E-03

Table 4.4 Numerical solution of the state-space first-order ODEs for an undamped system in free vibration.

3. Plot the discrete points (y, v) in the phase plane.

The numerical orbit is generated from the last two columns in **Table 4.4** using 4000 points, as shown in **Figure 4.12**, which is consistent with the elliptic orbit expected for a neutrally stable system.

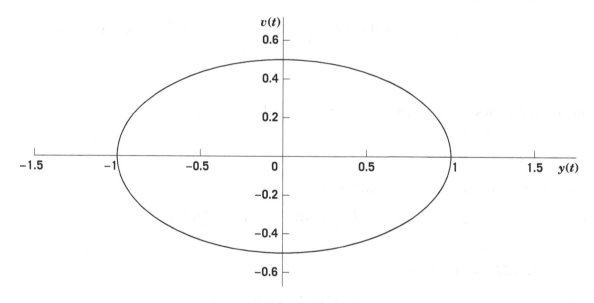

Figure 4.12 Phase-plane orbit of an undamped system in free vibration generated numerically from the state-space first-order ODEs.

Numerical orbit of an underdamped system

The numerical solution of the orbit of an underdamped system is found from Eqs. (4.52) and (4.53). The formulation is analogous to Eqs. (4.56) and (4.57), as summarized in **Table 4.5**, for $m = 1, 2, 3, ..., M$.

ODE	Discrete equation	
$\dfrac{dy}{dt} = v$	$y_m = y_{m-1} + v_{m-1}\Delta t$	(4.58)
$\dfrac{dv}{dt} = -2\zeta\omega_n v - \omega_n^2 y$	$v_m = v_{m-1} - \left(2\zeta\omega_n v_{m-1} + \omega_n^2 y_{m-1}\right)\Delta t$	(4.59)

Table 4.5 Discrete scheme for an underdamped system represented by state-space ODEs.

Example 4.11

Consider the discrete equations for the state-space variables for an underdamped system, Eqs. (4.58) and (4.59). The natural frequency is 1 rad/s, and the damping coefficient is 0.5. The ICs are $y(0) = y_o = 1$ and $\dfrac{dy(0)}{dt} = v_o = 0$.

1. Choose $\Delta t = \pi\left(10^{-3}\right)s$ and show that $\Delta t << p$, where p is the period.
2. Report t, $y(t)$, and $v(t)$ for $m = 1$ and 2.
3. Run the simulation using 4000 points and plot the discrete points (y, v) in the phase plane.

Solution:

1. Choose $\Delta t = \pi\left(10^{-3}\right)s$ and show that $\Delta t << p$, where p is the period.

From Chapter 2, the period of an underdamped system is given by $p = \dfrac{2\pi}{\omega_d}$, where

$\omega_d = \omega_n\sqrt{1-\zeta^2}$. Given $\omega_n = 1\dfrac{rad}{s}$ and $\zeta = \dfrac{1}{2}$ yields $\omega_d = 0.867\dfrac{rad}{s}$ and $p = 7.26s$. The ratio

$\dfrac{\Delta t}{p} = 4.3\left(10^{-4}\right)$. Thus, $\Delta t << p$.

2. Report t, $y(t)$ and $v(t)$ for $m = 1$ and 2.

For $m = 1$, $\Delta t = \pi\left(10^{-3}\right)s$.

From Eq. (4.58), $y_1 = y_o + v_o\Delta t = 1$, with $v_o = 0$.

From Eq. (4.59), $v_1 = v_o - \left(v_o + y_o\right)\Delta t = -\pi\left(10^{-3}\right)$.

For $m = 2$, $t_2 = 2\Delta t = 2\pi\left(10^{-3}\right)s$.

From Eq. (4.58), $y_2 = y_1 + v_1 \Delta t = 1 - 0.7854(10^{-3})\pi(10^{-3}) \simeq 1$.

From Eq. (4.59), $v_2 = v_1 - (v_1 + y_1)\Delta t = -\pi(10^{-3}) - \left[-\pi(10^{-3}) + 1\right]\pi(10^{-3}) = -6.2733(10^{-3})$, and

so on and so forth. The first ten points, including the ICs, are summarized in **Table 4.6**.

m	Δt	$y(t)$	$v(t)$
IC	0	1	0
1	3.1416E-03	1.0000E+00	−3.1416E-03
2	6.2832E-03	9.9999E-01	−6.2733E-03
3	9.4248E-03	9.9997E-01	−9.3952E-03
4	1.2566E-02	9.9994E-01	−1.2507E-02
5	1.5708E-02	9.9990E-01	−1.5609E-02
6	1.8850E-02	9.9985E-01	−1.8702E-02
7	2.1991E-02	9.9979E-01	−2.1784E-02
8	2.5133E-02	9.9973E-01	−2.4856E-02
9	2.8274E-02	9.9965E-01	−2.7919E-02

Table 4.6 Numerical solution of the state-space first-order ODEs
for an underdamped system in free vibration.

2. Run the simulation using 4000 points and plot the discrete points (y, v) in the phase plane.
The numerical orbit is generated from the last two columns in **Table 4.6** using 4000 points, as
shown in **Figure 4.13**.

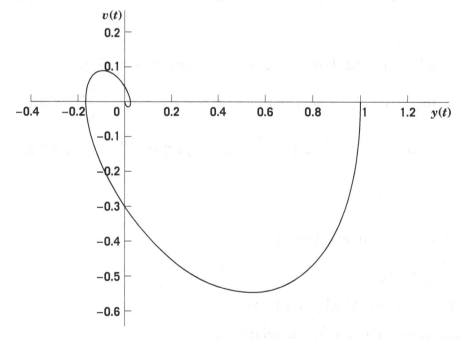

Figure 4.13 Phase-plane orbit generated numerically from the state-space first-order ODEs
for an underdamped system in free vibration.

For reference, the underdamped displacement is generated from the third column of **Table 4.6**, as shown in **Figure 4.14**, which is the expected behavior of such a system in free vibration.

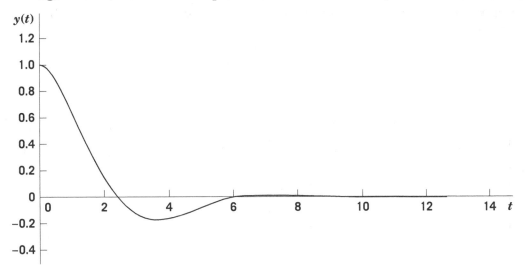

Figure 4.14 The displacement of an underdamped system in free vibration generated numerically from the state-space first-order ODEs.

Next you will learn about a system of second-order ODEs, also known as multi-degree-of-freedom systems, and see that it is sufficient to keep the formulation as coupled second-order ODEs rather than transforming the equations to a system of first-order ODEs.

4.6 Second-Order Systems

In Chapter 2, you studied second-order ODEs of the form $\dfrac{d^2 y}{dt^2} + a_1 \dfrac{dy}{dt} + a_0 y = g(t)$ that can be

rewritten as $\dfrac{d^2 y}{dt^2} = ay + b\dfrac{dy}{dt} + g(t)$ where $a = -a_0$ and $b = -a_1$ are constants. The foregoing ODE can be extended to model engineering problems as a system of coupled second-order ODEs with n dependent variables, $y_1, y_2, \ldots, y_n$,

$$\frac{d^2 y_1}{dt^2} = a_{11}y_1 + a_{12}y_2 + \cdots + a_{1n}y_n + b_{11}\frac{dy_1}{dt} + b_{12}\frac{dy_2}{dt} + \cdots + b_{1n}\frac{dy_n}{dt} + g_1(t)$$

$$\frac{d^2 y_2}{dt^2} = a_{21}y_1 + a_{22}y_2 + \cdots + a_{2n}y_n + b_{21}\frac{dy_1}{dt} + b_{22}\frac{dy_2}{dt} + \cdots + b_{2n}\frac{dy_n}{dt} + g_2(t)$$

$$\vdots \qquad \qquad \vdots \qquad \qquad \qquad \qquad \vdots$$

(4.60)

$$\frac{d^2 y_n}{dt^2} = a_{n1}y_1 + a_{n2}y_2 + \cdots + a_{nn}y_n + b_{n1}\frac{dy_1}{dt} + b_{n2}\frac{dy_2}{dt} + \cdots + b_{nn}\frac{dy_n}{dt} + g_n(t)$$

where a_{ij} and b_{ij} for $i = 1, 2, \ldots, n$ and $j = 1, 2, \ldots, n$ are constant coefficients and $g_i(t)$ are forcing functions. The system of ODEs in Eq. (4.60) is coupled because all the dependent variables appear in all equations. As in first-order ODEs, the system of Eqs. (4.60) can be written in matrix form as

$$\frac{d^2 Y}{dt^2} = AY + B\frac{dY}{dt} + G \tag{4.61}$$

where

$$Y = \begin{bmatrix} y_1 \\ y_2 \\ \cdot \\ \cdot \\ \cdot \\ \cdot \\ y_n \end{bmatrix}, A = \begin{bmatrix} a_{11} & a_{12} & \cdot & \cdot & \cdot & a_{1n} \\ a_{21} & a_{22} & \cdot & \cdot & \cdot & a_{2n} \\ \cdot & & & & & \cdot \\ \cdot & & & & & \cdot \\ \cdot & & & & & \cdot \\ a_{n1} & a_{n2} & \cdot & \cdot & \cdot & a_{nn} \end{bmatrix}, B = \begin{bmatrix} b_{11} & b_{12} & \cdot & \cdot & \cdot & b_{1n} \\ b_{21} & b_{22} & \cdot & \cdot & \cdot & b_{2n} \\ \cdot & & & & & \cdot \\ \cdot & & & & & \cdot \\ \cdot & & & & & \cdot \\ b_{n1} & b_{n2} & \cdot & \cdot & \cdot & b_{nn} \end{bmatrix} \text{ and } G = \begin{bmatrix} g_1(t) \\ g_2(t) \\ \cdot \\ \cdot \\ \cdot \\ g_n(t) \end{bmatrix}$$

The ICs for the system of equations are given by the two-column matrices

$$\begin{bmatrix} y_1(t_o) \\ y_2(t_o) \\ \cdot \\ \cdot \\ \cdot \\ y_n(t_o) \end{bmatrix} \text{ and } \begin{bmatrix} \dfrac{dy_1(t_o)}{dt} \\ \dfrac{dy_2(t_o)}{dt} \\ \cdot \\ \cdot \\ \cdot \\ \dfrac{dy_n(t_o)}{dt} \end{bmatrix}.$$

By analogy to $\dfrac{d^2 y}{dt^2} + 2\zeta\omega_n \dfrac{dy}{dt} + \omega_n^2 y = g(t)$ or $\dfrac{d^2 y}{dt^2} = -\omega_n^2 y - 2\zeta\omega_n \dfrac{dy}{dt} + g(t)$, the elements of matrix A in Eq. (4.61) can be interpreted to contain frequency information about the system, whereas matrix B elements contain damping information. The system of equations given by Eq. (4.61) is homogeneous when $G = Z$, and nonhomogeneous when $G \neq Z$, where Z is the zero matrix. Furthermore, the system is undamped if matrix B is also the zero matrix, so that Eq. (4.61) reduces to

$$\frac{d^2 Y}{dt^2} = AY \tag{4.62}$$

In engineering systems, Eq. (4.62) in itself is quite useful. For example, consider the two-story building in **Figure 4.1**, as part of a preliminary design in a seismic area with earthquake frequencies falling into a certain range. By modeling the building in free vibration, that is, $G = Z$, and no damping, that is, $B = Z$, it is possible to obtain the building natural frequencies from the eigenvalues of Eq. (4.62), and the building lateral displacements from the corresponding eigenvectors, which are discussed in the next section. If the natural frequencies fall outside the earthquake frequency range, the building will not resonate with the earthquake catastrophically. Thus, the solution of the homogeneous and undamped system of second-order ODEs given by Eq. (4.62), as an eigenvalue problem, is sufficient to predict whether or not the earthquake forcing frequency would cause building resonance. Thus, there is no need to solve a system of nonhomogeneous ODEs for this particular purpose.

4.7 Eigenvalues of Homogeneous and Linear System of Second-Order ODEs

Eigenvalues and eigenvectors

Following the same approach used in coupled first-order ODEs, propose an exponential solution $Y = Ve^{\sqrt{\lambda}t}$ where V is the set of n eigenvectors v_1, v_2, ..., v_n that corresponds to the set of eigenvalues λ_1, λ_2, ..., λ_n of the $n \times n$ matrix A. The choice of $\sqrt{\lambda}$ in the exponent is for mathematical convenience so that the eigenvalue problem formulation is identical to that covered earlier. Substituting $Y = Ve^{\sqrt{\lambda}t}$ and its second derivative, $\dfrac{d^2Y}{dt^2} = \lambda Ve^{\sqrt{\lambda}t}$ in Eq. (4.62) yields

$\lambda Ve^{\sqrt{\lambda}t} = AVe^{\sqrt{\lambda}t}$ or the classical eigenvalue formulation $\lambda V = AV$ after dividing both sides of the equation by $e^{\sqrt{\lambda}t}$.

Example 4.12

Consider the following system of second-order homogeneous ODEs,

$$\frac{d^2y_1}{dt^2} - y_2 = 0$$

$$\frac{d^2y_2}{dt^2} + 4y_1 + 5y_2 = 0$$

(4.63)

1. Cast the system of ODEs in the form of Eq. (4.62).
2. Find the eigenvalues.
3. Find the normalized eigenvectors corresponding to the eigenvalues.
4. Find the general solution for $y_1(t)$ and $y_2(t)$.

Solution:

1. Cast the system of ODEs in the form of Eq. (4.62).

Rewrite Eq. (4.63) as $\dfrac{d^2 y_1}{dt^2} = (0)y_1 + y_2$ or in matrix form, $\begin{bmatrix} \dfrac{d^2 y_1}{dt^2} \\ \dfrac{d^2 y_2}{dt^2} \end{bmatrix} = \begin{bmatrix} 0 & 1 \\ -4 & -5 \end{bmatrix}\begin{bmatrix} y_1 \\ y_2 \end{bmatrix}$ so that

$$A = \begin{bmatrix} 0 & 1 \\ -4 & -5 \end{bmatrix}.$$

2. Find the eigenvalues.

The eigenvalues of matrix A are found from $|A - \lambda I| = 0$, or $\begin{vmatrix} -\lambda & 1 \\ -4 & -5-\lambda \end{vmatrix} = 0$ that yields the

characteristic equation $\lambda^2 + 5\lambda + 4 = 0$ with roots $\lambda_1 = -1$ and $\lambda_2 = -4$.

3. Find the normalized eigenvectors corresponding to the eigenvalues.

The eigenvalues are found from Eq. (4.20), which is expanded to

$$\begin{bmatrix} 0-\lambda & 1 \\ -4 & -5-\lambda \end{bmatrix}\begin{bmatrix} v_1 \\ v_2 \end{bmatrix} = 0 \tag{4.64}$$

where v_1 and v_2 are the elements of the eigenvector V.

For $\lambda_1 = -1$, Eq. (4.64) yields $\begin{bmatrix} 1 & 1 \\ -4 & -4 \end{bmatrix}\begin{bmatrix} v_1 \\ v_2 \end{bmatrix} = 0$ whose rows degenerate to $v_1 + v_2 = 0$, or

$v_1 = -v_2$. Thus, the eigenvector is $V_1 = \begin{bmatrix} -1 \\ 1 \end{bmatrix}$. The normalized eigenvector, $\hat{v}_1$ is found by

dividing V_1 by its magnitude, that is, $\hat{v}_1 = \dfrac{1}{\sqrt{2}}\begin{bmatrix} -1 \\ 1 \end{bmatrix} = \begin{bmatrix} -\dfrac{1}{\sqrt{2}} \\ \dfrac{1}{\sqrt{2}} \end{bmatrix}.$

Similarly, for $\lambda_2 = -4$, Eq. (4.64) yields $\begin{bmatrix} 4 & 1 \\ -4 & -1 \end{bmatrix}\begin{bmatrix} v_1 \\ v_2 \end{bmatrix} = 0$ whose rows degenerate to

$4v_1 + v_2 = 0$, or $4v_1 = -v_2$. Thus, $V_2 = \begin{bmatrix} -1 \\ 4 \end{bmatrix}$ and $\hat{v}_2 = \begin{bmatrix} -\dfrac{1}{\sqrt{17}} \\ \dfrac{4}{\sqrt{17}} \end{bmatrix}.$

The eigenvalues and normalized eigenvectors can also be found using CAS.

4. Find the general solution for $y_1(t)$ and $y_2(t)$.

The general solution is formed from the proposed general solution, $Y = Ve^{\sqrt{\lambda}t}$, using the principle of linear superposition, which yields $Y = C_1\hat{v}_1 e^{\sqrt{\lambda_1}t} + C_2\hat{v}_2 e^{\sqrt{\lambda_2}t}$, or

$$\begin{bmatrix} y_1(t) \\ y_2(t) \end{bmatrix} = \begin{bmatrix} -\dfrac{1}{\sqrt{2}} \\ \dfrac{1}{\sqrt{2}} \end{bmatrix} C_1 e^{\sqrt{-1}t} + \begin{bmatrix} -\dfrac{1}{\sqrt{17}} \\ \dfrac{4}{\sqrt{17}} \end{bmatrix} C_2 e^{\sqrt{-4}t} \tag{4.65}$$

Using Euler's formula, write $C_1 e^{\sqrt{-1}t}$ as $C_1 e^{it} = C_1[\cos(t) + i\sin(t)]$, which can be written in terms of new constants as $C_1 e^{it} = a\cos(t) + b\sin(t) = c_1\cos(t + \phi_1)$, based on Eqs. (1.67)–(1.72). Similarly, write $C_2 e^{\sqrt{-4}t} = C_2 e^{2it} = C_2[\cos(2t) + i\sin(2t)] = c_2\cos(2t + \phi_2)$ so that upon expansion, Eq. (4.65) yields

$$y_1(t) = -\frac{1}{\sqrt{2}}c_1\cos(t + \phi_1) - \frac{1}{\sqrt{17}}c_2\cos(2t + \phi_2) \tag{4.66}$$

and

$$y_2(t) = \frac{1}{\sqrt{2}}c_1\cos(t + \phi_1) + \frac{4}{\sqrt{17}}c_2\cos(2t + \phi_2) \tag{4.67}$$

which are general solutions

Example 4.12 provides valuable lessons for any engineering system, electrical, mechanical, chemical, and the like, which is modeled by two coupled second-order ODEs using the form of Eq. (4.62), for example.

$$\begin{bmatrix} \dfrac{d^2 y_1}{dt^2} \\ \dfrac{d^2 y_2}{dt^2} \end{bmatrix} = \begin{bmatrix} a_{11} & a_{12} \\ a_{21} & a_{22} \end{bmatrix} \begin{bmatrix} y_1 \\ y_2 \end{bmatrix} \tag{4.68}$$

such that the eigenvalues are negative, that is, $\lambda_1 < 0$ and $\lambda_2 < 0$. The first lesson is that the general solution of Eq. (4.68) can be written as

$$Y(t) = \hat{v}_1 c_1\cos(\omega_1 t + \phi_1) + \hat{v}_2 c_2\cos(\omega_2 t + \phi_2) \tag{4.69}$$

where $\hat{v}_1$ and $\hat{v}_2$ are the normalized eigenvectors that correspond to the eigenvalues λ_1 and λ_2 obtained from $|A - \lambda I| = 0$. The second lesson is that $\lambda_1 < 0$ and $\lambda_2 < 0$ yield an oscillating solution with natural frequencies $\omega_1 = \sqrt{|\lambda_1|}$ and $\omega_2 = \sqrt{|\lambda_2|}$. Third, for definiteness, order the frequencies such that $\omega_1 < \omega_2$; that is, ω_1 is the low frequency, and ω_2 is the high frequency. If the system starts with ICs such that $c_2 = 0$, as you may demonstrate in a homework problem, then Eq. (4.69) yields $Y = \hat{v}_1 c_1 \cos(\omega_1 t + \phi_1)$, or

$$y_1(t) = v_1 c_1 \cos(\omega_1 t + \phi_1) \tag{4.70}$$

and

$$y_2(t) = v_2 c_1 \cos(\omega_1 t + \phi_1) \tag{4.71}$$

where v_1 and v_2 are the elements of the normalized eigenvalue, $\hat{v}_1$. Equations (4.70) and (4.71) are called low-mode oscillations at the low frequency ω_1. $y_1(t)$, and $y_2(t)$ are called the low-mode shapes. The ratio of the two shapes is $\dfrac{y_1(t)}{y_2(t)} = \dfrac{v_1}{v_2}$, or $y_1(t) = y_2(t)\dfrac{v_1}{v_2}$, which means that $y_1(t)$ is scaled relative to $y_2(t)$ by the ratio of v_1 and v_2.

If, on the other hand, the system starts with ICs such that $c_1 = 0$, then Eq. (4.69) yields $Y = \hat{v}_2 c_2 \cos(\omega_2 t + \phi_2)$, or

$$y_1(t) = v_1 c_2 \cos(\omega_2 t + \phi_2) \tag{4.72}$$

and

$$y_2(t) = v_2 c_2 \cos(\omega_2 t + \phi_2) \tag{4.73}$$

where v_1 and v_2 are the elements of the normalized eigenvalue, $\hat{v}_2$. Equations (4.72) and (4.73) are called high-mode oscillations at the high frequency ω_2. $y_1(t)$ and $y_2(t)$ are called the high-mode shapes. Once again $y_1(t) = y_2(t)\dfrac{v_1}{v_2}$. It is possible also that for some ICs, neither c_1 nor c_2 is zero, so that the system oscillates with a combination of low-mode and high-mode shapes.

Modeling of two-degree of freedom, mass-spring system

Consider an oscillating system that consists of two masses, m_1 and m_2, two springs with stiffnesses k_1 and k_2, and two-degree-of-freedom for displacement, $x_1(t)$ and $x_2(t)$, as shown in **Figure 4.15**. This model is typical for the two-story building described in **Figure 4.1**, which may be subject to earthquake modeled by $x_g(t)$.

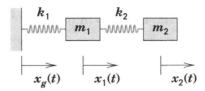

Figure 4.15 A two-degree-of-freedom oscillating system.

In order to set the stage for the analysis of a two-degree-of-freedom system, consider first a one-degree-of-freedom system.

One-degree-of-freedom with fixed support

Consider the mass, m, and spring with stiffness, k, as shown in **Figure 4.16**. This spring is fixed at the left end so that $x_g(t) = 0$, and the mass has one-degree-of-freedom, $x(t)$.

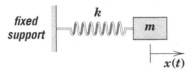

Figure 4.16 Mass and spring with one end fixed.

The analysis proceeds by considering the free body diagram of the mass, as shown in **Figure 4.17**. The mass is assumed at this instant of time to be accelerating to the right, which is consistent with the direction of coordinate x in **Figure 4.16**. In this case, the spring restoring force, F, is opposite to the assumed motion; that is, it acts to the left when the motion is to the right.

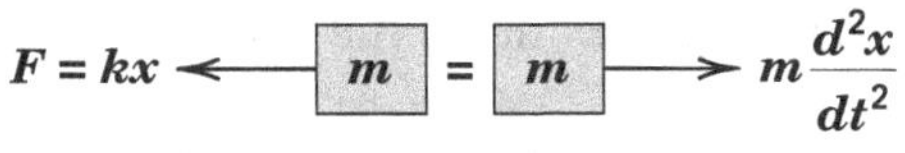

$$F = kx \longleftarrow \boxed{m} = \boxed{m} \longrightarrow m\frac{d^2x}{dt^2}$$

Figure 4.17 Free body diagram of mass m.

From Newton's law of motion

$$-F = m\frac{d^2x}{dt^2} \qquad (4.74)$$

The displacement of the spring is relative to fixed support, $x_g(t) = 0$, which degenerates to x. From Hooke's law for a spring

$$F = kx \qquad (4.75)$$

Substituting Eq. (4.75) into Eq. (4.74) yields the governing ODE $-kx = m\frac{d^2x}{dt^2}$, or

$$\frac{d^2x}{dt^2} + \frac{k}{m}x = 0 \qquad (4.76)$$

Equation (4.76) is the classical ODE for a system in free vibration without damping, which was covered in Chapter 2.

One-degree-of-freedom with moving support

Consider the mass, m, and spring with stiffness, k, as shown in **Figure 4.18**. This spring is attached to a moving support with displacement, $x_g(t)$. The mass still has one-degree-of-freedom, $x(t)$.

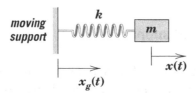

Figure 4.18 Mass and spring attached to a moving support.

The free body diagram of the mass, as shown in **Figure 4.19**, is similar to that in **Figure 4.17**. The only difference is the effect of the moving support. Whereas the displacement of the spring in **Figure 4.16** was relative to a fixed support, here the spring displacement is relative to the moving support; that is, $x - x_g$.

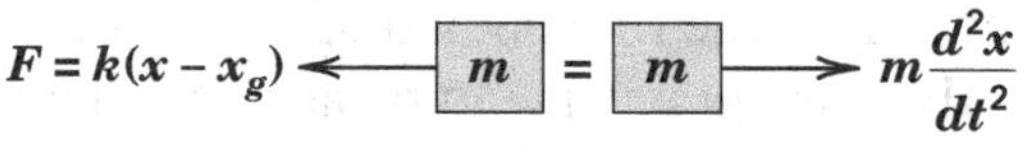

$$F = k(x - x_g) \longleftarrow \boxed{m} = \boxed{m} \longrightarrow m\frac{d^2x}{dt^2}$$

Figure 4.19 Free body diagram of mass, m.

Newton's law of motion, Eq. (4.74), applies here; the governing ODE is $-F = m\dfrac{d^2x}{dt^2}$. For a relative displacement, $x - x_g$, Hooke's law for a spring is given by

$$F = k\left(x - x_g\right) \tag{4.77}$$

So that the governing ODE is $-k\left(x - x_g\right) = m\dfrac{d^2x}{dt^2}$, or

$$\frac{d^2x}{dt^2} + \frac{k}{m}x = \frac{k}{m}x_g \tag{4.78}$$

Equation (4.78) is the classical ODE for a system in forced vibration without damping, which was covered in Chapter 2. The forcing function is the term on the right-hand side of the equation.

Two-degree-of-freedom with moving support

Consider two masses, m_1 and m_2, with displacements, x_1 and x_2, respectively; two springs with stiffnesses, k_1 and k_1; and the lateral displacement of the moving support, $x_g(t)$, as shown in **Figure 4.20**.

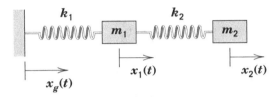

Figure 4.20 Mass-spring model.

Consider the free body diagram of the mass on the right, m_2, as shown in **Figure 4.21**. Here, the spring displacement is x_2 relative to x_1; that is, $x_2 - x_1$, so that by Hooke's law, the spring force, designated F_{21}, is given by

$$F_{21} = k_2 \left(x_2 - x_1 \right) \tag{4.79}$$

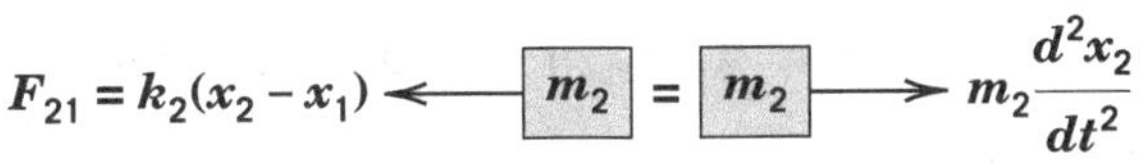

$$F_{21} = k_2(x_2 - x_1) \longleftarrow \boxed{m_2} = \boxed{m_2} \longrightarrow m_2 \frac{d^2 x_2}{dt^2}$$

Figure 4.21 Free body diagram of mass m_2.

From Newton's law of motion

$$-F_{21} = m_2 \frac{d^2 x_2}{dt^2} \tag{4.80}$$

Substituting Eq. (4.79) into Eq. (4.80) yields the first governing ODE for mass m_2,

$$-k_2 \left(x_2 - x_1 \right) = m_2 \frac{d^2 x_2}{dt^2} \tag{4.81}$$

Next, consider the free body diagram of mass m_1, as shown in **Figure 4.22**.

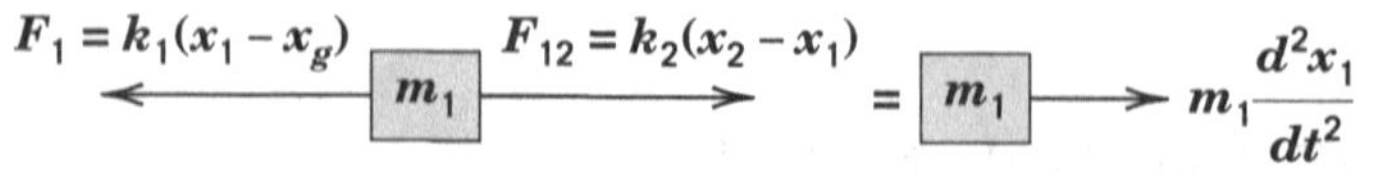

Figure 4.22 Free body diagram of mass m_1.

Here, m_1 experiences two forces. One force is the spring displacement relative to the lateral displacement of the support, which is given by Hooke's law as

$$F_1 = k_1 \left(x_1 - x_g \right) \tag{4.82}$$

The second force, designated F_{12}, must satisfy Newton's third law: for every action there is an equal and opposite reaction. Since F_{21} acts to the left, as shown in **Figure 4.21**, F_{12} acts to the right, as shown in **Figure 4.22**. Applying Newton's law of motion yields

$$-F_1 + F_{12} = m_1 \frac{d^2 x_1}{dt^2} \tag{4.83}$$

Substituting Eqs. (4.82) and $F_{12} = k_2 \left(x_2 - x_1 \right)$ into Eq. (4.83) yields the second governing ODE for mass m_1

$$-k_1 \left(x_1 - x_g \right) + k_2 \left(x_2 - x_1 \right) = m_1 \frac{d^2 x_1}{dt^2} \tag{4.84}$$

Equations (4.81) and (4.84) can be rewritten as

$$
\begin{aligned}
m_1 \frac{d^2 x_1}{dt^2} &= -\left(k_1 + k_2 \right) x_1 + k_2 x_2 + k_1 x_g \\
m_2 \frac{d^2 x_2}{dt^2} &= k_2 x_1 - k_2 x_2
\end{aligned}
\tag{4.85}
$$

or in matrix form

$$\frac{d^2}{dt^2} \begin{bmatrix} x_1 \\ x_2 \end{bmatrix} = \begin{bmatrix} -\dfrac{k_1 + k_2}{m_1} & \dfrac{k_2}{m_1} \\ \dfrac{k_2}{m_2} & -\dfrac{k_2}{m_2} \end{bmatrix} \begin{bmatrix} x_1 \\ x_2 \end{bmatrix} + \begin{bmatrix} \dfrac{k_1 x_g}{m_1} \\ 0 \end{bmatrix} \tag{4.86}$$

Equation (4.86) constitutes a system of linear, nonhomogeneous, and coupled system of second-order ODEs with constant coefficients. The system is nonhomogeneous by virtue of $\begin{bmatrix} \dfrac{k_1 x_g}{m_1} \\ 0 \end{bmatrix}$ on the right-hand side. For the purpose of finding the natural frequencies using eigenvalues, it is sufficient to solve the homogeneous part of Eq. (4.86); that is,

$$\frac{d^2}{dt^2}\begin{bmatrix} x_1 \\ x_2 \end{bmatrix} = \begin{bmatrix} -\dfrac{k_1+k_2}{m_1} & \dfrac{k_2}{m_1} \\ \dfrac{k_2}{m_2} & -\dfrac{k_2}{m_2} \end{bmatrix}\begin{bmatrix} x_1 \\ x_2 \end{bmatrix} \qquad (4.87)$$

Once you find the natural frequencies from Eq. (4.87), you can infer from an engineering point of view that if the support $x_g(t)$ is excited, say by an earthquake, with a frequency that happened to match one of the eigenvalues of the system, then resonance occurs, which is destructive.

Equation (4.87) is in the form of Eq. (4.62), where $A = \begin{bmatrix} -\dfrac{k_1+k_2}{m_1} & \dfrac{k_2}{m_1} \\ \dfrac{k_2}{m_2} & -\dfrac{k_2}{m_2} \end{bmatrix}$. The eigenvalues are found from matrix A.

Example 4.13 Building Vibration

An engineer wants to find the natural frequencies of the two-story building described in **Figure 4.1**, in order to determine if they are within the earthquake frequency range, $0.05\dfrac{rad}{s}$ to $1\dfrac{rad}{s}$, at that location. The building lateral displacement is modeled by Eq. (4.87).

1. Determine the eigenvalues and normalized eigenvectors.
2. Determine if the low-mode and high-mode natural frequencies of the building fall within the range of the earthquake oscillation frequency.
3. Find expressions for the low-mode and high-mode shapes of the lateral displacements, $x_1(t)$ and $x_2(t)$.

Data: $m_1 = m_2 = 10^5\, kg$; stiffnesses: $k_1 = 10^7\,\dfrac{N}{m}$ and $k_2 = 2k_1$.

Solution:

1. Determine the eigenvalues and eigenvectors.

Substituting the values of mass and stiffness in Eq. (4.87) yields $\dfrac{d^2}{dt^2}\begin{bmatrix} x_1 \\ x_2 \end{bmatrix} = \begin{bmatrix} -300 & 200 \\ 200 & -200 \end{bmatrix}\begin{bmatrix} x_1 \\ x_2 \end{bmatrix}$.

Thus $A = \begin{bmatrix} -300 & 200 \\ 200 & -200 \end{bmatrix}$ whose eigenvalues are found from $\begin{vmatrix} -300-\lambda & 200 \\ 200 & -200-\lambda \end{vmatrix} = 0$, either

manually following the steps in Example 4.12, or using CAS, which yields $\lambda_1 = -43.85$ and $\lambda_2 = -456.16$. The corresponding normalized eigenvectors are found using CAS, which are

$$\hat{v}_1 = \begin{bmatrix} 0.6154 \\ 0.7882 \end{bmatrix} \text{ and } \hat{v}_2 = \begin{bmatrix} 0.7882 \\ -0.6154 \end{bmatrix}.$$

2. Determine if the low-mode and high-mode natural frequencies of the building fall within the range of the earthquake oscillation frequency.

The low-mode and high-mode natural frequencies are given by $\omega_1 = \sqrt{|\lambda_1|} = 6.62\dfrac{rad}{s}$ and

$\omega_2 = \sqrt{|\lambda_2|} = 21.36\dfrac{rad}{s}$. Since both frequencies are greater than $1\dfrac{rad}{s}$, the building is outside the range of the earthquake oscillation frequencies and should not resonate seismically with increasing amplitude.

3. Find expressions for the low-mode and high-mode shapes of the lateral displacements, $x_1(t)$ and $x_2(t)$.

The displacements $x_1(t)$ and $x_2(t)$ for the low-mode frequency are given by similar expression

to Eqs. (4.70) and (4.71) using v_1 and v_2 of the normalized eigenvalue, $\hat{v}_1 = \begin{bmatrix} 0.6154 \\ 0.7882 \end{bmatrix}$; that is,

$$x_1(t) = 0.6154c_1 \cos(6.62t + \phi_1)$$
$$x_2(t) = 0.7882c_1 \cos(6.62t + \phi_1)$$

so that $x_1(t) = x_2(t)\dfrac{v_1}{v_2}$, or $x_1(t) = 0.781x_2(t)$. Simply put, if $x_2(t)$ displacement is one unit,

for example, 1 cm to the right, then $x_1(t)$ displacement is 0.781 cm also to the right, and vice versa. These low-mode displacements can be plotted by setting $c_1 = 1$ without loss of generality,

as shown in **Figure 4.23**. Here, both floors oscillate at the same frequency $\omega_1 = 6.62\dfrac{rad}{s}$ and

phase angle, ϕ_1, but with amplitudes, $0.6154c_1$ for the first floor and $0.7882c_1$ for the second floor.

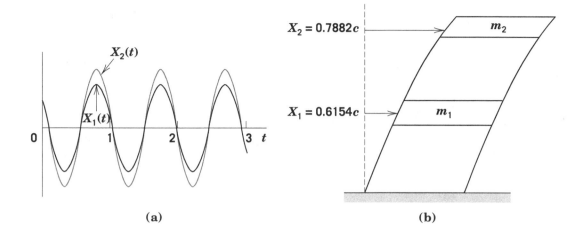

(a)   **(b)**

Figure 4.23 Two-story building displacement at $\omega_1 = 6.62\dfrac{rad}{s}$: (a) $x_1(t)$ and $x_2(t)$ in harmonic

motion; and (b) building low-mode shape.

Similarly, the displacements $x_1(t)$ and $x_2(t)$ for the high-mode frequency are found with

$\hat{v}_2 = \begin{bmatrix} 0.7882 \\ -0.6154 \end{bmatrix}$; that is,

$$x_1(t) = 0.7882c_2 \cos\left(21.36t + \phi_2\right)$$
$$x_2(t) = -0.6154c_2 \cos\left(21.36t + \phi_2\right)$$

so that $x_1(t) = x_2(t)\dfrac{v_1}{v_2}$, or $x_1(t) = -1.28x_2(t)$. Here, if $x_2(t)$ displacement is one unit, for

example, 1 *cm* to the right, then $x_1(t)$ displacement is 1.28 *cm* to the left, and vice versa. Here,

both floors oscillate at the same frequency $\omega_2 = 21.36\dfrac{rad}{s}$ and phase angle, ϕ_2, but in out-of-

phase motion, as shown in **Figure 4.24**.

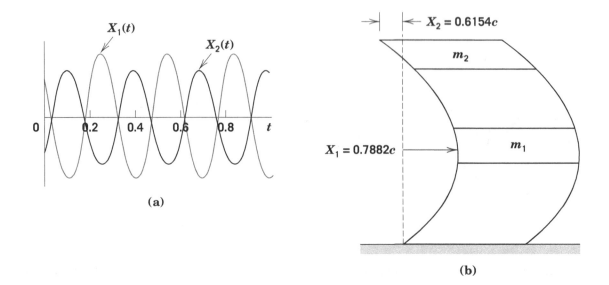

(a)

(b)

Figure 4.24 Two-story building displacement at $\omega_2 = 21.36\dfrac{rad}{s}$: (a) $x_1(t)$ out-of-phase with $x_2(t)$; and (b) building high-mode shape.

In this engineering example you learned how a set of coupled, homogeneous, linear, and constant coefficient ODEs is formulated as an eigenvalue problem, where the eigenvalues are related to the natural frequencies of the building and the eigenvectors are related to the shape the building will take at each natural frequency.

HOMEWORK PROBLEMS FOR CHAPTER FOUR

Problems for Section 4.2

For Problems 1–4, define $v = \dfrac{dy}{dt}$.

a. Cast each ODE in the form of Eq. (4.2) using the state-space variables v and y.
b. Identify Y, A, and G.

1. $2\dfrac{d^2 y}{dt^2} = 3$;
2. $2\dfrac{d^2 y}{dt^2} + \dfrac{dy}{dt} = -2y$;
3. $2\dfrac{d^2 y}{dt^2} - 3y - 1 = 0$;
4. $\dfrac{d^2 y}{dt^2} + y = 0$

For Problems 5–8, define the intermediate variables $y_3 = \dfrac{dy_1}{dt}$ and $y_4 = \dfrac{dy_2}{dt}$.

a. Cast each ODE in the form of Eq. (4.2) using state-space variables.
b. Identify Y, A, and G.

5.
$$\dfrac{d^2 y_1}{dt^2} = 1$$
$$\dfrac{d^2 y_2}{dt^2} = -1$$

6.
$$4\dfrac{d^2 y_1}{dt^2} + 3\dfrac{dy_1}{dt} + 2\dfrac{dy_2}{dt} + y_1 + 2y_2 = 3$$
$$\dfrac{d^2 y_2}{dt^2} = 0$$

7.
$$\dfrac{d^2 y_1}{dt^2} + \dfrac{dy_1}{dt} + y_2 + 5 = 0$$
$$\dfrac{d^2 y_2}{dt^2} - \dfrac{dy_1}{dt} + y_2 = 0$$

8.
$$\dfrac{d^2 y_1}{dt^2} + \dfrac{dy_1}{dt} + \dfrac{dy_2}{dt} + y_1 + y_2 = 1$$
$$\dfrac{d^2 y_2}{dt^2} + \dfrac{dy_2}{dt} - \dfrac{dy_1}{dt} + y_2 - y_1 = 0$$

Problems for Section 4.3

For Problems 9–14,

a. Write the given ODE as a system of coupled first-order ODEs.
b. Find the eigenvalues and determine whether the ODE describes a neutrally stable, asymptotically stable, or unstable system.

9. $2\dfrac{d^2 y}{dt^2} - 2y = 0$;
10. $2\dfrac{d^2 y}{dt^2} + 5\dfrac{dy}{dt} + 2y = 0$;
11. $2\dfrac{d^2 y}{dt^2} + 5\dfrac{dy}{dt} + 2y = 0$;

12. $2\dfrac{d^2y}{dt^2}+5\dfrac{dy}{dt}+2y=0$;

13. $\dfrac{d^2y_1}{dt^2}=0$;
$\dfrac{d^2y_2}{dt^2}=0$

14. $\dfrac{d^2y_1}{dt^2}+\dfrac{dy_1}{dt}+y_2=0$
$\dfrac{d^2y_2}{dt^2}-\dfrac{dy_1}{dt}+y_2=0$

For Problems 15–18, consider the Lorenz equations, originally developed by Edward Lorenz for atmospheric modeling. The following equations are a linear version of a system of first-order ODEs near the equilibrium point (0,0,0):

$$\frac{dx}{dt}=\sigma y-\sigma x$$

$$\frac{dy}{dt}=rx-y$$

$$\frac{dz}{dt}=-bz$$

where $b=\dfrac{8}{3}$, $\sigma=10$, $r>0$ and $r\neq1$.

a. Find the eigenvalues and determine whether the ODEs describe a neutrally stable, asymptotically stable, or unstable system.

15. $r=0.999$;   16. $r=1.001$;   17. $r=0.001$;   18. $r=2$.

19. Revisit Example 2.6 about the hinge design problem from Chapter 2, as shown in **Figure 4.25**. The angular displacement, $\theta(t)$, is described by the second-order ODE, $\dfrac{d^2\theta}{dt^2}+\dfrac{c_d}{mL^2}\dfrac{d\theta}{dt}+\dfrac{g}{L}\theta=0$, where $m=30kg$, $L=1m$, $c_d=53.59\dfrac{N.m}{s}$ and $g=9.81\dfrac{m}{s^2}$.

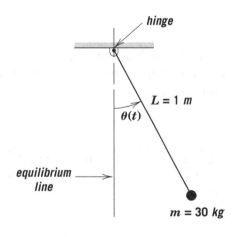

Figure 4.25 Rod, mass, and swing model.

a. Write the given ODE as a system of coupled first-order ODEs.
b. Find the eigenvalues and determine whether the ODE describes a neutrally stable, asymptotically stable, or unstable system.

20. Revisit Problem 19 where the hinge friction coefficient is neglected so that the equation of motion is $\dfrac{d^2\theta}{dt^2}+\dfrac{g}{L}\theta=0$, where $g=9.81\dfrac{m}{s^2}$.

a. Find the pendulum length, L, in m so that $|\lambda_1|=|\lambda_2|=p$.

21. Consider the emergency water supply tower, as shown in **Figure 4.26a**, which consists of a spherical reservoir on top of a slender beam. Under the influence of an excitation force F, for example, wind, or seismic activity, the tower may displace about its equilibrium position, as shown in **Figure 4.26b**. Suppose the tower is modeled as a single-degree-of-freedom with the reservoir as a point mass, m, at the tip of a slender beam, with an effective stiffness k, as shown in **Figure 4.26c**. The mass is $5\times10^5\,kg$.

a. Find the oscillation period of the system in s and the stiffness in N/m so that $|\lambda_1|=|\lambda_2|=\pi$.

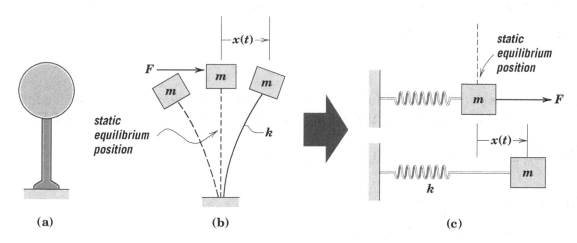

Figure 4.26 Water tower: (a) equilibrium position; (b) lateral displacement; and (c) equivalent mass-spring model.

For Problems 22–26, consider the radio tuner circuit, as shown in **Figure 4.3**, without the forcing voltage. The circuit is modeled as a second-order ODE in terms of charge q,

$$\frac{d^2q}{dt^2}+a_1\frac{dq}{dt}+a_0q=0\,,\text{ where } a_1=\frac{R}{L} \text{ and } a_0=\frac{1}{LC}.$$

a. Define $i=\dfrac{dq}{dt}$ and transform the second-order ODE to a system of two coupled first-order ODEs with $i(t)$ and $q(t)$ as the state-space variables.

b. Find a symbolic expression for the eigenvalues, λ_1 and λ_2, in terms of C, L, and R.

c. Classify the system as neutrally stable, asymptotically stable, or unstable.

22. $C = 10^{-10}\, farad$, $L = 10^4\, henry$, $R = 2(10^7)\, ohm$.

23. $C = 10^{-10}\, farad$, $L = 10^4\, henry$, $R = 2(10^5)\, ohm$.

24. $C = 10^{-6}\, farad$, $L = 10^3\, henry$, $R = 7(10^4)\, ohm$.

25. $C = 10^{-6}\, farad$, $L = 10^3\, henry$, $R = 6(10^4)\, ohm$.

26. $C = 10^{-6}\, farad$, $L = 10^3\, henry$, $R = 0\, ohm$.

Problems for Section 4.4

27. Consider the free vibration of a single-degree-of-freedom, undamped system, as shown in **Figure 4.2**, with ICs $y(0) = 1$ and $\dfrac{dy(0)}{dt} = 0$.

a. Find the orbit equation of this system from the displacement, $y(t)$ and velocity $v(t)$.

b. Plot the orbit in a phase plane for $\omega_n = \dfrac{1}{2}$.

28. Consider the electrical circuit for a radio tuner, as shown in **Figure 4.3**, without the forcing voltage $V_S(t)$ and resistor, so that the circuit model is reduced to the second-order ODE

$\dfrac{d^2 i}{dt^2} + \dfrac{1}{LC} i = 0$, where $L = 5(10^4)\, henry$, and $C = 52.75(10^{-12})\, farad$. The ICs are

$i(0) = 10^{-3}\, amp$ and $\dfrac{di(0)}{dt} = 0$.

a. Find the orbit equation of this system.

b. Plot the orbit in a phase plane and determine if the circuit is neutrally stable, asymptotically stable, or unstable.

Problems for Section 4.5

For Problems 29–31, define $v = \dfrac{dy}{dt}$ and set up the necessary first-order discrete equations.

a. Report t, $y(t)$ and $v(t)$ for $m = 1$ and 2.

b. Run the simulation in a spreadsheet using M points and plot the discrete points (y, v) in the phase plane.

c. Determine from the phase plane if the ODE is neutrally stable, asymptotically stable, or unstable.

29. $6\dfrac{d^2 y}{dt^2} + y = 0$; $y(0) = 0$, $v(0) = 1$; $\Delta t = \pi\left(10^{-3}\right) s$; $M = 10000$.

30. $3\dfrac{d^2 y}{dt^2} + \dfrac{dy}{dt} + y = 0$; $y(0) = 0$, $v(0) = 1$; $\Delta t = \pi\left(10^{-3}\right) s$; $M = 10000$.

31. $\dfrac{d^2 y}{dt^2} - \dfrac{dy}{dt} + y = 0$; $y(0) = 1$, $v(0) = 1$; $\Delta t = \pi\left(10^{-3}\right) s$; $M = 5000$.

32. Consider the radio tuner circuit, as shown in **Figure 4.3**, without a forcing voltage. The circuit is modeled as a second-order ODE in terms of charge q, $\dfrac{d^2 q}{dt^2} + \dfrac{R}{L}\dfrac{dq}{dt} + \dfrac{1}{LC} q = 0$, where $C = 10^{-6}\, farad$, $L = 10^3\, henry$, and $R = 3\left(10^4\right) ohm$.

a. Define $i = \dfrac{dq}{dt}$ and set up the necessary first-order discrete equations. The ICs are $q(0) = 0$ and $i(0) = 25\, amp$.

b. Report t, $q(t)$ and $i(t)$ for $m = 1$ and 2 using $\Delta t = 10^{-3} s$.

c. Run the simulation in a spreadsheet using $M = 1000$ and plot the discrete points (q, i) in the phase plane.

d. Determine from the phase plane if the circuit is neutrally stable, asymptotically stable, or unstable.

Problems for Sections 4.6–4.7

33. Consider the general solution $Y = C_1 \hat{v}_1 e^{\sqrt{\lambda_1} t} + C_2 \hat{v}_2 e^{\sqrt{\lambda_2} t}$ found in Example 4.12 prior to Eq. (4.65). For definiteness, designate λ_1 and λ_2 as the low-mode and high-mode eigenvalues, respectively. Thus, $\hat{v}_1$ and $\hat{v}_2$ are the low-mode and high-mode eigenvectors, respectively. Expanding the foregoing matrix yields

$$\begin{bmatrix} y_1(t) \\ y_2(t) \end{bmatrix} = \begin{bmatrix} v_{1l} \\ v_{2l} \end{bmatrix} C_1 e^{\sqrt{\lambda_1} t} + \begin{bmatrix} v_{1h} \\ v_{2h} \end{bmatrix} C_2 e^{\sqrt{\lambda_2} t} \qquad (4.88)$$

where the subscripts l and h stand for low and high, respectively.

a. Show that when the dependent variables are initialized with, $\hat{v}_1$, the low-mode eigenvector,

that is when $\begin{bmatrix} y_1(0) \\ y_2(0) \end{bmatrix} = \begin{bmatrix} v_{1l} \\ v_{2l} \end{bmatrix}$, then $C_2 = 0$ so that Eq. (4.88) degenerates to the low-mode

general solution. Conversely, you can show that $C_1 = 0$ if $\begin{bmatrix} y_1(0) \\ y_2(0) \end{bmatrix} = \begin{bmatrix} v_{1h} \\ v_{2h} \end{bmatrix}$.

For Problems 34–36,

a. Cast the system of ODEs in the form of Eq. (4.68).
b. Find the eigenvalues.
c. Find the normalized eigenvectors corresponding to the eigenvalues.
d. Find the general solution for $y_1(t)$ and $y_2(t)$ with the IC $\begin{bmatrix} y_1(0) \\ y_2(0) \end{bmatrix} = \begin{bmatrix} v_{1h} \\ v_{2h} \end{bmatrix}$, where v_{1h} and

v_{2h} are the components of the high-mode normalized eigenvector, $\hat{v}_2$.

34. $\begin{aligned} &\dfrac{d^2 y_1}{dt^2} + y_1 - y_2 = 0 \\ &\dfrac{d^2 y_2}{dt^2} + 2y_2 = 0 \end{aligned}$;
35. $\begin{aligned} &\dfrac{d^2 y_1}{dt^2} + 4y_1 = 0 \\ &\dfrac{d^2 y_2}{dt^2} - y_1 + 5y_2 = 0 \end{aligned}$;
36. $\begin{aligned} &\dfrac{d^2 y_1}{dt^2} + 4y_1 - y_2 = 0 \\ &\dfrac{d^2 y_2}{dt^2} - y_1 + 4y_2 = 0 \end{aligned}$.

37. An engineer wants to find the natural frequencies of the two-story building described in **Figure 4.1**, in order to determine if they are within the earthquake frequency range, $0.05\dfrac{rad}{s}$ to

$1\dfrac{rad}{s}$ at that location. The building's lateral displacement is modeled by Eq. (4.87).

a. Determine the eigenvalues and normalized eigenvectors.
b. Determine if the low-mode and high-mode natural frequencies of the building fall within the range of the earthquake oscillation frequency.
c. Find expressions for the low-mode and high-mode shapes of the lateral displacements, $x_1(t)$ and $x_2(t)$.
d. Indicate which mode is in-phase or out-of-phase.

Data: $m_1 = m_2 = 10^5 \, kg$; stiffnesses: $k_1 = 10k_2$ and $k_2 = 10^5 \dfrac{N}{m}$.

38. Consider the three-story building, as shown in **Figure 4.27**, which is modeled as a three-degree-of-freedom system.

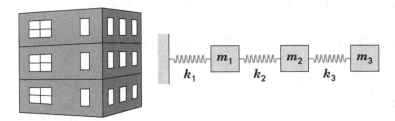

Figure 4.27 Three-story building and its equivalent mass-spring system.

The free body diagram for each mass is given in **Figure 4.28** so that you can easily apply Newton's law of motion in three ODEs.

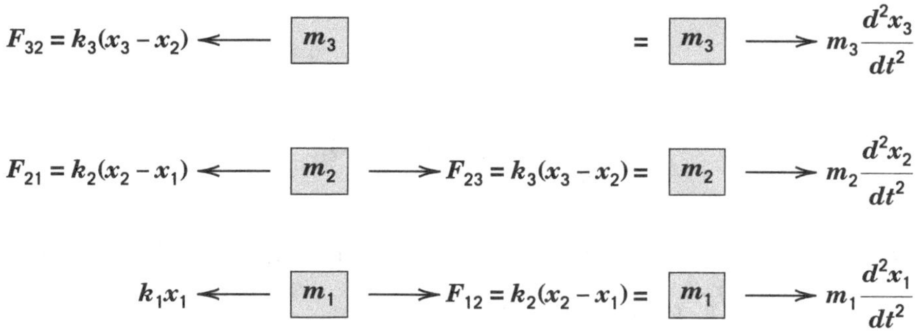

$$F_{32} = k_3(x_3 - x_2) \longleftarrow \boxed{m_3} \qquad = \boxed{m_3} \longrightarrow m_3\frac{d^2x_3}{dt^2}$$

$$F_{21} = k_2(x_2 - x_1) \longleftarrow \boxed{m_2} \longrightarrow F_{23} = k_3(x_3 - x_2) = \boxed{m_2} \longrightarrow m_2\frac{d^2x_2}{dt^2}$$

$$k_1x_1 \longleftarrow \boxed{m_1} \longrightarrow F_{12} = k_2(x_2 - x_1) = \boxed{m_1} \longrightarrow m_1\frac{d^2x_1}{dt^2}$$

Figure 4.28 Free body diagram of each floor.

a. Cast the system of three coupled second-order ODEs in the form of Eq. (4.62).
b. Determine the eigenvalues and normalized eigenvectors.
c. Determine if any of three natural frequencies of the building fall within the range of the earthquake oscillation frequency, $0.05\frac{rad}{s}$ to $1\frac{rad}{s}$.

Data: $m_1 = m_2 = m_3 = 10^5 kg$; stiffnesses: $k_1 = 10^7 \frac{N}{m}$, $k_2 = 2k_1$ and $k_3 = k_1$.

39. The rotating crankshaft of an engine is attached to a flywheel and cam gear, as shown in **Figure 4.29a**. The spinning shaft may resonate at a critical frequency accompanied with excessive vibration that leads to failure.

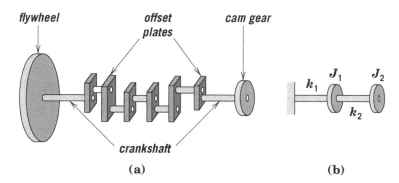

(a)   (b)

Figure 4.29 Partial engine components: (a) crankshaft attached to a flywheel and cam gear; and (b) equivalent disk-rotational spring system.

A model of the crankshaft is shown in **Figure 4.29b** using a rotational disk-spring system in order to determine the crankshaft torsional vibration frequencies. The left shaft in **Figure 4.29b** models the crankshaft leftmost part in **Figure 4.29a**, with a rotational spring stiffness k_1. The right shaft in **Figure 4.29b** models the rightmost part of the crankshaft with a rotational spring stiffness k_2. The middle disk in **Figure 4.29b** models the offset plates with a mass moment of inertia J_1, whereas the right disk models the cam gear with a mass moment of inertia J_2. The governing equations of motion in rotation are given by

$$J_1 \frac{d^2\theta_1}{dt^2} + k_1\theta_1 + k_2\left(\theta_1 - \theta_2\right) = 0$$

$$J_2 \frac{d^2\theta_2}{dt^2} + k_2\left(\theta_2 - \theta_1\right) = 0$$

where θ_1 and θ_2 are the angular displacements of the crankshaft as labeled in **Figure 4.29a**.

Data: $J_1 = 0.04 kg.m^2$; $J_2 = \frac{J_1}{2}$; $k_1 = 0.5\left(10^5\right) N.m$; $k_2 = 2k_1$.

a. Cast this system of second-order ODEs in the form of Eq. (4.62).
b. Determine the eigenvalues and normalized eigenvectors.
c. Determine the low-mode and high-mode natural frequencies.
d. Find the critical frequency that causes the crankshaft to resonate.

CHAPTER FIVE

LAPLACE TRANSFORM

5.1 Introduction

In Chapters 1, 2, and 4, engineering systems were modeled by first-order and second-order ordinary differential equations (ODEs), and solutions were found in the time domain, or simply t-domain. Furthermore, in Chapter 4, the systems of ODEs were limited to homogeneous equations. In principle, a system of nonhomogeneous equations as formulated by Eqs. (4.2) and (4.61), where the G matrix is nonzero, can be solved in the t-domain for a given set of initial conditions (ICs). However, such solutions can be found expeditiously by the Laplace transform method. In this method, the ODE is first transformed from the t-domain to an algebraic equation in the frequency domain, which is designated the s-domain; second, the solution is found in that domain; and third, the solution is transformed back to the t-domain.

In the Laplace transform method, the ICs are integrated in the algebraic equation; hence no additional steps are required to fix constants of integration as in the t-domain. Furthermore, the homogeneous and particular solutions are found simultaneously, unlike the method used in Chapter 2, for example, where those solutions were found separately and then added together. Coupled with Computer-Algebra Systems (CAS), you can push the envelope with systems of coupled ODEs in the s-domain more efficiently than the t-domain. That said, the Laplace transform method requires linear ODEs with constant coefficients and existence of the Laplace transform of forcing functions. For a general nth–order ODE

$$\frac{d^n y}{dt^n} + a_{n-1}\frac{d^{n-1} y}{dt^{n-1}} + a_{n-2}\frac{d^{n-2} y}{dt^{n-2}} + \cdots + a_1\frac{dy}{dt} + a_0 y = r(t)$$

the dependent variable $y(t)$ is defined as the response or output of the system, and $r(t)$ is defined as the forcing function or input to the system. The concept of a system, forcing function, and response was introduced in Chapter 1, as shown in the block diagram in **Figure 5.1**, as a representation of ODEs. The block diagram should remind you about three important components: the governing ODE that models a system; the forcing function or input that causes an effect on the system via the ODE; and the response or output that is a measure of the system's action, mechanically, electrically, or chemically, to name a few areas of interest to engineers. The response is the solution of the ODE.

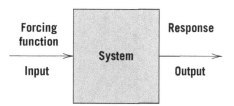

Figure 5.1 A block diagram of a system with a forcing function and a response.

5.2 Basic Concepts

The Laplace transform is useful for solving ODEs in time and space; however, in this guide only the t-domain is considered. The following steps are used to solve ODEs by the Laplace transform:

1. Transform the ODE from a differential equation in $y(t)$ in the t-domain to an algebraic equation in $Y(s)$ in the s-domain.
2. Find the Laplace transform of the forcing function, $R(s)$.
3. Substitute $R(s)$ and the ICs, for example, $y(0)$ and $\dfrac{dy(0)}{dt}$ into the algebraic equation of $Y(s)$.
4. Take the inverse Laplace transform of $Y(s)$ using tables for simple problems, or CAS for algebraically cumbersome problems in order to find the desired response in the t-domain, $y(t)$.

$R(s)$ and $Y(s)$ are defined as the system forcing function and response, respectively, in the s-domain, as shown by the block diagram in **Figure 5.2**.

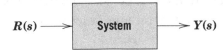

Figure 5.2 Block diagram in the s-domain with $R(s)$ as the forcing function and $Y(s)$ as the response.

5.3 Forcing Functions

The following forcing functions are frequently encountered in engineering systems:

-   unit step function, commonly known as the Heaviside function;
-   ramp function;
-   Dirac delta function;
-   sinusoid function.

Unit step or Heaviside function

The unit step or Heaviside function $u(t-a)$ is a piecewise continuous function that is zero in the interval $-\infty < t < a$, and then jumps to a value of one in the interval $a \le t < \infty$. For clarity, the variable t_0 is used at the jump so that the unit step function is defined as

$$u(t-t_0) = \begin{cases} 0 & t < t_0 \\ 1 & t \ge t_0 \end{cases} \tag{5.1}$$

as shown in **Figure 5.3**.

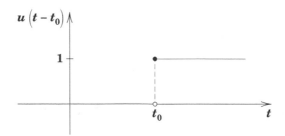

Figure 5.3 A unit step function at $t = t_0$.

Think of the unit step function as a signal that is "off" for $t < t_0$, and "on" for $t \ge t_0$. An excellent analogy is the electric switch to turn a light bulb on at $t = t_0$, which was off for $t < t_0$. Quite often, the term *delayed* is used in engineering for $t_0 > 0$, to indicate that $r(t)$ takes place at a later time $t = t_0$ relative to $t = 0$. In general, a step function $r(t)$ with magnitude $|r_m|$ can be expressed in terms of the unit step function $u(t-t_0)$ as

$$r(t) = r_m u(t-t_0) \tag{5.2}$$

$r(t)$ may shift above or below the horizontal axis depending on whether $r_m > 0$ or $r_m < 0$, respectively, as shown in **Figure 5.4**.

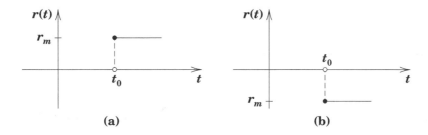

Figure 5.4 A step function $r(t)$ for: (a) $r_m > 0$; and (b) $r_m < 0$.

A step function is an idealization. In reality, the function will have a rise time, as shown in **Figure 5.5**. In engineering applications, however, when the rise time is small, the function is modeled as a step function. As a matter of fact, electrical devices, known as function generators, are capable of generating nearly ideal step functions with extremely small rise times.

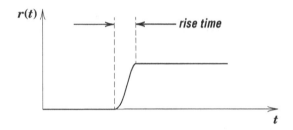

Figure 5.5 Real unit step function with rise time.

Example 5.1

1. Plot the step functions $r_1(t) = u(t-1)$, $r_2(t) = -u(t-2)$, and $r(t) = r_1(t) + r_2(t)$.
2. Interpret the result.

Solution:

1. Plot the step functions $r_1(t) = u(t-1)$, $r_2(t) = -u(t-2)$, and $r(t) = r_1(t) + r_2(t)$.

Using the definition of the unit step function, Eq. (5.1), $r_1(t) = u(t-1) = \begin{cases} 0 & t < 1 \\ 1 & t \geq 1 \end{cases}$,

as shown in **Figure 5.6a**.

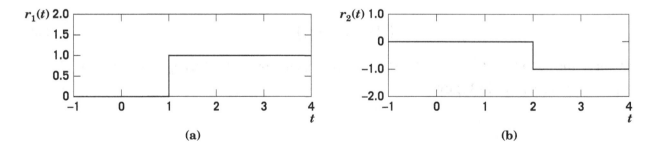

(a)   (b)

Figure 5.6 Step functions: (a) $r_1(t) = u(t-1)$; and (b) $r_2(t) = -u(t-2)$.

$r_2(t) = -u(t-2) = \begin{cases} 0 & t < 2 \\ -1 & t \geq 2 \end{cases}$ shifts from 0 to -1 at $t_0 = 2$, as shown in **Figure 5.6b**.

$r(t) = r_1(t) + r_2(t) = u(t-1) + \left[-u(t-2)\right]$, which is a linear superposition of **Figure 5.6a** and **Figure 5.6b**, is shown in **Figure 5.7a** and **Figure 5.7b**.

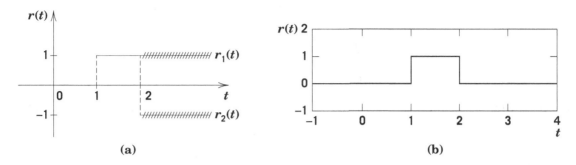

Figure 5.7 Step functions: (a) linear superposition of **Figure 5.6a** and **Figure 5.6b**; and (b) the resultant function $r(t)$.

2. Interpret the result.

It can be seen from **Figure 5.7a** that adding the two functions $r_1(t)$ and $r_2(t)$ results in the hatched lines canceling each other out for $t \geq 2$. Thus, the resulting function $r(t)$ is a finite step function over the interval $1 \leq t < 2$, where $r(t) = \begin{cases} 0 & t < 1 \\ 1 & 1 \leq t < 2, \\ 0 & t \geq 2 \end{cases}$

which you can demonstrate numerically, as shown in **Table 5.1**. $r(t)$ is known as rectangular pulse.

Time interval	$u(t-1)$	$-u(t-2)$	$r(t) = u(t-1) - u(t-2)$
$t < 1$	0; off for $t < 1$	0; off for $t < 1$	0
$1 \leq t < 2$	1; on for $t \geq 1$	0; off for $t < 2$	1
$t \geq 2$	1; on for $t \geq 1$	-1; on for $t \geq 2$	0

Table 5.1 Evaluation of the function $r(t) = u(t-1) - u(t-2)$.

Ramp function

A ramp function is modeled as a straight line. A ramp function $r(t)$ with slope m, which passes through $r(t_0)$, is given by $r(t) - r(t_0) = m(t - t_0)$, or

$$r(t) = m(t - t_0) + r(t_0) \tag{5.3}$$

Multiple piecewise functions

In general, a number of N piecewise functions, $r_0(t)$, $r_1(t)$, ..., $r_{N-1}(t)$, over the interval $t_0 \le t < t_N$ as shown in **Figure 5.8**, can be expressed compactly using step functions as

$$r(t) = \sum_{i=0}^{N-1} r_i(t) \left[u(t-t_i) - u(t-t_{i+1}) \right] \qquad (5.4)$$

If in addition there is a function, $r_N(t)$ that runs indefinitely in the interval $t_N \le t < \infty$, simply add $r_N(t)u(t-t_N)$ to Eq. (5.4). Notice that the difference of step functions in Eq. (5.4), $u(t-t_i) - u(t-t_{i+1})$ is a rectangular pulse over the interval $t_i \le t < t_{i+1}$.

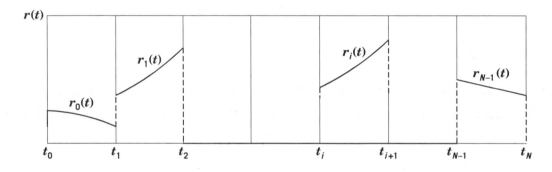

Figure 5.8 General piecewise functions expressed compactly using step functions.

Example 5.2 Piecewise Volumetric Flow Rate

In the soft-drink mixing process, as shown in Figure 1.1 in Chapter 1, the inlet volumetric flow rate $\dot{Q}_i$ was increased suddenly from zero to $0.1 \dfrac{m^3}{s}$ at $t=0$, as shown in

Figure 5.9. After one hour, production demand required a sudden drop in $\dot{Q}_i$ to $0.05 \dfrac{m^3}{s}$. Then after three hours from the initial time, production demand required a sudden increase in $\dot{Q}_i$ to $0.15 \dfrac{m^3}{s}$.

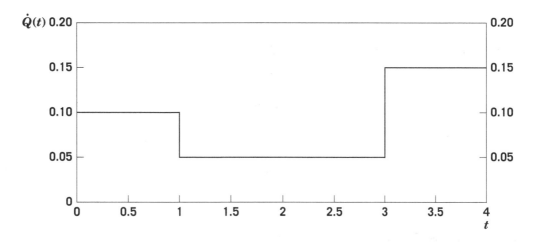

Figure 5.9 The inlet volumetric flow rate $\dot{Q}_i$ at different time intervals.

1. Model $\dot{Q}_i(t)$ using step functions, where t is in hours.

Solution:

Equation (5.4) is used over the interval $0 \le t < 3$. For the step function that runs indefinitely over the interval $t \ge 3$, add $0.15u(t-3)$ to the solution. Thus,

$$\dot{Q}(t) = 0.1\left[u(t) - u(t-1)\right] + 0.05\left[u(t-1) - u(t-3)\right] + 0.15u(t-3)$$

that reduces to $\dot{Q}_i = 0.1u(t) - 0.05u(t-1) + 0.1u(t-3)$

Example 5.3 Simulation of a Valve Angle

Consider the soft-drink production system, as shown in Figure 1.1 in Chapter 1. For a particular batch, the valve angle setting, $\theta(t)$, is controlled remotely, where t is in minutes. Over the interval $0 \le t < 1$, the valve is opened at a constant rate of 10 degrees/min to allow a certain volume flow rate of water in the tank. At $t = 1$, the valve is turned off rapidly to allow water and syrup to mix well. Then, at $t = 2$, the valve is turned on rapidly to $10°$ that remains constant over the interval $2 \le t < 3$. Finally, the valve is closed at a constant rate from $\theta(3) = 10°$ to $\theta(3.25) = 0°$ in order to discharge the batch to another tank in the plant.

1. Express $\theta(t)$ as a piecewise function and plot it over the interval $0 \le t < \infty$.
2. Express $\theta(t)$ compactly using Eq. (5.4).

Solution:

1. Express $\theta(t)$ as a piecewise function and plot it over the interval $0 \le t < \infty$.

Over the interval $0 \le t < 3$, $\theta(t) = \begin{cases} 10t & 0 \le t < 1 \\ 0^o & 1 \le t < 2 \\ 10^o & 2 \le t < 3 \end{cases}$.

Over the interval $3 \le t \le 3.25$, the line passes through $\theta(3) = 10$. Thus, $\theta(t) - 10 = m(t-3)$,

or $\theta(t) = m(t-3) + 10$ based on Eq. (5.3). Using $\theta(3.25) = 0^o$ yields the slope

$m = \dfrac{0-10}{3.25-3} = -40 \dfrac{\text{degrees}}{\text{min}}$. Therefore, $\theta(t) = -40(t-3) + 10 = -40t + 130$ over the interval

$3 \le t \le 3.25$. Over the interval $t \ge 3.25$, $\theta(t) = 0$.

$\theta(t)$ is shown in **Figure 5.10** over the interval $0 \le t < +\infty$.

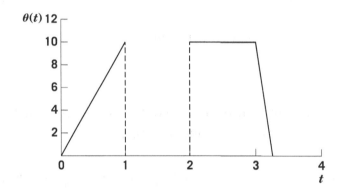

Figure 5.10 The valve angle over the time interval $0 \le t < +\infty$.

2. Express $\theta(t)$ compactly using Eq. (5.4).

Using Eq. (5.4), $\theta(t)$ can be expressed compactly as

$$\theta(t) = 10t\left[u(t) - u(t-1)\right] + 10\left[u(t-2) - u(t-3)\right] + (-40t + 130)\left[u(t-3) - u(t-3.25)\right]$$

This example illustrates how to express several piecewise continuous functions compactly. Depending on the value of t, some step functions are either turned on or turned off, yielding the required value of the function at that time.

■

Dirac delta function

The Dirac delta function models impulsive excitations that act over a very short time, Δt. Essentially, this function is modeled as a spike as $\Delta t \to 0$. One way to understand how the spike develops in a limiting sense is by considering a rectangular pulse over a time interval, $\Delta t = t_2 - t_1$, as shown in **Figure 5.11a**.

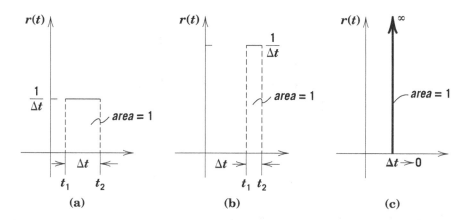

Figure 5.11 Rectangular pulse for progressively decreasing time intervals: (a) unity area rectangular pulse with finite Δt; (b) unity area tall rectangular pulse as Δt decreases; and (c) unity area Dirac delta function spiking to infinity as Δt goes to zero.

The pulse magnitude is defined in a special way as $r(t) = \dfrac{1}{\Delta t}$ over Δt, so that the area under the curve is unity; that is, $\displaystyle\int_{-\infty}^{\infty} r(t)\,dt = \int_{t_1}^{t_2} \dfrac{1}{\Delta t}\,dt = \dfrac{1}{\Delta t}\int_{t_1}^{t_2} dt = \dfrac{t_2 - t_1}{\Delta t} = 1$.

As Δt becomes smaller, $r(t)$ becomes taller, as shown in **Figure 5.11b**, such that the area under the curve remains equal to one. In the limit $\Delta t \to 0$, $r(t) = \displaystyle\lim_{\Delta t \to 0} \dfrac{1}{\Delta t} \to \infty$, which is shown in **Figure 5.11c** as a spike running toward infinity such that the area remains equal to one. The limiting case of $r(t) \to \infty$ as $\Delta t \to 0$ such that $\displaystyle\int_{-\infty}^{\infty} r(t)\,dt = 1$ is the basis for the formal definition of the unit Dirac delta function, $\delta(t - t_0)$ at $t = t_0$,

$$\delta(t - t_0) = \begin{cases} 0 & t \neq t_0 \\ \infty & t = t_0 \end{cases} \tag{5.5}$$

as shown in **Figure 5.12a**, such that the area under the curve is equal to one, that is, $\displaystyle\int_{-\infty}^{+\infty} \delta(t - t_0)\,dt = 1$. The dimension of $\delta(t - t_0)$ is 1/time by virtue of the definition of $r(t)$, which implies that the foregoing integral is dimensionless. It is obvious that when $t_0 = 0$, the unit Dirac delta function is a spike at the origin, as shown in **Figure 5.12b**. Equation (5.5) is not a piecewise continuous function by the nature of its unboundedness at $t = t_0$. For that reason $\delta(t - t_0)$ is classified as a singular function.

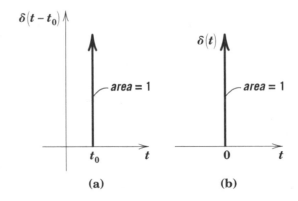

(a)   (b)

Figure 5.12 Unit Dirac delta function: (a) $\delta(t-t_0)$, $t_0 \neq 0$; and (b) $\delta(t)$, $t_0 = 0$.

The Dirac delta function in a mechanical system

In mechanical systems, the Dirac delta function is used to model impulsive excitations such as the rapid impact of an aircraft landing gear on the runway, as shown by the impulsive force, $F(t)$ in **Figure 5.13a**. At the landing instance, $t = t_0$, $F(t)$ grows rapidly to a maximum, as shown in **Figure 5.13b**. Then, the impact force decays rapidly to zero at time $t = t_1$. This impulsive excitation over the short time interval $\Delta t = t_1 - t_0$, causes tremendous stress on the landing gear.

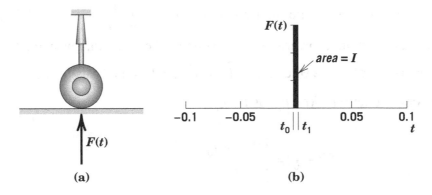

(a)   (b)

Figure 5.13 An impulsive excitation: (a) impulsive force on landing gear; and (b) representation of the force as a spike over a very short time interval.

From mechanics, the impulse is defined as the area under the curve in **Figure 5.13b**,

$$I = \int_{t_0}^{t_1} F(t)\,dt \tag{5.6}$$

Let $F(t)$ be modeled as a rectangular pulse with amplitude, F_{imp}, where the subscript "imp" is used to emphasize impulse. As $\Delta t \rightarrow 0$, $F(t)$ approaches a spike such that the area is always equal to the impulse, I, so that Eq. (5.6) yields $I = \lim\limits_{\Delta t \rightarrow 0}(F_{imp}\Delta t)$, or

$$F_{imp} = I \lim_{\Delta t \rightarrow 0}\left(\frac{1}{\Delta t}\right) \tag{5.7}$$

However, $\lim\limits_{\Delta t \rightarrow 0}\left(\dfrac{1}{\Delta t}\right)$ is the Dirac delta function so that Eq. (5.7) is written as

$$F_{imp} = I(0)\delta(t) \tag{5.8}$$

by assuming that the landing gear impact takes place at $t_0 = 0$, where $I(0)$ is the impulse magnitude at $t_0 = 0$. Equation (5.8) is interpreted by saying that an impulsive force can be modeled as the product of an impulse with given magnitude and the Dirac delta function. As mentioned earlier, note that the dimension of $\delta(t)$ is 1/time, say $1/s$. In addition the dimension of I in SI is $N.s$ based on its definition in Eq. (5.6). Thus, the units on the right-hand side of Eq. (5.8) $N.s(1/s)$ yields N, which is the force unit as expected.

The Dirac delta function in an electrical system

In Section 1.4 in Chapter 1, you may have explored RC circuits as signal filters using first-order ODEs. If you consider the simplest circuit consisting of only an uncharged capacitor across a voltage source, as shown in **Figure 5.14a**, and apply a voltage step at $t = 0$, as shown in **Figure 5.14b**, via a switch, what electrical phenomenon would be described by the Dirac delta function?

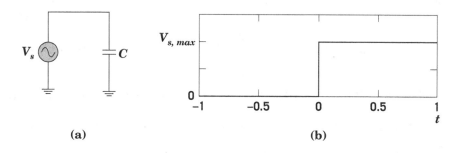

(a)   (b)

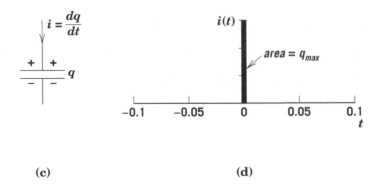

(c)

(c)   (d)

Figure 5.14 Modeling impulsive current: (a) simple capacitor-voltage circuit; (b) applied step voltage at $t = 0$; (c) capacitor current and charge; and (d) impulsive current at $t = 0$.

This question can be answered using several laws and definitions. First, for the simple circuit shown in **Figure 5.14a**, Kirchhoff's voltage loop law yields $V_C = V_S$, where V_C is the voltage across the capacitor. Second, the maximum charge, q_{max}, that can accumulate on the capacitor, as shown in **Figure 5.14c**, is by definition $q_{max} = CV_C$, or in this case, $q_{max} = CV_S$, where C is the capacitance. Third, the charge buildup on the capacitor, from $q = 0$ to $q = q_{max}$, depends on its rate via the definition of current, $i(t) = \dfrac{dq}{dt}$ that yields by integration

$$q = \int_0^t i(t') dt' \tag{5.9}$$

where t' is a dummy variable of integration.

Fourth, from Ohm's law, the current through the capacitor is given by $i(t) = C\dfrac{dV_C}{dt}$ or $i(t) = C\dfrac{dV_S}{dt}$. Thus, $i(t)$ is proportional to $\dfrac{dV_S}{dt}$, the slope of $V_S(t)$. From **Figure 5.14b**, the slope is zero except at the step jump, where $\dfrac{dV_S}{dt} \to \infty$ for a vertical line, so that $i(t)$ spikes at $t = 0$, as shown in **Figure 5.14d**. Let $i(t)$ be modeled as a rectangular pulse with amplitude, i_{imp}. As $\Delta t \to 0$, $i(t)$ approaches a spike such that the area is always equal to q_{max}, so that Eq. (5.9) yields $q_{max} = \lim\limits_{\Delta t \to 0} \left(i_{imp} \Delta t \right)$, or

$$i_{imp} = q_{max} \lim_{\Delta t \to 0} \left(\frac{1}{\Delta t} \right) \tag{5.10}$$

However, $\lim\limits_{\Delta t \to 0}\left(\dfrac{1}{\Delta t}\right)$ is the Dirac delta function so that Eq. (5.10) is written as

$$i_{imp} = q_{max}(0)\delta(t) \qquad (5.11)$$

where $q_{max}(0)$ is the maximum charge at $t_0 = 0$. Equation (5.11) is interpreted by saying that for a step voltage across a capacitor, an impulsive current can be modeled as the product of maximum charge and the Dirac delta function. Here, the units on the right-hand side of Eq. (5.11) are *coulomb*(1/s) that is *ampere* by definition, which is the current unit as expected.

The previous analysis of impulsive phenomenon was limited to two simple mechanical and electrical systems in order to demonstrate the relationship between the Dirac delta function and impulsive excitations such as force and current. What if you are interested in applying an impulsive voltage, $V_{s,imp}$, to an RLC circuit, for example, the radio tuner circuit covered in Chapter 2, as shown in **Figure 5.15a**? Is it possible to derive a relationship between $V_{s,imp}$ and $\delta(t)$ analogous to that between i_{imp} and $\delta(t)$ in Eq. (5.11)?

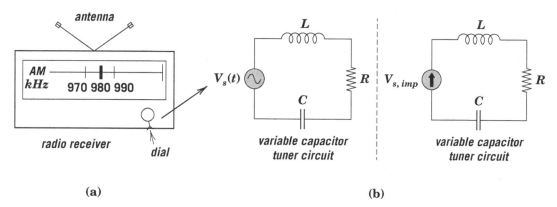

Figure 5.15 Radio tuner circuit: (a) sinusoidal voltage source; and (b) impulsive voltage source.

Deriving a relationship between $V_{s,imp}$ and $\delta(t)$ from first principles as in Eq. (5.11) may not be possible with several electrical elements at once, for example, RLC. The alternative approach is to model the impulse as a limiting process of a rectangular pulse. For example, suppose you apply a rectangular voltage pulse of amplitude, V_{max} over a very short time interval Δt. The area under the curve is $A = V_{max}.\Delta t$ or $V_{max} = A\dfrac{1}{\Delta t}$. As $\Delta t \to 0$, the voltage spikes such that area A remains constant. Thus, from $V_{max} = A\dfrac{1}{\Delta t}$ you can define the impulsive source voltage as

$$V_{s,imp} = A(0)\delta(t) \qquad (5.12)$$

where $A(0) = V_{max}.\Delta t$ is the area of the impulsive source voltage at $t_0 = 0$. Here, the units on the right-hand side of Eq. (5.12) are *volts.s*(1/s) that yield *volts*, which is the voltage unit as expected.

The impulsive source voltage is depicted by the symbol ⬆, as shown in **Figure 5.15b**, in contrast to the actual sinusoidal source voltage as depicted by the symbol 〰 in **Figure 5.15a**.

Applying an impulsive voltage as a forcing function to an RLC circuit or even more complicated circuits using Eq. (5.12) simulates the free response of the system, which is useful for parameters identification such as damping, the damped period, and duration of transient, to name a few important quantities.

You can generalize the foregoing concepts to any impulsive function, $r_{imp}(t)$ at time t_0 as

$$r_{imp}(t) = A(t_0)\delta(t - t_0) \tag{5.13}$$

where $A(t_0)$ is a generalized impulse, which is the area of a rectangular pulse applied over a very short time interval Δt. Examples of impulsive function in electrical and mechanical systems are summarized in **Table 5.2**. The Dirac delta function is not limited to electrical and mechanical systems only; it also appears in many other areas, for example, physics, chemical engineering, and probability theory, to name a few areas.

Generalized impulse (area) $A(t_0)$	Impulsive function $r_{imp}(t)$
Force impulse: $A(t_0) = F_{max}.\Delta t$	Force: $F_{imp}(t) = A(t_0)\delta(t - t_o)$, N
Voltage impulse: $A(t_0) = V_{max}.\Delta t$	voltage: $V_{imp}(t) = A(t_0)\delta(t - t_o)$, volt
Current impulse: $A(t_0) = i_{max}.\Delta t$	current: $i_{imp}(t) = A(t_0)\delta(t - t_o)$, ampere

Table 5.2 Examples of impulsive functions in mechanical and electrical systems.

The following integral is one of the useful properties of the Dirac delta function,

$$\int_{-\infty}^{+\infty} f(t)\delta(t - t_0)dt = f(t_0) \tag{5.14}$$

Example 5.4 Simulation of an Impact Force

Concrete piers are tested periodically to evaluate their strength which can deteriorate over time due to environmental conditions such as salt water, acid rain, and seismic activities. The test is done by striking the top of the pier with an electronic hammer connected to a load cell, as shown in **Figure 5.16a**. The impact causes a compression wave to oscillate up and down, thereby causing the pier to vibrate. From this information, the engineer can determine the strength of concrete as you will learn in Chapter 7, Discrete Fourier Transform. In this test the impact force $F(t)$ is measured as a function of time t, as shown in **Figure 5.16b**.

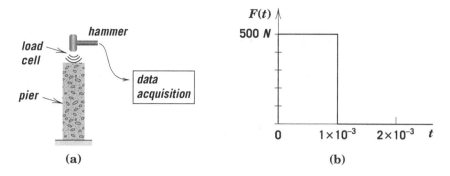

Figure 5.16 Impulsive test of concrete piers: (a) hammer induces an impact force; and (b) impact force as a rectangular pulse over a time interval.

1. Model $F(t)$ as a rectangular force pulse.
2. Model $F(t)$ as an impulsive force if the time interval approaches zero such that the impulse remains constant.

Solution:

1. Using Eq. (5.4), $F(t) = 500\left[u(t) - u(t - 10^{-3})\right]N$ from **Figure 5.16b**.

2. The impulse is $A = 500N\left(10^{-3}s\right) = 0.5N.s$ as defined by the area under the force curve in **Figure 5.16b**. As $\Delta t \to 0$, the impulse is concentrated at $t = 0$ with magnitude $A(0) = 0.5N.s$ so that from **Table 5.2** $F_{imp} = 0.5\delta(t)N$.

∎

Sinusoid functions

Sinusoid functions should be reviewed in Chapters 1 and 2, which were used to model various systems.

5.4 Laplace Transform

The Laplace transform of a piecewise continuous function $f(t)$ in the t-domain $0 \le t < \infty$, denoted $\mathcal{L}\left[f(t)\right]$ or $F(s)$, is defined as

$$\mathcal{L}\left[f(t)\right] = F(s) = \int_0^\infty f(t)e^{-st}dt \qquad (5.15)$$

provided the integral exists, where s is a parameter. Conversely, the inverse Laplace transform of $F(s)$ is $f(t) = \mathcal{L}^{-1}\left[F(s)\right]$

Example 5.5

Revisit the simulation of an impact force in Example 5.4.

1. Find the Laplace transform of $F(t) = 500 \left[u(t) - u(t - 10^{-3}) \right] N$ by direct integration using Eq. (5.15).
2. Find the Laplace transform of $F_{imp} = 0.5\delta(t)$.

Solution

1. Find the Laplace transform of $F(t) = 500 \left[u(t) - u(t - 10^{-3}) \right] N$ by direct integration using Eq. (5.15).

Using Eq. (5.15), the Laplace transform of $F(t) = 500 \left[u(t) - u(t - 10^{-3}) \right] N$ by direct integration is $\mathscr{L}\left[F(t) \right] = F(s) = \int_0^\infty 500 \left[u(t) - u(t - 10^{-3}) \right] e^{-st} dt$, or

$F(s) = 500 \int_0^\infty u(t) e^{-st} dt - 500 \int_{10^{-3}}^\infty u(t - 10^{-3}) e^{-st} dt$. The lower limit in the second integral is changed using the property of a step function.

Evaluating the first integral yields $\int_0^\infty \overset{=1}{\cancel{u(t)}} e^{-st} dt = -\frac{1}{s} e^{-st} \Big|_0^\infty = \frac{1}{s}$.

Evaluating the second integral yields $\int_{10^{-3}}^\infty \overset{=1}{\cancel{u(t - 10^{-3})}} e^{-st} dt = -\frac{1}{s} e^{-st} \Big|_{10^{-3}}^\infty = \frac{1}{s} e^{-10^{-3} s}$.

Thus, $F(s) = 500 \left(\frac{1}{s} - \frac{1}{s} e^{-10^{-3} s} \right) = 500 \frac{1 - e^{-10^{-3} s}}{s}$.

2. Find the Laplace transform of $F_{imp} = 0.5\delta(t)$.

The Laplace transform of $F_{imp} = 0.5\delta(t)$ is $\mathscr{L}\left[F_{imp} \right] = F_{imp}(s) = 0.5 \int_0^\infty \delta(t) e^{-st} dt$. The Dirac delta function is zero over the interval $-\infty < t < 0$, so that $\int_0^\infty \delta(t) e^{-st} dt = e^{-s(0)} = 1$ using Eq. (5.14). Thus, $F_{imp}(s) = 0.5$.

Example 5.6 Laplace Transform of a Valve Angle

Revisit the water tank problem in Example 1.7 in Chapter 1. The valve angle is given by the piecewise continuous function

$$\theta(t) = \begin{cases} \dfrac{t}{6} & 0 \le t < 60 \\ 10 & t \ge 60 \end{cases} \tag{5.16}$$

1. Find $\theta(s)$ by direct integration using Eq. (5.15).
2. Find $\theta(s)$ using the shifting property of a step function.

Solution:

1. Find $\theta(s)$ by direct integration using Eq. (5.15).

   Substituting Eq. (5.16) in Eq. (5.15) yields

$$\theta(s) = \int_0^\infty \theta(t)e^{-st}dt = \int_0^{60} \frac{t}{6}e^{-st}dt + \int_{60}^\infty 10e^{-st}dt \tag{5.17}$$

Integrating by parts, the first integral on the right-hand side yields

$$\int_0^{60} \underbrace{\frac{t}{6}}_{u}\underbrace{e^{-st}dt}_{dv} = \underbrace{\frac{t}{6}}_{u}\underbrace{\frac{e^{-st}}{-s}}_{v}\Bigg|_0^{60} - \int_0^{60} \underbrace{\frac{e^{-st}}{-s}}_{v}\underbrace{\frac{1}{6}dt}_{du} = -10\frac{e^{-60s}}{s} - \frac{e^{-st}}{6s^2}\Bigg|_0^{60} = -10\frac{e^{-60s}}{s} - \frac{e^{-60s}}{6s^2} + \frac{1}{6s^2}.$$

The second integral yields $\int_{60}^\infty 10e^{-st}dt = \dfrac{10e^{-st}}{-s}\Bigg|_{60}^\infty = \dfrac{10e^{-60s}}{s}$.

You may also evaluate $\int_0^{60} \dfrac{t}{6}e^{-st}dt$ using CAS. Substituting the results in Eq. (5.17) yields

$$\theta(s) = -10\frac{e^{-60s}}{s} - \frac{e^{-60s}}{6s^2} + \frac{1}{6s^2} + \frac{10e^{-60s}}{s} = \frac{1-e^{-60s}}{6s^2}.$$

2. Find $\theta(s)$ using the shifting property of a step function.

   Express Eq. (5.16) compactly using Eq. (5.4), $\theta(t) = \dfrac{t}{6}[u(t) - u(t-60)] + 10u(t-60)$ so that Laplace transform

$$\theta(s) = \frac{1}{6}\mathscr{L}[tu(t)] - \frac{1}{6}\mathscr{L}[tu(t-60)] + 10\,\mathscr{L}[u(t-60)] \tag{5.18}$$

The shifting property of a step function is given by

$$\mathscr{L}\big[f(t-a)u(t-a)\big] = e^{-as}\,\mathscr{L}\big[f(t)\big] \qquad (5.19)$$

In order to apply the shifting property of a step function, rewrite the middle Laplace term in Eq. (5.18) as

$$\mathscr{L}\big[tu(t-60)\big] = \mathscr{L}\big[(t-60+60)u(t-60)\big] = \mathscr{L}\big[(t-60)u(t-60)\big] + 60\,\mathscr{L}\big[u(t-60)\big]$$

so that

$$\theta(s) = \frac{1}{6}\mathscr{L}\big[tu(t)\big] - \frac{1}{6}\mathscr{L}\big[(t-60)u(t-60)\big] \qquad (5.20)$$

Applying the shifting property, Eq. (5.19), to each term in Eq. (5.20) yields

$$\mathscr{L}\big[tu(t)\big] = e^{-s(0)}\,\mathscr{L}\big[t\big] = e^{-s(0)}\frac{1}{s^2} = \frac{1}{s^2}\,.$$

$$\mathscr{L}\big[(t-60)u(t-60)\big] = e^{-60s}\,\mathscr{L}\big[t\big] = e^{-60s}\frac{1}{s^2}\,.$$

Substituting the results in Eq. (5.20) yields $\theta(s) = \dfrac{1}{6}\dfrac{1}{s^2} - \dfrac{1}{6}\dfrac{e^{-60s}}{s^2} = \dfrac{1-e^{-60s}}{6s^2}\,.$

In the next two sections, you will solve first- and second-order ODEs by the Laplace transform.

5.5 Laplace Transform of First-Order ODEs

Given a first-order linear ODE in engineering form, $\tau\dfrac{dy}{dt} + y = f(t)$ over the interval $t>0$ with IC $y(0)$, its Laplace transform is given by

$$Y(s) = \frac{F(s) + \tau\,y(0)}{\tau s + 1} \qquad (5.21)$$

where τ is the time constant.

Example 5.7 The Water Tank Problem Solved by The Laplace Transform

Revisit the fluid level in a tank, $h(t)$, in Example 1.7, Chapter 1, which is governed by

$$\tau \frac{dh}{dt} + h = f(t), \text{ where the forcing function is } f(t) = \left(\frac{k_v}{k_o}\right)\theta \text{ and } \theta(t) = \begin{cases} \dfrac{t}{6} & 0 \le t < 60 \\ 10 & t \ge 60 \end{cases}.$$

The IC is $h(0) = 0$. Furthermore, $\dfrac{k_v}{k_o} = 0.197$ and $\tau = 60$.

1. Solve the problem by the Laplace transform method.

Solution:

The Laplace transform of the ODE is given by Eq. (5.21),

$$H(s) = \frac{F(s) + \tau h(0)}{\tau s + 1} \tag{5.22}$$

The Laplace transform of $f(t) = \left(\dfrac{k_v}{k_o}\right)\theta$ is $F(s) = \dfrac{k_v}{k_o}\theta(s)$, where $\theta(s) = \dfrac{1 - e^{-60s}}{6s^2}$ is found in

Example 5.6. Substituting $F(s)$, $h(0) = 0$, and $\tau = 60$ in Eq. (5.22) yields

$$H(s) = \frac{1}{6}\frac{k_v}{k_o}\frac{1 - e^{-60s}}{s^2(60s + 1)} \tag{5.23}$$

The response $h(t)$ in the t-domain is found by taking the inverse Laplace transform of $H(s)$. Using CAS, the inverse Laplace transform of Eq. (5.23) yields

$$h(t) = \frac{1}{6}\frac{k_v}{k_o}\left[\left(120 - t - 60e^{-\frac{t-60}{60}}\right)u(t - 60) + t - 60 + 60e^{-\frac{t}{60}}\right] \tag{5.24}$$

Equation (5.24) is the same result as that found in Example 1.7, except it is in a compact form that is typical in CAS. To show that Eq. (5.24) is the same result, substitute $u(t - 60)$ by its appropriate value as follows:

For $t < 60$, $u(t - 60) = 0$ so that Eq. (5.24) yields $h(t) = \dfrac{1}{6}\dfrac{k_v}{k_o}\left(t - 60 + 60e^{-\frac{t}{60}}\right)$. Using

$\dfrac{k_v}{k_o} = 0.197$ yields $h(t) = 1.97\left(e^{-\frac{t}{60}} - 1\right) + 0.0328t$.

For $t \geq 60$, $u(t-60)=1$ so that Eq. (5.24) yields $h(t)=10\dfrac{k_v}{k_o}\left[1-e^{-\frac{t-60}{60}}+e^{-\frac{t}{60}}\right]$, or

$h(t)=10\dfrac{k_v}{k_o}\left[1-e^{-\frac{t}{60}}(1-e)\right]$, which reduces to $-3.38e^{-\frac{t}{60}}+1.97$.

This example demonstrates the advantage of the Laplace transform in solving engineering problems with piecewise continuous forcing function. Notice that the solution here is solved in one step using CAS, whereas in Example 1.7 the solution was found in two steps.

■

5.6 Laplace Transform of Second-Order ODEs

Given a second-order linear ODE in engineering form, $\dfrac{d^2y}{dt^2}+2\zeta\omega_n\dfrac{dy}{dt}+\omega_n^2 y=f(t)$ over the

interval $t>0$, or for simplicity $\dfrac{d^2y}{dt^2}+a_1\dfrac{dy}{dt}+a_0 y=f(t)$ where $a_1=2\zeta\omega_n$ and $a_0=\omega_n^2$, with

ICs $y(0)$ and $\dfrac{dy(0)}{dt}$, its Laplace transform is given by

$$Y(s)=\dfrac{F(s)+(s+a_1)y(0)+\dfrac{dy(0)}{dt}}{s^2+a_1 s+a_0} \tag{5.25}$$

Example 5.8 Transient Response of a Damped RLC Circuit

Consider the radio tuner circuit, as shown in **Figure 5.15**. The current $i(t)$ is governed by the second-order ODE

$$\dfrac{d^2 i}{dt^2}+a_1\dfrac{di}{dt}+a_0 i=\dfrac{1}{L}\dfrac{dV_s}{dt} \tag{5.26}$$

where $a_0=\dfrac{1}{LC}$ and $a_1=\dfrac{R}{L}$. The ICs are $i(0)=0$ and $\dfrac{di(0)}{dt}=0$. In this example you shall

model $V_s(t)$ in Eq. (5.26) as a Dirac delta function as shown in **Figure 5.15b**. As explained in the paragraph following Eq. (5.12), an impulsive voltage applied to an RLC circuit simulates the free response of the system; that is, $i(t)$ is expected to decay to zero. Take $C=\dfrac{5}{\pi^2}\left(10^{-10}\right)$

farad, $L=5\left(10^{-4}\right)$ *henry* and $R=100\ ohm$.

1. Model the forcing voltage $V_s(t)$ as an impulsive voltage at $t_0 = 0$ with a maximum amplitude of 1.5 *volts*, which occurs over a time interval $\Delta t = 10 \times 10^{-6} s$.
2. Solve for $i(t)$ using the Laplace transform.
3. Estimate how long it takes the transient current amplitude to reach 1% of its initial response, $i(0^+)$.

Solution:

1. Model the forcing voltage $V_s(t)$ as an impulsive voltage at $t_0 = 0$ with a maximum amplitude of 1.5 *volts*, which occur over a time interval $\Delta t = 10 \times 10^{-6} s$.

From **Table 5.2**, $V_s(t) = A(0)\delta(t)$, where $A(0) = V_{max}.\Delta t$, so that

$$V_s(t) = 15 \times 10^{-6}\delta(t) \tag{5.27}$$

2. Solve for $i(t)$ using the Laplace transform.

The forcing function is $f(t) = \dfrac{1}{L}\dfrac{dV_s}{dt}$. Using Eq. (5.27) and $L = 5(10^{-4})\ henry$ yields

$f(t) = 0.03\dfrac{d\delta}{dt}$. Its Laplace transform is $F(s) = 0.03\ \mathscr{L}\left[\dfrac{d\delta}{dt}\right]$. Using CAS, $\mathscr{L}\left[\dfrac{d\delta}{dt}\right] = s$ so that $F(s) = 0.03s$.

Furthermore, using the values of C and L yields $a_0 = \dfrac{1}{LC} = \dfrac{1}{5(10^{-4})\dfrac{5}{\pi^2}(10^{-10})} = 4\pi^2(10^{12})$, and

$a_1 = \dfrac{100}{5(10^{-4})} = 2(10^5)$. Substituting $F(s)$, a_0, a_1, $i(0) = 0$ and $\dfrac{di(0)}{dt} = 0$ in Eq. (5.25) yields

the current Laplace transform, $I(s)$,

$$I(s) = \frac{0.03s}{s^2 + 2(10^5)s + 4\pi^2(10^{12})} \tag{5.28}$$

Using CAS, the inverse Laplace transform of Eq. (5.28) is

$$i(t) = \frac{0.03e^{-10^5 t}}{\sqrt{400\pi^2 - 1}}\left[\sqrt{400\pi^2 - 1}\cos\left(\sqrt{400\pi^2 - 1}\times10^5 t\right) - \sin\left(\sqrt{400\pi^2 - 1}\times10^5 t\right)\right] \tag{5.29}$$

5-21

If you set $t = 0$ in Eq. (5.29) you get $i(0) = 0.03$ that appears to be conflicting with the IC $i(0) = 0$. By the nature of the impulsive function, you should note that $i(0) = 0.03$ is the response immediately after the impulsive voltage is applied, which should be written as $i(0^+) = 0.03$. The same is true for the first derivative; that is, $\dfrac{di(0)}{dt}$ found from Eq. (5.29) is $\dfrac{di(0^+)}{dt}$.

3. Estimate how long it takes the transient current amplitude to reach 1% of its initial response, $i(0^+)$.

The response $i(t)$ given by Eq. (5.29) is shown in **Figure 5.17**, which decays from $i(0^+) = 0.03$ *ampere* to approximately zero near $t = 5 \times 10^{-5} s$.

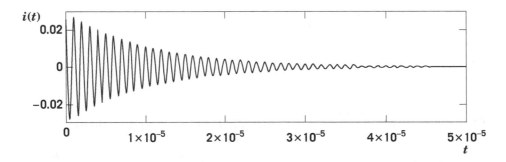

Figure 5.17 Transient current response due to an impulsive voltage.

In order to determine the time when the amplitude reaches 1% of $i(0^+)$, or 0.0003 *ampere*, plot $i(t)$ over the interval $4 \times 10^{-5} \le t \le 5 \times 10^{-5}$, as well as a constant line at 0.0003, as shown in **Figure 5.18**. From the graph, it can be seen that the amplitude reaches 1% of $i(0^+)$ at $t = 4.6 \times 10^{-5} s$.

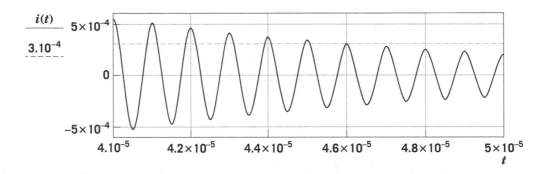

Figure 5.18 Transient current response over the interval $4 \times 10^{-5} \le t \le 5 \times 10^{-5}$.

In the next section, you will solve a system of coupled first-order ODEs using the Laplace transform.

5.7 Laplace Transform of First-Order Coupled ODEs

A first-order coupled system of n ODEs has the form

$$\frac{dY}{dt} = AY + G \qquad (5.30)$$

where $Y = \begin{bmatrix} y_1(t) \\ y_2(t) \\ \cdot \\ \cdot \\ \cdot \\ y_n(t) \end{bmatrix}$ is the dependent variable column matrix, $A = \begin{bmatrix} a_{11} & a_{12} & \cdot & \cdot & a_{1n} \\ a_{21} & a_{22} & \cdot & \cdot & a_{2n} \\ \cdot & & \cdot & & \\ \cdot & & & \cdot & \\ \cdot & & & & \cdot \\ a_{n1} & a_{n2} & \cdot & \cdot & a_{nn} \end{bmatrix}$ is

the coefficient matrix and $G = \begin{bmatrix} g_1 \\ g_2 \\ \cdot \\ \cdot \\ \cdot \\ g_n \end{bmatrix}$ is the forcing function column matrix. The ICs for the

system of equations are given by the column matrix $Y(0) = \begin{bmatrix} y_1(0) \\ y_2(0) \\ \cdot \\ \cdot \\ \cdot \\ y_n(0) \end{bmatrix}$

The Laplace transform of Eq. (5.30) is

$$sY(s) - Y(0) = AY(s) + G(s)$$

or $\left[sI\text{-}A \right] Y(s) = Y(0) + G(s)$, so that

$$Y(s) = [sI - A]^{-1} \left[Y(0) + G(s) \right] \qquad (5.31)$$

where I is the identity matrix. The solution in the t-domain, $Y(t)$, is found by the inverse Laplace transform of Eq. (5.31) using CAS. The Laplace transform method shall be used to solve two applied engineering problem in the area of heat transfer and audio system design.

Modeling of tubular heat exchanger as a system of coupled first-order ODEs

Consider the flow of a fluid at temperature $T_f(t)$ inside a tube at temperature, $T_w(t)$ as part of a normal operation in a power plant, as shown in **Figure 5.19a**. The tube surrounding is at temperature, T_s. As part of a plan for plant maintenance shutdown, it is required to stop the fluid flow and estimate the time it takes $T_f(t)$ to reach T_s within 5 °C from some ICs $T_f(0)$ and $T_w(0)$. The time history of $T_f(t)$ can be predicted from the conservation of energy.

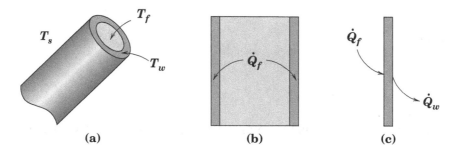

(a)   (b)   (c)

Figure 5.19 Tubular heat exchanger: (a) physical component; (b) fluid heat rate balanced by the rate of its internal energy decrease; and (c) energy balance on the cylindrical wall.

As the fluid cools down, conservation of energy requires the rate of fluid internal energy decrease, $-(mc)_f \dfrac{dT_f}{dt}$ to balance the fluid rate of heat transfer to the tube, $\dot{Q}_f$, as shown in **Figure 5.19b**, where m and c are the mass and specific heat of the fluid, respectively. Thus,

$$\dot{Q}_f = -(mc)_f \frac{dT_f}{dt} \tag{5.32}$$

Similarly, conservation of energy requires the fluid rate of heat transfer to the tube, $\dot{Q}_f$ to balance the rate of wall internal energy increase, $(mc)_w \dfrac{dT_w}{dt}$, and any wall heat transfer rate to the surrounding, $\dot{Q}_w$, as shown in **Figure 5.19c**, where m and c are the mass and specific heat of the wall, respectively. Thus,

$$\dot{Q}_f = (mc)_w \frac{dT_w}{dt} + \dot{Q}_w \qquad (5.33)$$

Furthermore, Newton's law of cooling states that the rate of heat transfer is proportional to temperature change. For heat transfer from the fluid to the wall, such that $T_f > T_w$,

$$\dot{Q}_f = (hA)_f (T_f - T_w) \qquad (5.34)$$

and for heat transfer from the wall to the surrounding, such that $T_w > T_s$

$$\dot{Q}_w = (hA)_w (T_w - T_s) \qquad (5.35)$$

where h and A in Eq. (5.34) are the fluid heat transfer coefficient and tube internal surface area, respectively, and h and A in Eq. (5.35) are the surrounding air heat transfer coefficient and tube external surface area, respectively. Substituting Eqs. (5.34) and (5.35) in Eqs. (5.32) and (5.33) yields

$$\frac{dT_f}{dt} = -\left(\frac{hA}{mc}\right)_f T_f + \left(\frac{hA}{mc}\right)_f T_w$$

$$\frac{dT_w}{dt} = \frac{(hA)_f}{(mc)_w} T_f - \frac{(hA)_f + (hA)_w}{(mc)_w} T_w + \left(\frac{hA}{mc}\right)_w T_s \qquad (5.36)$$

For clarity, The ODEs in Eq. (5.36) are recast as

$$\frac{dT_f}{dt} = a_{11}T_f + a_{12}T_w + g_1$$

$$\frac{dT_w}{dt} = a_{21}T_f + a_{22}T_w + g_2 \qquad (5.37)$$

where $a_{11} = -\left(\frac{hA}{mc}\right)_f$, $a_{12} = \left(\frac{hA}{mc}\right)_f$ $a_{21} = \frac{(hA)_f}{(mc)_w}$, $a_{22} = -\frac{(hA)_f + (hA)_w}{(mc)_w}$, $g_1 = 0$ and

$g_2 = \left(\frac{hA}{mc}\right)_w T_s$.

Equation (5.37) is a first-order system of linear, nonhomogeneous, and coupled ODEs with constant coefficients. The unknown variables are $T_f(t)$ and $T_w(t)$, which have the form of

Eq. (5.30), where $Y = \begin{bmatrix} T_f \\ T_w \end{bmatrix}$, $A = \begin{bmatrix} a_{11} & a_{12} \\ a_{21} & a_{22} \end{bmatrix}$ and $G = \begin{bmatrix} g_1 \\ g_2 \end{bmatrix}$. Its solution is found by the Laplace transform method with ICs $Y(0) = \begin{bmatrix} T_f(0) \\ T_w(0) \end{bmatrix}$.

Example 5.9 Tubular Heat Exchanger

1. Solve the system of coupled ODEs given by Eq. (5.37) by the Laplace transform method to find the instantaneous temperatures of the fluid and wall.

2. Plot $T_f(t)$ and $T_w(t)$ versus t in s.

3. Determine how long it will take the fluid temperature to reach $(T_S + 5)\,^\circ C$ once the unit is shut down.

Data: $T_s = 20^\circ C$; $T_f(0) = T_w(0) = 100^\circ C$; $(hA)_f = (hA)_w = 50 \dfrac{W}{^\circ C}$; $(mc)_f = 10^6 \dfrac{Joules}{^\circ C}$; and $(mc)_w = 2(10^6)\dfrac{Joules}{^\circ C}$.

Solution:

1. Solve the system of coupled ODEs given by Eq. (5.37) by the Laplace transform method to find the instantaneous temperatures of the fluid and wall.

From the given data, $a_{11} = -\left(\dfrac{hA}{mc}\right)_f = -5(10^{-5})$, $a_{12} = \left(\dfrac{hA}{mc}\right)_f = 5(10^{-5})$

$a_{21} = \dfrac{(hA)_f}{(mc)_w} = 2.5(10^{-5})$, $a_{22} = -\dfrac{(hA)_f + (hA)_w}{(mc)_w} = -5(10^{-5})$, and $g_2 = \left(\dfrac{hA}{mc}\right)_w T_s = 5(10^{-4})$.

Substituting the foregoing values in Eq. (5.37) yields

$$\frac{dT_f}{dt} = -5(10^{-5})T_f + 5(10^{-5})T_w$$

$$\frac{dT_w}{dt} = 2.5(10^{-5})T_f - 5(10^{-5})T_w + 5(10^{-4})$$

or in matrix form,

$$\frac{d}{dt}\begin{bmatrix} T_f \\ T_w \end{bmatrix} = \begin{bmatrix} -5(10^{-5}) & 5(10^{-5}) \\ 2.5(10^{-5}) & -5(10^{-5}) \end{bmatrix}\begin{bmatrix} T_f \\ T_w \end{bmatrix} + \begin{bmatrix} 0 \\ 5(10^{-4}) \end{bmatrix} \tag{5.38}$$

The Laplace transform of Eq. (5.38) is given by Eq. (5.31), where

$$A = \begin{bmatrix} a_{11} & a_{12} \\ a_{21} & a_{22} \end{bmatrix} = 10^{-5} \begin{bmatrix} -5 & 5 \\ 2.5 & -5 \end{bmatrix}, \ Y(0) = \begin{bmatrix} T_f(0) \\ T_w(0) \end{bmatrix} = \begin{bmatrix} 100 \\ 100 \end{bmatrix} \text{ and } G(s) = \mathscr{L} \begin{bmatrix} 0 \\ 5(10^{-4}) \end{bmatrix}.$$

Using Laplace transform tables or CAS, $G(s) = \mathscr{L} \begin{bmatrix} 0 \\ 5(10^{-4}) \end{bmatrix} = \begin{bmatrix} 0 \\ \dfrac{5(10^{-4})}{s} \end{bmatrix}$, so that

$$[Y(0) + G(s)] = \begin{bmatrix} 100 \\ 100 + \dfrac{5(10^{-4})}{s} \end{bmatrix} \tag{5.39}$$

Next, evaluate matrix $[sI - A]$: $[sI - A] = \begin{bmatrix} s - a_{11} & -a_{12} \\ -a_{21} & s - a_{22} \end{bmatrix} = \begin{bmatrix} s + 5(10^{-5}) & -5(10^{-5}) \\ -2.5(10^{-5}) & s + 5(10^{-5}) \end{bmatrix}.$

Using CAS to find the inverse of $[sI - A]$ yields

$$[sI - A]^{-1} = \frac{1}{2(10^4)s^2 + 2s + 2.5(10^{-5})} \begin{bmatrix} 2(10^4)s + 1 & 1 \\ \dfrac{1}{2} & 2(10^4)s + 1 \end{bmatrix} \tag{5.40}$$

Substituting Eqs. (5.39) and (5.40) in Eq. (5.31) yields

$$\begin{bmatrix} T_f(s) \\ T_w(s) \end{bmatrix} = \frac{1}{2(10^4)s^2 + 2s + 2.5(10^{-5})} \begin{bmatrix} 2(10^4)s + 1 & 1 \\ \dfrac{1}{2} & 2(10^4)s + 1 \end{bmatrix} \begin{bmatrix} 100 \\ 100 + \dfrac{5(10^{-4})}{s} \end{bmatrix} \tag{5.41}$$

Processing the matrix product on the right-hand side of Eq. (5.41) using CAS yields

$$T_f(s) = 20 \frac{4(10^9)s^2 + 4(10^5)s + 1}{8(10^8)s^3 + 8(10^4)s^2 + s} \tag{5.42}$$

and

$$T_w(s) = 20 \frac{4(10^9)s^2 + 3.2(10^5)s + 1}{8(10^8)s^3 + 8(10^4)s^2 + s} \tag{5.43}$$

Taking the inverse Laplace transform of Eqs. (5.42) and (5.43) using CAS yields

$$T_f(t) = 20 + 80e^{-5\left(10^{-5}\right)t}\left[\cosh(at) + \sqrt{2}\sinh(at)\right] \tag{5.44}$$

and

$$T_w(t) = 20 + 80e^{-5\left(10^{-5}\right)t}\left[\cosh(at) + \frac{\sqrt{2}}{2}\sinh(at)\right] \tag{5.45}$$

where $a = 2.5\sqrt{2}\left(10^{-5}\right)$.

2. Plot $T_f(t)$ and $T_w(t)$ versus t in s.

$T_f(t)$ and $T_w(t)$ are plotted in **Figure 5.20**.

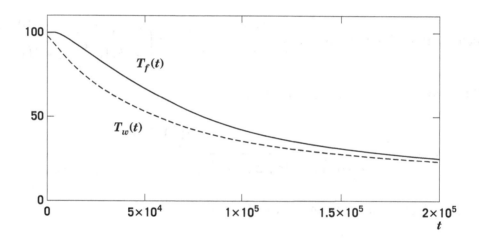

Figure 5.20 Fluid and wall temperatures versus time.

3. Determine how long it will take for the fluid temperature to reach $\left(T_S + 5\right){}^{\circ}C$ once the unit is shut down.

The time can be estimated graphically or by substituting $25\,{}^{\circ}C$ in Eq. (5.44); that is, $25 = 20 + 80e^{-5\left(10^{-5}\right)t}\left[\cosh(at) + \sqrt{2}\sinh(at)\right]$, and solving for t using CAS, which yields $t = 2.022\left(10^{5}\right)s = 2.34\,days$. This is the maintenance shutdown waiting period from the time the tubular heat exchanger operation is halted.

Stereo system with woofer

Consider a speaker system consisting of woofer, midrange, and tweeter speakers, as shown in **Figure 5.21**.

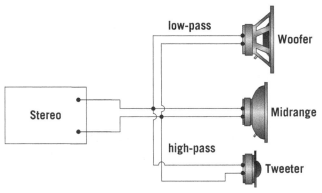

Figure 5.21 Stereo system with woofer, midrange, and tweeter speakers.

The system will have to reproduce sound in the audible frequency range of $20Hz - 20kHz$. The woofer, midrange, and tweeter speakers are designed to reproduce sounds in the $20Hz - 1.4kHz$, $1kHz - 10kHz$, and $6kHz - 20kHz$ range, respectively. For that purpose, a low-pass filter is used in the woofer to filter frequencies greater than $1.4kHz$; a band-pass filter is used in the midrange speaker to exclude frequencies less than $1kHz$ and greater than $10kHz$; and a high-pass filter is used in the tweeter to filter frequencies less than $6kHz$ (see Chapter 1, Section 1.4, for more information on filters). A filter circuit can be modeled as a two-branch circuit. For example, the woofer filter circuit is modeled as two-branches, as shown in **Figure 5.22**.

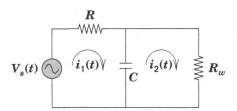

Figure 5.22 Low-pass circuit for the woofer filter.

Applying Kirchhoff's voltage loop law to the left branch of the circuit leads to

$$i_1 R + \frac{1}{C} \int (i_1 - i_2) dt = V_s(t) \tag{5.46}$$

where $i_1 R$ is the voltage across the filter resistor, and $\frac{1}{C} \int (i_1 - i_2) dt$ is the voltage across the filter capacitor. R is selected appropriately to filter frequencies greater than $1.4kHz$ for the woofer.

Applying Kirchhoff's voltage loop law to the right branch of the circuit leads to

$$i_2 R_w + \frac{1}{C} \int (i_2 - i_1) dt = 0 \tag{5.47}$$

where R_w is the woofer resistance, and $i_2 R_w$ is the voltage across the woofer. To eliminate the integrals, take the derivatives of Eqs. (5.46) and (5.47), which yield

$$\frac{di_1}{dt} = \frac{1}{RC}(i_2 - i_1) + \frac{1}{R}\frac{dV_s}{dt}$$
$$\frac{di_2}{dt} = \frac{1}{R_w C}(i_1 - i_2)$$

(5.48)

or in matrix form as

$$\frac{d}{dt}\begin{bmatrix} i_1 \\ i_2 \end{bmatrix} = \begin{bmatrix} a_{11} & a_{12} \\ a_{21} & a_{22} \end{bmatrix}\begin{bmatrix} i_1 \\ i_2 \end{bmatrix} + \begin{bmatrix} \dfrac{1}{R}\dfrac{dV_s}{dt} \\ 0 \end{bmatrix}$$

(5.49)

where $a_{11} = -\dfrac{1}{RC}$; $a_{12} = \dfrac{1}{RC}$, $a_{21} = \dfrac{1}{R_w C}$ and $a_{22} = -\dfrac{1}{R_w C}$. Eq. (5.49) is a linear,

nonhomogeneous and coupled first-order system of ODEs with constant coefficients, which has

the form of Eq. (5.30), where $Y = \begin{bmatrix} i_1 \\ i_2 \end{bmatrix}$, $A = \begin{bmatrix} a_{11} & a_{12} \\ a_{21} & a_{22} \end{bmatrix}$ and $G = \begin{bmatrix} \dfrac{1}{R}\dfrac{dV_s}{dt} \\ 0 \end{bmatrix}$. Its solution is found

by the Laplace transform method with ICs $Y(0) = \begin{bmatrix} i_1(0) \\ i_2(0) \end{bmatrix}$.

Example 5.10 Woofer Loudspeaker

Consider the woofer low-pass filter circuit in **Figure 5.22**, which should filter frequencies greater than $1.4 kHz$. Consider a test voltage $V(t) = A\sin(\omega t)$ to simulate a signal from the stereo.

Data: $A = 1\ volt$, $R = 0.1 ohm$, $C = 0.001137\ farad$, $R_w = 8 ohms$.

Recall from Figure 1.22, Chapter 1, that for a low-pass filter $\omega \ll \omega_c$, where $\omega_c = \dfrac{1}{\tau} = \dfrac{1}{RC}$

is the cutoff frequency. The choice of $R = 0.1 ohm$ and $C = 0.001137\ farad$ yields

$\omega_c = 8795\dfrac{rad}{s}$, so that $f_c = \dfrac{\omega_c}{2\pi} = 1400 Hz$ as required for the woofer. In order to demonstrate the validity of this model:

1. Solve Eq. (5.49) by the Laplace transform method with ICs $Y(0) = \begin{bmatrix} i_1(0) \\ i_2(0) \end{bmatrix} = \begin{bmatrix} 0 \\ 0 \end{bmatrix}$.

2. Find the amplitude of the steady oscillating current i_s from $i_2(t)$.

3. Find the amplitude of the steady voltage across the woofer using $V_w = i_s R_w$.

4. Plot V_w versus $\dfrac{\omega}{\omega_c}$ to validate the model consistently with Figure 1.22, Chapter 1.

Solution:

1. Solve Eq. (5.49) by the Laplace transform method with ICs $Y(0) = \begin{bmatrix} i_1(0) \\ i_2(0) \end{bmatrix} = \begin{bmatrix} 0 \\ 0 \end{bmatrix}$.

The Laplace transform of Eq. (5.49) is given by Eq. (5.31), where

$$A = \begin{bmatrix} a_{11} & a_{12} \\ a_{21} & a_{22} \end{bmatrix} = \begin{bmatrix} -\dfrac{1}{RC} & \dfrac{1}{RC} \\ \dfrac{1}{R_wC} & -\dfrac{1}{R_wC} \end{bmatrix}.$$

For simplicity, write $A = \begin{bmatrix} -a & a \\ b & -b \end{bmatrix}$, where $a = \dfrac{1}{RC}$ and $b = \dfrac{1}{R_wC}$.

Matrix G is $G = \begin{bmatrix} \dfrac{1}{R}\dfrac{dV_s}{dt} \\ 0 \end{bmatrix} = \begin{bmatrix} \dfrac{1}{R}\omega\cos(\omega t) \\ 0 \end{bmatrix}$. Its Laplace transform is $G(s) = \begin{bmatrix} \dfrac{s\omega}{R(s^2+\omega^2)} \\ 0 \end{bmatrix}$, so

that

$$\left[Y(0) + G(s) \right] = \begin{bmatrix} \dfrac{s\omega}{R(s^2+\omega^2)} \\ 0 \end{bmatrix} \tag{5.50}$$

Next, evaluate matrix $[sI - A]$: $[sI - A] = \begin{bmatrix} s - a_{11} & -a_{12} \\ -a_{21} & s - a_{22} \end{bmatrix} = \begin{bmatrix} s+a & -a \\ -b & s+b \end{bmatrix}$.

Using CAS to find the inverse of $[sI - A]$ yields

$$[sI - A]^{-1} = \frac{1}{s(s+a+b)}\begin{bmatrix} s+b & a \\ b & s+a \end{bmatrix} \tag{5.51}$$

Substituting Eqs. (5.50) and (5.51) in Eq. (5.31) yields

$$\begin{bmatrix} I_1(s) \\ I_2(s) \end{bmatrix} = \frac{1}{s(s+a+b)}\begin{bmatrix} s+b & a \\ b & s+a \end{bmatrix}\begin{bmatrix} \dfrac{s\omega}{R(s^2+\omega^2)} \\ 0 \end{bmatrix} \tag{5.52}$$

Processing the matrix product on the right-hand side of Eq. (5.52) using CAS yields

$$I_1(s) = \frac{\omega(s+b)}{R(s^2+\omega^2)(s+a+b)} \tag{5.53}$$

$$I_2(s) = \frac{\omega b}{R(s^2 + \omega^2)(s + a + b)} \qquad (5.54)$$

The voltage through the woofer speaker is of interest; thus apply the inverse Laplace transform only to Eq. (5.54). Using CAS,

$$i_2(t) = \frac{\omega b}{R\left[\omega^2 + (a+b)^2\right]}\left[\underbrace{e^{-(a+b)t}}_{\text{transient}} + \frac{a+b}{\omega}\sin(\omega t) - \cos(\omega t)\right] \qquad (5.55)$$

2. Find the amplitude of the steady oscillating current i_s from $i_2(t)$.

The transient in Eq. (5.55) is given by the exponential term with negative exponent because $a + b > 0$. Therefore, the steady oscillating current is

$$i_2(t) = \frac{b}{R\left[\omega^2 + (a+b)^2\right]}\left[(a+b)\sin(\omega t) - \omega\cos(\omega t)\right] \qquad (5.56)$$

Based on Eqs. (1.68–1.69) in Chapter 1, the amplitude of the trigonometric term between brackets in Eq. (5.56) is $\sqrt{\omega^2 + (a+b)^2}$. Thus, the amplitude of the steady oscillating current is

$$i_s = \frac{b\sqrt{\omega^2 + (a+b)^2}}{R\left[\omega^2 + (a+b)^2\right]} = \frac{b}{R\sqrt{\omega^2 + (a+b)^2}}.$$

3. Find the amplitude of the steady voltage across the woofer using $V_w = i_s R_w$.

The amplitude of the voltage through the woofer is

$$V_w = \frac{bR_w}{R\sqrt{\omega^2 + (a+b)^2}} \qquad (5.57)$$

4. Plot V_w versus $\dfrac{\omega}{\omega_c}$ to validate the model consistent with Figure 1.22, Chapter 1.

Introduce $\dfrac{\omega_c^2}{\omega_c^2}$ in the denominator of Eq. (5.57), that is, $V_w = \dfrac{bR_w}{R\sqrt{\omega^2\dfrac{\omega_c^2}{\omega_c^2} + (a+b)^2}}$, or

$$V_w(X) = \frac{bR_w}{R\sqrt{X^2\omega_c^2 + (a+b)^2}} \qquad (5.58)$$

where $X = \dfrac{\omega}{\omega_c}$. Plotting $V_w(X)$ versus X on a semi-log scale with $R = 0.1$, $R_w = 8$, $a = 8795$, $b = 109.94$, and $\omega_c = 8795$ is shown in **Figure 5.23**, which is consistent with the behavior of a low-pass filter as shown in Figure 1.22, Chapter 1, and validates the given data and circuit model.

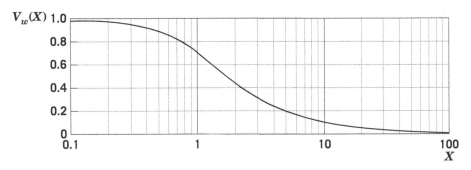

Figure 5.23 Woofer voltage as part of a low-pass filter limiting input frequencies to less than 1.4 *kHz*.

In this problem you learned how to develop a model for a low-pass filter in a stereo, which consisted of linear, nonhomogeneous, and coupled ODEs with constant coefficients. The system was solved using the Laplace transform to show how input frequencies to the woofer greater than 1.4 *kHz* are filtered.

The Laplace transforms of single and system of ODEs are summarized in **Table 5.3**.

ODE type	ODE	Laplace transform
Single first-order	$\tau \dfrac{dy}{dt} + y = f(t)$	$Y(s) = \dfrac{F(s) + \tau\, y(0)}{\tau s + 1}$
Single second-order	$\dfrac{d^2 y}{dt^2} + 2\zeta\omega_n \dfrac{dy}{dt} + \omega_n^2 y = f(t)$ or $\dfrac{d^2 y}{dt^2} + a_1 \dfrac{dy}{dt} + a_0 y(t) = r(t)$ $a_1 = 2\zeta\omega_n$ and $a_0 = \omega_n^2$	$Y(s) = \dfrac{R(s) + (s + a_1)y(0) + \dfrac{dy(0)}{dt}}{s^2 + a_1 s + a_0}$
System of first-order	$\dfrac{d\mathrm{Y}}{dt} = \mathrm{A}\mathrm{Y} + \mathrm{G}$	$\mathrm{Y}(s) = \left[s\mathrm{I} - \mathrm{A} \right]^{-1} \left[\mathrm{Y}(0) - \mathrm{G}(s) \right]$

Table 5.3 Summary of Laplace transform of single and system of ODEs.

HOMEWORK PROBLEMS FOR CHAPTER FIVE

Problems for Section 5.3

For Problems 1–3, $\theta(t)$ describes the valve angle to control the flow into a tank, as shown in Figure 1.3, Chapter 1.

a. Express the valve angle, $\theta(t)$ in degrees as a piecewise function.
b. Express $\theta(t)$ compactly using Eq. (5.4).
c. Plot the compact expression using CAS over the interval $0 \leq t < \infty$.

1.

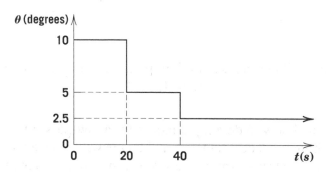

Figure 5.24 Valve angle as a piecewise function for Problem 1.

2.

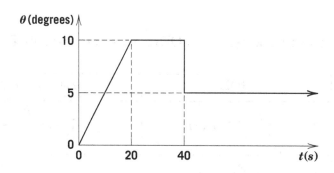

Figure 5.25 Valve angle as a piecewise function for Problem 2.

3.

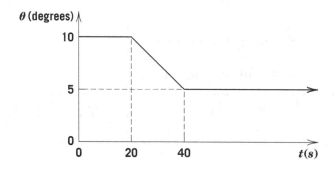

Figure 5.26 Valve angle as a piecewise function for Problem 3.

4. In a pharmaceutical application, vials are filled with a chemical compound by positioning them on a rotary table driven by a motor, as shown in **Figure 5.27**. The process cycle begins from rest with the motor turning uniformly at an angular acceleration $\alpha = 1$ *revolution / s^2* until the table reaches an angular speed of $\omega = 0.25$ *revolution / s*. Then, the motor decelerates to zero speed at an acceleration $\alpha = -1$ *revolution / s^2*. Finally, the system dwells for 1 *sec*; that is, no motion takes place during this time interval, while a vial is being filled with the chemical compound. Afterward, the cycle is repeated.

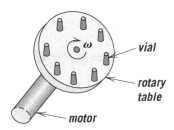

Figure 5.27 Vials positioned on a rotary table driven by a motor.

a. Express $\alpha(t)$ in *revolution / s^2* as a piecewise function over one cycle.

b. Express $\alpha(t)$ in *revolution / s^2* compactly using Eq. (5.4) over one cycle.

c. Express $\omega(t)$ in *revolution / s* as a piecewise function over one cycle.

d. Express $\omega(t)$ in *revolution / s* compactly using Eq. (5.4) over one cycle.

e. Plot the compact expressions of $\alpha(t)$ and $\omega(t)$ using CAS over one cycle.

5. In reference to the simulation of an impact force in Example 5.4, consider the case where the engineer applies the impact force $F(t)$ twice. The first time, a 500 *N* rectangular pulse is applied at $t_0 = 0$ and lasts 5 *ms*. Shortly, a second 250 *N* rectangular pulse is applied at $t_1 = 1s$ and lasts 5 *ms*, as shown in **Figure 5.28**.

Figure 5.28 Double rectangular pulse applied to a concrete pier.

a. Express $F(t)$ as a rectangular force pulse over the interval $0 \le t < \infty$ using Eq. (5.4), where t is in s.

b. Model $F(t)$ compactly using the Dirac delta function by assuming that $\Delta t \to 0$ for each rectangular pulse while its impulse remains constant. The first function takes place at $t_0 = 0$, and the second function at $t_1 = 1s$.

6. The response of structures to seismic ground displacement $x_g(t)$ is studied by modeling ground motion by a rectangular acceleration or velocity pulse. Consider the case of a $1.5\frac{m}{s}$ rectangular velocity pulse, $v(t)$, which takes place at $t_0 = 0$ and lasts $0.5s$.

a. Express $v(t)$ compactly using Eq. (5.4).

b. Find an expression for acceleration while considering the double discontinuity in $v(t)$. Here Δt does not go to zero.

c. Find a compact expression for the ground displacement, $x_g(t)$.

Problems for Section 5.4

For Problems 7–9, see Example 5.6.

a. Find the Laplace transform $\theta(s)$ by direct integration using Eq. (5.15).

b. Express $\theta(t)$ compactly using Eq. (5.4) and find $\theta(s)$ using the shifting property of a step function. Obviously, the expression should match that found in part a.

c. Take the inverse Laplace transform of $\theta(s)$ using CAS.

d. Validate your expression by plotting $\theta(t)$ as found in part a or b to reproduce the graph as given in the problem.

7. Problem 1 and **Figure 5.24**.
8. Problem 2 and **Figure 5.25**.
9. Problem 3 and **Figure 5.26**.

Problems for Section 5.5

For Problems 10–12, revisit the fluid level in a tank, $h(t)$, in Example 1.7, Chapter 1, which is governed by $\tau\frac{dh}{dt} + h = f(t)$, where the forcing function is $f(t) = \left(\frac{k_v}{k_o}\right)\theta$. The IC is $h(0) = 0$.

Furthermore, $\frac{k_v}{k_o} = 0.2$ and $\tau = 60$.

a. Solve the problem by the Laplace transform method using Eq. (5.21).
b. Plot the compact expression of $h(t)$ using CAS over the interval $0 \le t < \infty$.

10. $\theta(t)$ as shown in **Figure 5.24**.

11. $\theta(t)$ as shown in **Figure 5.25**.

12. $\theta(t)$ as shown in **Figure 5.26**.

13. Consider a low-pass filter governed by $\tau \dfrac{dV_o}{dt} + V_o = V(t)$, where $\tau = 0.02s$, and
$V(t) = 2\sin(0.5\pi t) + \sin(120\pi t)$. The IC is $V_o(0) = 0$.

a. Solve for the output $V_o(t)$ using the Laplace transform.

14. Consider a low-pass filter governed by $\tau \dfrac{dV_o}{dt} + V_o = V(t)$, where $\tau = 0.02s$. The IC is
$V_o(0) = 0$.

a. Model the forcing voltage $V(t)$ as impulsive at $t_0 = 0$ with a maximum amplitude of 2 *volts*, which occurs over a time interval $\Delta t = 10 \times 10^{-6}s$.
b. Solve for the output $V_o(t)$ using the Laplace transform.
c. Plot $V_o(t)$ over the interval $0 \le t < 0.15s$.
d. Estimate how long it takes the transient voltage amplitude to reach 0.5% of its initial response, $V_o(0^+)$.

15. Consider a high-pass filter governed by $\tau \dfrac{dV_o}{dt} + V_o = \tau \dfrac{dV}{dt}$, where $\tau = 0.02s$, and
$V(t) = 2\sin(0.5\pi t) + \sin(120\pi t)$. The IC is $V_o(0) = 0$.

a. Solve for the output $V_o(t)$ using the Laplace transform.

16. Consider a high-pass filter governed by $\tau \dfrac{dV_o}{dt} + V_o = \tau \dfrac{dV}{dt}$, where $\tau = 0.02s$. The IC is
$V_o(0) = 0$.

a. Model the forcing voltage $V(t)$ as a rectangular pulse at $t_0 = 0$ with a maximum amplitude of 2 *volts*, which occurs over a time interval $\Delta t = 0.2s$.
b. Solve for the output $V_o(t)$ using the Laplace transform.
c. Plot $V_o(t)$ over the interval $0 \le t < 0.2s$.
d. Estimate how long it takes the voltage to drop by 63.21 % from its initial response, $V_o(0^+)$.

Problems for Section 5.6

17. Revisit the transient response of a damped RLC circuit in Example 5.8. In lieu of the current $i(t)$ governed by Eq. (5.26), you can analyze the voltage response across the capacitor, $V_C(t)$. Substituting $i(t) = C\dfrac{dV_C}{dt}$, given by Ohm's law across the capacitor, in Eq. (5.26) yields

$$\frac{d^2V_C}{dt^2} + a_1\frac{dV_C}{dt} + a_0 V_C = a_0 V_s \qquad (5.59)$$

where $a_0 = \dfrac{1}{LC}$ and $a_1 = \dfrac{R}{L}$. The ICs are $i(0) = 0$ and $\dfrac{di(0)}{dt} = 0$. Take $C = \dfrac{5}{\pi^2}\left(10^{-10}\right)$ farad, $L = 5\left(10^{-4}\right)$ henry and $R = 100$ ohm.

a. Model the forcing voltage $V_s(t)$ as an impulsive voltage at $t_0 = 0$ with a maximum amplitude of 2 volts, which occurs over a time interval $\Delta t = 10^{-6}\, s$.
b. Solve for $V_C(t)$ using the Laplace transform using Eq. (5.25).
c. Plot $V_C(t)$ over the interval $0 \le t < 5\times10^{-5}\, s$ in order to visualize the overall response.
d. Plot $V_C(t)$ over the interval $0 \le t < 1\times10^{-5}\, s$ and report the first maximum value of the voltage response, $V_{C,max}$.
e. Estimate how long it takes the transient voltage amplitude to reach 1% of its initial maximum response, $V_{C,max}$.

18. Consider a car whose shock absorber on one side failed, which reduces the system to a mass-spring system. The car travels on an uneven road modeled by the tire path, $y_t = A\cos\left(\dfrac{2\pi x}{\lambda}\right)$, as shown in **Figure 5.29**, where the road wavelength and amplitude are $50m$ and $0.01\, m$, respectively. y_t can be converted to a temporal sinusoid using the car speed— see Eq. (2.97–2.99)— so that the sprung mass is governed by the second-order ODE

$$\frac{d^2y}{dt^2} + \frac{k}{m}y = \frac{k}{m}A\cos(\omega t) \qquad (5.60)$$

where $\omega = \dfrac{2\pi v}{\lambda}$. The car velocity is 45 mph (20 m/s). The sprung mass is 225 kg and the spring stiffness is $87\left(10^3\right)\dfrac{N}{m}$.

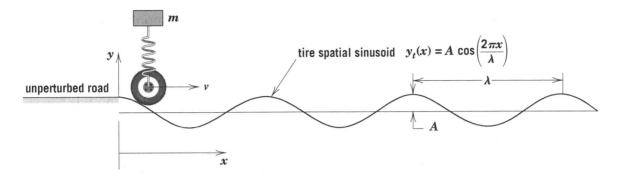

Figure 5.29 Spring-mass system moving at velocity v on a sinusoidal road.

a. Solve Eq. (5.60) for the sprung mass response, $y(t)$ by the Laplace transform using Eq. (5.25). The ICs are $y(0) = 0$ and $\dfrac{dy(0)}{dt} = 0$.

b. Plot $y(t)$ over the interval $0 \leq t < 10s$, and comment on its amplitude as far as safety is concerned.

19. Repeat Problem 18 where the car travels over a step modeled by the tire path

$$y_t = A\left[u(x) - u(x - \lambda)\right] \tag{5.61}$$

as shown in **Figure 5.30a**, where $A = 0.1\,m$ is the height and $\lambda = 1m$ is the width of the step. Equation (5.61) is converted to time as $y_t = A\left[u(t) - u\left(t - \dfrac{\lambda}{v}\right)\right]$, as shown in **Figure 5.30b**.

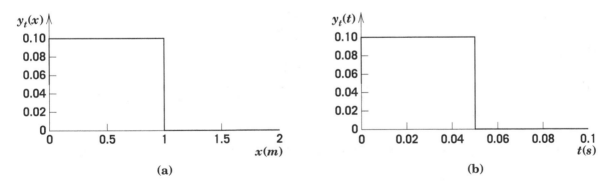

Figure 5.30 Step model: (a) spatial step model; and (b) temporal step model.

In this case, the sprung mass is governed by the second-order ODE

$$\frac{d^2y}{dt^2} + \frac{k}{m}y = \frac{k}{m}A\left[u(t) - u\left(t - \frac{\lambda}{v}\right)\right] \tag{5.62}$$

The car velocity is 45 *mph* (20 *m/s*). The sprung mass is 225 *kg* and the spring stiffness is $87(10^3)\dfrac{N}{m}$.

a. Solve Eq. (5.62) for the sprung mass response, $y(t)$ by the Laplace transform using Eq. (5.25). The ICs are $y(0) = 0$ and $\dfrac{dy(0)}{dt} = 0$.

b. Plot $y(t)$ over the interval $0 \le t < 1s$, and comment on its amplitude as far as safety is concerned.

20. Repeat Problem 18 where the car travels over a hump modeled by the tire path

$$y_t = \frac{A}{2}\left[1 - \cos\left(\frac{2\pi x}{\lambda}\right)\right]\left[u(x) - u(x - \lambda)\right] \tag{5.63}$$

as shown in **Figure 5.31a**, where $A = 0.1\ m$ is the height and $\lambda = 1m$ is the width of the hump. The rectangular spatial pulse, $\left[u(x) - u(x - \lambda)\right]$, is necessary in order to terminate the hump at $\lambda = 0.1m$. Equation (5.63) is converted to time as $y_t = \frac{A}{2}\left[1 - \cos(\omega t)\right]\left[u(t) - u\left(t - \frac{\lambda}{v}\right)\right]$, as shown in **Figure 5.31b**, where $\omega = \dfrac{2\pi v}{\lambda}$.

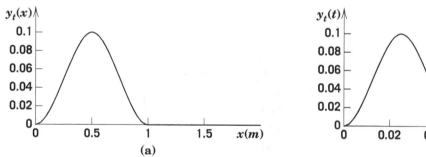 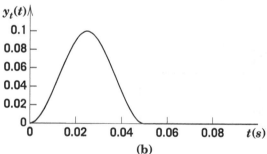

(a)   (b)

Figure 5.31 Hump model: (a) spatial hump model; and (b) temporal hump model.

In this case the sprung mass is governed by the second-order ODE:

$$\frac{d^2 y}{dt^2} + \frac{k}{m}y = \frac{k}{m}\frac{A}{2}\left[1 - \cos(\omega t)\right]\left[u(t) - u\left(t - \frac{\lambda}{v}\right)\right] \tag{5.64}$$

The car velocity is 45 *mph* (20 *m/s*). The sprung mass is 225 *kg* and the spring stiffness is $87\left(10^3\right)\dfrac{N}{m}$.

a. Solve Eq. (5.64) for the sprung mass response, $y(t)$ by the Laplace transform using Eq. (5.25). The ICs are $y(0)=0$ and $\dfrac{dy(0)}{dt}=0$.

b. Plot $y(t)$ over the interval $0 \le t < 1s$, and comment on its amplitude as far as safety is concerned.

21. In a sheet metal manufacturing company, a strip of metal, as shown in **Figure 5.32** , is used to make aluminum cans. The sheet metal drum is driven by a motor. The voltage motor, $V_a(t)$, is regulated in order to accelerate the sheet metal uniformly, then move it at uniform velocity, and finally decelerate it uniformly to rest. This control strategy is modeled by the applied voltage to the motor as a forcing function, as shown in **Figure 5.32b**. At $t=3s$, the sheet metal is cut by a hydraulic ram. The system dwells for 0.5s; that is, no motion takes place during that time before the cycle is repeated.

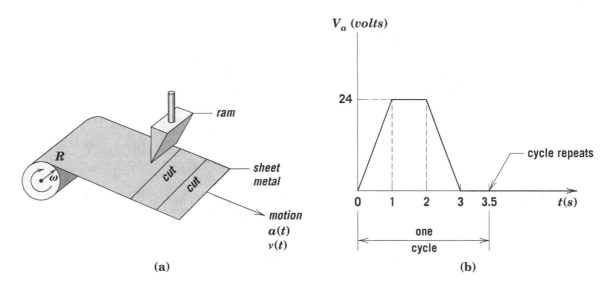

Figure 5.32 Aluminum cans manufacturing: (a) motor driving sheet metal drum; and (b) motor piecewise control voltage over 3.5-*s* time interval.

The motor angular speed is modeled by the ODE $\dfrac{d^2\omega}{dt^2}+600\dfrac{d\omega}{dt}+90\times10^3\omega = 90\times10^3 V_a(t)$, where $V_a(t)$ is the applied voltage in *volts*, as a piecewise function.

a. Express $V_a(t)$ compactly over the interval $0 \le t \le 3.5s$ using Eq. (5.4).

b. Find the Laplace transform, $V_a(s)$, using CAS.

c. Find the angular speed of the motor over the interval $0 \le t \le 3.5s$ by the Laplace transform using Eq. (5.25), subject to the following ICs: the initial angular speed and acceleration are
$$\omega(0)=0 \ rad/s, \text{ and } \frac{d\omega(0)}{dt}=3000\frac{rad}{s^2}, \text{ respectively.}$$

d. Plot $\omega(t)$ over the time interval $0 \le t \le 3.5s$.

22. Read Problem 16 in Chapter 2 first for detailed information about buoy analysis.

A helicopter company that specializes in retrieving cylindrical buoys from the ocean simulated the response of buoys in a lab with different wave amplitudes and frequencies.

Relative to the equilibrium position with zero initial displacement and speed, as shown in **Figure 5.33a**, the buoy is set to oscillate vertically by inducing a wave of height $w(t)$ in a test tank, as shown in **Figure 5.33b**. In this case, the instantaneous upward buoyancy force is given by $F_b = \rho A\big[d-y+w(t)\big]g$.

The wave is simulated as a sinusoid given by $w(t)=H\sin(\omega t)$, where H and ω are the wave amplitude and frequency, respectively.

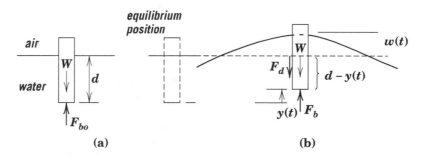

Figure 5.33 Model of cylindrical buoy: (a) buoy equilibrium position; and (b) buoy subject to wave amplitude.

a. Substitute F_b, $W=mg$, $m=\rho A d$, and $F_d = c_d\frac{dy}{dt}$ in Newton's law of motion
$$F_b - W - F_d = ma = m\frac{d^2y}{dt^2},$$ and derive the governing ODE for the instantaneous displacement of the buoy, $y(t)$.

b. Cast the ODE in the engineering form $\frac{d^2y}{dt^2}+2\zeta\omega_n\frac{dy}{dt}+\omega_n^2 y = f(t)$. Define the natural frequency, ω_n, the damping factor, ζ, and the forcing function, $f(t)$.

c. For negligible damping coefficient, c_d, find the undamped forced response, $y(t)$ by the Laplace transform method with zero ICs. The buoy cross sectional area is $A = 0.01m^2$, and its equilibrium depth is $d = 0.5m$. The water density is $\rho = 1000 \dfrac{kg}{m^3}$. The wave height is

$H = 0.05m$. The forcing frequency is $\omega = \dfrac{\omega_n}{2}$ and $g = 10 \dfrac{m}{s^2}$.

d. Plot on the same graph the response $y(t)$ and the forcing function $f(t)$ over the domain $t \geq 0$.

Problems for Section 5.7

For Problems 23–24, consider the RLC circuit as shown in **Figure 5.34** and **Figure 5.35**. The instantaneous current $i(t)$ is described by Kirchhoff's voltage loop law as

$$L\frac{di}{dt} + Ri + v = v_S \tag{5.65}$$

where v is the instantaneous voltage across the capacitor. By Ohm's law, the current across the capacitor is given by

$$i = C\frac{dv}{dt} \tag{5.66}$$

a. Cast Eqs. (5.65) and (5.66) as a system of coupled first-order ODEs in the form of Eq. (5.30), where the vector Y(t) is defined as $Y(t) = \begin{bmatrix} i(t) \\ v(t) \end{bmatrix}$.

b. Find $i(t)$ and $v(t)$ using the Laplace transform. The ICs are $i(0) = 0$ and $v(0) = 0$.

c. Plot $i(t)$ and $v(t)$ over the domain $t \geq 0$.

23. $L = 10 \, henries$, $R = 1000 \, ohms$, $v_s = 100 \, volts$, and $C = 4\left(10^{-6}\right) farad$.

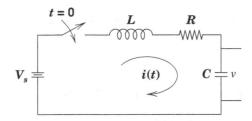

Figure 5.34 RLC circuit activated by a constant-voltage source.

24. $L = 1$ henry, $R = 100$ ohms, $v_s = 100\cos\left(100\pi t + \dfrac{\pi}{6}\right)$ volts, and $C = 4\left(10^{-6}\right)$ farad.

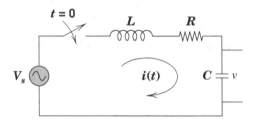

Figure 5.35 RLC circuit activated by a time-varying voltage source.

For Problems 25–27, consider the flow in two cascaded tanks as part of a chemical filtration process. The instantaneous fluid heights in the tanks are shown in **Figure 5.36**, which are modeled by the following coupled first-order ODEs,

$$4\frac{dh_1}{dt} = -0.04\left(h_1 - h_2\right) + 0.2\theta$$

$$4\frac{dh_2}{dt} = -0.3h_2 + 0.04\left(h_1 - h_2\right)$$

(5.67)

where h_1, h_2 are the instantaneous heights in m, and θ is the valve angle in degrees.

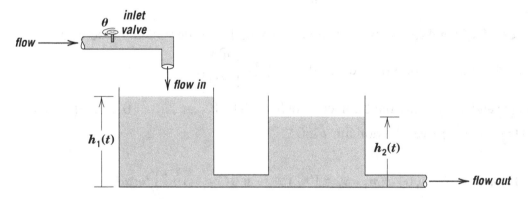

Figure 5.36 Two tanks in series used in a filtration process.

a. Find the Laplace transform of the piecewise forcing function $\theta(t)$.
b. Cast the system of Eq. (5.67) in the form of Eq. (5.30), where the vector Y(t) is defined as $Y(t) = \begin{bmatrix} h_1(t) \\ h_2(t) \end{bmatrix}$.
c. Find $h_1(t)$ and $h_2(t)$ using the Laplace transform. The ICs are $h_1(0) = h_2(0) = 0$.
d. Plot $h_1(t)$ and $h_2(t)$ over the time domain $t \geq 0$.
e. Estimate from the graphs the heights at steady-state.p

25. The valve is stepped from 0 to 60° as shown in **Figure 5.37**.

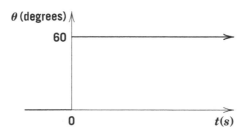

Figure 5.37 Valve setting for Eq. (5.67).

26. The valve is opened as shown in **Figure 5.38**.

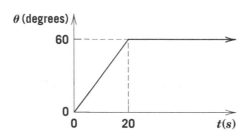

Figure 5.38 Valve setting for Eq. (5.67).

27. The valve is opened as shown in **Figure 5.39**.

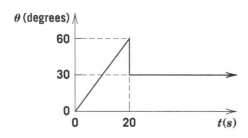

Figure 5.39 Valve setting for Eq. (5.67).

28. Higher-order ODEs can be solved as a system of first-order ODEs. Consider the angular displacement, $\theta\,(rad)$ of a motor that controls the motion of an elevator, as shown in **Figure 5.40**, which is described by

$$\frac{d^3\theta}{dt^3} + 2\zeta\omega_n\frac{d^2\theta}{dt^2} + \omega_n^2\frac{d\theta}{dt} = K\omega_n^2 V_m \tag{5.68}$$

where $\omega_n = 100\pi \dfrac{rad}{s}$, $\zeta = 1$, $V_m = 200 \ volts$ is the input voltage to the motor, and

$K = 1\left(\dfrac{1}{s.volt}\right)$ is a constant.

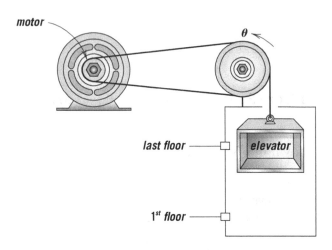

Figure 5.40 Elevator motion controlled by an external motor.

Define the angular speed as

$$\omega = \frac{d\theta}{dt} \tag{5.69}$$

in *rad/s*, and the angular acceleration as

$$\alpha = \frac{d\omega}{dt} \tag{5.70}$$

in *rad/s²*

a. Define $Y(t) = \begin{bmatrix} \theta(t) \\ \omega(t) \\ \alpha(t) \end{bmatrix}$ and cast Eqs. (5.68)–(5.70) as a system of coupled first-order ODEs in the form of Eq. (5.30).

b. Find Y(t) using the Laplace transform. The ICs are $\begin{bmatrix} \theta(0) \\ \omega(0) \\ \alpha(0) \end{bmatrix} = \begin{bmatrix} 0 \\ 0 \\ 1 \end{bmatrix}$.

c. Find the time to execute 10 revolutions, where t is in *s*. One revolution is equal to $2\pi \ rad$.

CHAPTER SIX

FOURIER SERIES AND CONTINUOUS FOURIER TRANSFORM

6.1 Introduction

Periodic phenomena, such as the rotation of the earth around the sun, the cycle of seasons, the occurrence of night and day, and heartbeat rhythms, to name a few examples, occur regularly in nature. There are also human-made periodic phenomena—for example, the wind turbine rotor designed to extract energy from wind, musical tones, and radio waves, to name a few cases of another set of endless periodic events. Periodicity is defined as the recurrence of an event in time or the repetitive pattern in space. That is, periodicity can be temporal as well as spatial. An example of temporal periodicity, simply called period, is the earth daily rotation around its axis, which is approximately 24 hours. An example of spatial periodicity is the repetitive pattern of cubic cells in a crystalline structure, which is on the order of 3 Angstrom, where 1 Angstrom = 10^{-10} m. The concept of wavelength as a measure of spatial periodicity was introduced briefly in Chapter 2; see Figure 2.28 and Eq. (2.97).

One of the ubiquitous periodic functions is the sine or cosine function. An example of a temporal sinusoid is in the area of fatigue testing of materials, as shown in **Figure 6.1a**. In this system, a test specimen mounted between two jaws is subject to a periodic force in time via an oscillating shaker. The purpose of the test is to determine how long it will take cracks to grow in a test specimen, as a function of the frequency of the shaker. An example of a periodic force is $f(t) = 10\sin(t)$, as shown in **Figure 6.1b**, where the force amplitude is 10 N and t is time in s. Think of negative values of time as past events relative to $t = 0$.

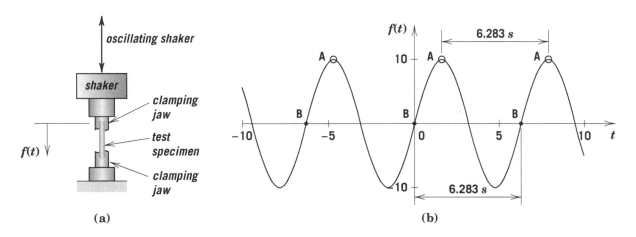

(a)   (b)

Figure 6.1 Fatigue test setup: (a) the test specimen subject to an oscillating force; and (b) the corresponding temporal sinusoid.

Given your familiarity with sine functions, you can easily identify periodicity by a repeating reference point, for example, point A or point B, as shown in the graph. You also should know from the basic form of a sinusoid, $f(t) = A\sin(\omega t)$, that the circular frequency is $\omega = 1\dfrac{rad}{s}$ in this case so that the frequency in *Hertz* is $f = \dfrac{\omega}{2\pi} = 0.159 Hz$, which yields a period $p = \dfrac{1}{f} = 2\pi = 6.283s$, as shown in **Figure 6.1b**. In the case of the test specimen, the crack phenomenon is characterized at a frequency $\omega = 1\dfrac{rad}{s}$. If the test is repeated at twice the frequency, for example, $f(t) = 10\sin(2t)$, the force function would occur at $\omega = 2\dfrac{rad}{s}$, with a period $p = \dfrac{1}{f} = \pi = 3.142s$, as shown by the top graph in **Figure 6.2**. You can see that by doubling the frequency, the occurrence of waves doubles in comparison to the first sine function, for example, two occurrences between the points labeled C in the top graph, in comparison to one occurrence between the points labeled B in the bottom graph. Notice also that the period is now halved.

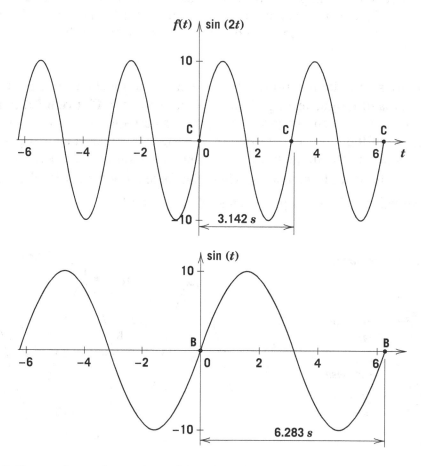

Figure 6.2 Comparison of one sinusoid with twice the frequency of another sinusoid.

Other periodic functions different from sinusoids appear extensively in engineering. For example, in the design of audio amplifiers, the audio signal should pass through a filter with minimum distortion. See Chapter 1, Section 1.4, for a review of mathematical modeling of filter circuits. An example of a periodic function is the repetitive step voltage $v_s(t)$ generated by a function generator as input to a low-pass filter, as shown in **Figure 6.3a**. In this case, $v_s(t)$ has magnitude of 20 *volts* and a period $p = 0.002s$ as shown in **Figure 6.3b**. The periodicity is shown by the repetition of point A per cycle. A typical cycle is defined over the interval $0 \leq t < p$ as a piecewise function

$$v_s(t) = \begin{cases} 20 & 0 \leq t < \dfrac{p}{2} \\ -20 & \dfrac{p}{2} \leq t < p \end{cases} \tag{6.1}$$

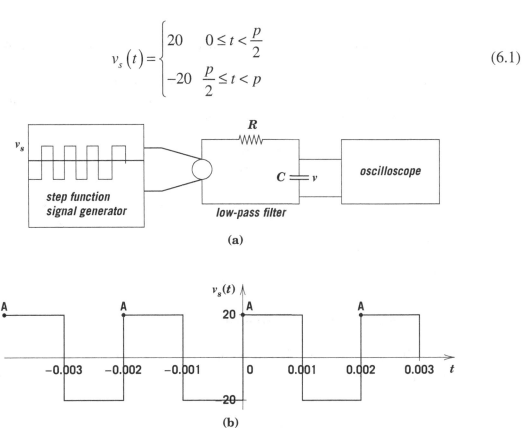

(a)

(b)

Figure 6.3 Low-pass filter with input periodic function: (a) step function wave generator, filter, and oscilloscope; and (b) input periodic voltage.

Mathematically, the response of the low-pass filter, $v(t)$, is obtained in the time domain by solving the first-order ODE

$$\tau \frac{dv}{dt} + v = v_s \tag{6.2}$$

with IC $v(0) = v_o$, subject to the periodic forcing function, $v_s(t)$ as depicted in **Figure 6.4**, where $\tau = RC$ is the circuit time constant.

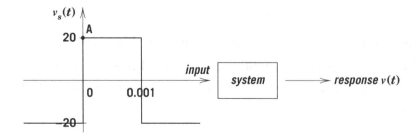

Figure 6.4 A block diagram of a system with a forcing function and a response.

Another example of a periodic function is that of electrical current in *ampere* as a function of time *t* in *s*, as shown in **Figure 6.5**. In some welding application such as thin aluminum, a triangular periodic current provides a fast weld with low heat loss.

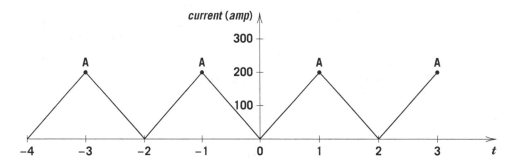

Figure 6.5 Periodic triangular current used in welding.

The periodicity is shown by the repetition of point A per cycle that has a period $p = 2s$. A typical current cycle is defined over the interval $0 \le t < p$ as a piecewise triangular function,

$$i(t) = \begin{cases} 200t & 0 \le t < \dfrac{p}{2} \\ -200(t-2) & \dfrac{p}{2} \le t < p \end{cases}$$ (6.3)

The list of periodic functions can go on and on. The point of this introduction is that the sine and cosine functions are the building blocks for any periodic and piecewise continuous functions like the repetitive step function in **Figure 6.3**, and the triangular function in **Figure 6.5**. By building blocks, we mean that a piecewise continuous and periodic function may be constructed from a linear combination of an infinite number of sine and cosine functions, which is the basis for Fourier series.

Thus, in this chapter you will be introduced to modeling piecewise continuous and periodic functions as Fourier series. Then you will learn how to apply a limiting process to a Fourier series in order to develop the Fourier transform, which transforms variables from the time domain to the frequency domain. The Fourier transform is a powerful tool in the analysis of nonperiodic functions. Furthermore, you will be introduced to the concept of bandwidth. In this chapter, unless otherwise specified, t stands for time, and the units are s for time, Hz for frequency f, and $\dfrac{rad}{s}$ for circular frequency ω.

6.2 Basic Concepts

Periodic functions

A periodic temporal function is defined as

$$f(t) = f(t \pm np) \tag{6.4}$$

for integer $n \geq 1$, where p is defined as the fundamental period of one cycle. The fundamental frequency is defined as

$$f_o = \frac{1}{p} \tag{6.5}$$

and the corresponding fundamental circular frequency as

$$\omega_o = 2\pi f_o = \frac{2\pi}{p} \tag{6.6}$$

Harmonic oscillation and the greatest common divisor

When an oscillation takes place with the fundamental circular frequency ω_o, for example, $f(t) = A_o \sin(\omega_o t)$, then ω_o is called the first harmonic. For any other oscillation at integer multiple frequency of ω_o, that is $\omega_o = n\omega$, ω is called the nth harmonic. Its corresponding harmonic sine function is $f(t) = A_n \sin(n\omega_o t)$. When several sinusoids with different frequencies are added together, they form a relatively complicated compounded waveform that has a fundamental frequency f_o as the greatest common divisor (GCD) of the frequencies that form the waveform. The harmonics that make up the waveform are multiple integers of the fundamental frequency of the waveform as illustrated in the following example.

Example 6.1

Consider the compounded waveform $g(t) = \sin(2\pi t) + \sin(6\pi t)$.

1. Find the fundamental circular frequency of $g(t)$, ω_o.
2. Express $g(t)$ in terms of ω_o, and identify its harmonics.

3. Define $f(t) = \sin(\omega_o t)$ as a reference function associated with the fundamental frequency, and demonstrate graphically that $g(t)$ and $f(t)$ have the same frequency.

Solution:

1. Find the fundamental circular frequency of $g(t)$, ω_o.

The compounded waveform $g(t)$ is composed of two functions: $\sin(2\pi t)$ and $\sin(6\pi t)$ with circular frequencies $2\pi \dfrac{rad}{s}$ and $6\pi \dfrac{rad}{s}$, or frequencies $1Hz$ and $3Hz$, respectively using $f = \dfrac{\omega}{2\pi}$. The GCD that divides $1Hz$ and $3Hz$ without a remainder is $1Hz$. Thus, the fundamental frequency of $g(t)$ is $f_o = 1Hz$ and its corresponding circular frequency is $\omega_o = 2\pi \dfrac{rad}{s}$.

2. Express $g(t)$ in terms of ω_o, and identify its harmonics.

Using $\omega_o = 2\pi \dfrac{rad}{s}$, $g(t) = \sin(2\pi t) + \sin(6\pi t)$ is rewritten as $g(t) = \sin(\omega_o t) + \sin(3\omega_o t)$. Thus, the first component that makes up $g(t)$ turned out to be a fundamental harmonic with frequency, ω_o. The second component of $g(t)$ is a third harmonic with frequency, $3\omega_o$.

3. Define $f(t) = \sin(\omega_o t)$ as a reference function associated with the fundamental frequency, and demonstrate graphically that $g(t)$ and $f(t)$ have the same frequency.

The reference function is $f(t) = \sin(\omega_o t) = \sin(2\pi t)$. As you can see in **Figure 6.6**, $g(t)$ has the same frequency of $f(t)$, as shown by the repeated occurrence of point A. The structure of $g(t)$ is accentuated by the thick line in the first cycle over the interval $0 \le t \le p$, where $p = 1s$ is the fundamental period. That is $g(t)$ has the same fundamental frequency ω_o as the reference function $f(t) = \sin(\omega_o t)$ shown as a dashed line. The third harmonic $\sin(3\omega_o t)$ inside $g(t)$ manifests itself as a higher frequency riding on the fundamental frequency in a form that resembles the letter "M" over the interval $0 \le t \le 0.5$ and "W" over the interval $0.5 \le t \le 1$.

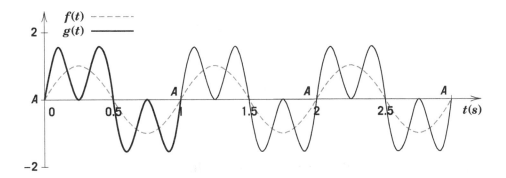

Figure 6.6 A compounded waveform $g(t) = \sin(\omega_o t) + \sin(3\omega_o t)$ that has the same fundamental frequency of $f(t) = \sin(\omega_o t)$.

In this Example, you learned how to find the fundamental frequency, ω_o of a compounded waveform using the GCD. Furthermore, in this particular example, ω_o and its sinusoid, $\sin(\omega_o t)$ appeared in the given function, $g(t)$. In a homework problem, you may explore that it is possible to construct a waveform from sinusoids with the fundamental harmonic appearing in none of the sinusoids that make up the waveform.

In the next section, you are introduced to the Fourier series as a compounded waveform that consists of an infinite number of sinusoids.

6.3 Fourier Series

The Fourier series of a periodic and piecewise continuous function $f(t)$ is defined in terms of a linear superposition of an infinite number of harmonics with a leading constant as

$$f(t) = a_o + \sum_{n=1}^{\infty} \left[a_n \cos(n\omega_o t) + b_n \sin(n\omega_o t) \right] \tag{6.7}$$

or, by substituting Eq. (6.6) in Eq. (6.7),

$$f(t) = a_o + \sum_{n=1}^{\infty} \left[a_n \cos\left(\frac{n\pi}{L} t\right) + b_n \sin\left(\frac{n\pi}{L} t\right) \right] \tag{6.8}$$

where $L = \dfrac{p}{2}$ is half of the fundamental period of $f(t)$, and the Fourier coefficients a_o, a_n, and b_n are given by

$$a_o = \frac{1}{2L} \int_{-L}^{L} f(t) \, dt \tag{6.9}$$

$$a_n = \frac{1}{L} \int_{-L}^{L} f(t) \cos\left(\frac{n\pi}{L} t\right) dt \qquad (6.10)$$

$$b_n = \frac{1}{L} \int_{-L}^{L} f(t) \sin\left(\frac{n\pi}{L} t\right) dt \qquad (6.11)$$

for integer $n \geq 1$.

Example 6.2 Audio Filters: Fourier Series of Input Voltage

Consider the input voltage $v_s(t)$, as shown in **Figure 6.3**, over the interval $-\infty < t < \infty$.

1. Find the Fourier series of $v_s(t)$.
2. Show graphically how well the addition of successive harmonics represents $v_s(t)$ over one period in the interval $0 \leq t \leq p$.

Solution:

1. Find the Fourier series of $v_s(t)$.

From **Figure 6.3**, the period of one cycle is $p = 0.002s$. Thus, $L = \frac{p}{2} = 0.001s$.

From Eq. (6.9), $a_o = \frac{1}{2L} \int_{-L}^{L} v_s(t) dt$. However, $v_s(t) = -20$ over $-L \leq t < 0$ and $v_s(t) = 20$ over $0 \leq t < L$, so that $a_o = \frac{1}{2L} \left[-20 \int_{-L}^{0} dt + 20 \int_{0}^{L} dt \right]$. Therefore, $a_o = \frac{1}{2L}(-20L + 20L) = 0$.

Similarly, from Eq. (6.10), $a_n = \frac{1}{L} \left[-20 \int_{-L}^{0} \cos\left(\frac{n\pi}{L} t\right) dt + 20 \int_{0}^{L} \cos\left(\frac{n\pi}{L} t\right) dt \right]$ that yields

$$a_n = \left(\frac{20}{n\pi}\right) \left[-\sin\left(\frac{n\pi}{L} t\right) \Big|_{-L}^{0} + \sin\left(\frac{n\pi}{L} t\right) \Big|_{0}^{L} \right] = \left(\frac{20}{n\pi}\right) \left[-\sin(n\pi) + \sin(n\pi) \right] = 0.$$

From Eq. (6.11), $b_n = \frac{1}{L} \left[-20 \int_{-L}^{0} \sin\left(\frac{n\pi}{L} t\right) dt + 20 \int_{0}^{L} \sin\left(\frac{n\pi}{L} t\right) dt \right]$ that yields

$$b_n = \frac{1}{L} \left[20 \left(\frac{L}{n\pi}\right) \cos\left(\frac{n\pi}{L} t\right) \Big|_{-L}^{0} - 20 \left(\frac{L}{n\pi}\right) \cos\left(\frac{n\pi}{L} t\right) \Big|_{0}^{L} \right], \text{ or}$$

$$b_n = \left(\frac{20}{n\pi}\right)\left[\cos(0) - \cos(n\pi) - \left(\cos(n\pi) - \cos(0)\right)\right] = 40\left(\frac{1 - \cos(n\pi)}{n\pi}\right).$$

Substituting the values of a_o, a_n, and b_n in Eq. (6.8) yields

$$v_s(t) = \frac{40}{\pi}\sum_{n=1}^{\infty}\frac{1 - \cos(n\pi)}{n}\sin\left(\frac{\pi n}{L}t\right) \tag{6.12}$$

2. Show graphically how well the addition of successive harmonics represents $v_s(t)$ over one period in the interval $0 \le t \le p$.

In practice, the infinite summation in Eq. (6.12) is truncated to a finite number of harmonics, for example,

$$v_s(t) = \frac{40}{\pi}\sum_{n=1}^{M}\frac{1 - \cos(n\pi)}{n}\sin\left(\frac{\pi n}{L}t\right) \tag{6.13}$$

for an integer $M \ge 1$. Notice that $1 - \cos(n\pi) = 0$ for even n, and $1 - \cos(n\pi) = 2$ for odd n. Thus,

$$v_s(t) = \frac{80}{\pi}\sum_{n=1}^{M}\frac{1}{n}\sin\left(\frac{\pi n}{L}t\right) \tag{6.14}$$

for an odd integer $M \ge 1$. For interpretation expediency, express Eq. (6.14) in terms of the fundamental frequency, f_o, Eq. (6.5), as

$$v_s(t) = \frac{80}{\pi}\sum_{n=1}^{M}\frac{1}{n}\sin\left(2\pi n f_o t\right) \tag{6.15}$$

where $f_o = \dfrac{1}{.002} = 500\,Hz$.

For $M = 1$,

$$v_s(t) = \underbrace{\frac{80}{\pi}\sin\left(2\pi f_o t\right)}_{\text{first harmonic,}\, f_o = 500\,Hz} \tag{6.16}$$

For $M = 3$,

$$v_s(t) = \frac{80}{\pi}\left(\underbrace{\sin\left(2\pi f_o t\right)}_{\text{first harmonic,}\, f_o = 500\,Hz} + \underbrace{\frac{1}{3}\sin\left[2\pi\left(3 f_o\right)t\right]}_{\text{third harmonic,}\, 3 f_o = 1500\,Hz}\right) \tag{6.17}$$

For $M = 5$, $v_s(t) = \dfrac{80}{\pi}\left(\underbrace{\sin\left(2\pi f_o t\right)}_{\text{first harmonic, } f_o = 500\,Hz} + \underbrace{\dfrac{1}{3}\sin\left[2\pi\left(3f_o\right)t\right]}_{\text{third harmonic, } 3f_o = 1500\,Hz} + \underbrace{\dfrac{1}{5}\sin\left[2\pi\left(5f_o\right)t\right]}_{\text{fifth harmonic, } 5f_o = 2500\,Hz}\right)$ (6.18)

and so on and so forth.

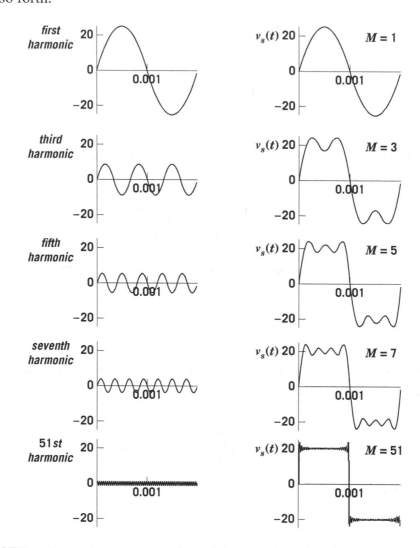

Figure 6.7 Fourier series representation of the step function described in **Figure 6.3**.

Equations (6.16)–(6.18) are shown in **Figure 6.7** along with the seventh and fifty-first harmonics. Several observations can be made. The first observation is that the period of the first or fundamental harmonic matches that of the given function, $v_s(t)$, that is $p = 0.002s$. However, the fundamental harmonic does not replicate the shape of the desired $v_s(t)$ as a step function.

Second, as the number of harmonics increases, the high frequency terms in each Fourier series appear as peaks and valleys riding on top of the fundamental harmonic. For example, the third harmonic, which consists of three cycles and six peaks and valleys, as shown on the left side of the figure, also manifests itself in the solution, $v_s(t)$ with six peaks and valleys riding on top of the fundamental harmonic. However, as illustrated in Example 6.1 using the GCD, all the compounded waveforms, given by the various Fourier expressions in Eqs. (6.16)–(6.18) for different values of M, have the same fundamental frequency, f_o, which appears as the leading term in each equation. Notice also the overshoot near the point where $v_s(t)$ is discontinuous, which is inherent in Fourier series at discontinuity points in the function, known as the Gibbs phenomenon. This is caused by attempting to reproduce a discontinuous function by a superposition of continuous harmonics. Third, notice the amplitudes of harmonics in the left figures, which decrease with higher frequencies as a result of the factor $1/n$ in the solution. Thus, the contribution of amplitude to the final waveform diminishes with higher frequencies.

∎

Lanczos sigma factor

Various methods can be used to minimize the Gibbs phenomenon overshoot. One of the methods is the Lanczos sigma factor method in which the factor $\dfrac{\sin\left(\dfrac{n\pi}{M}\right)}{\dfrac{n\pi}{M}}$ is introduced in the solution and the sum is taken to M-1. For example, rewriting Eq. (6.15) with the Lanczos sigma factor yields

$$v_s(t) = \frac{80}{\pi} \sum_{n=1}^{M-1} \frac{\sin\left(\dfrac{n\pi}{M}\right)}{\dfrac{n\pi}{M}} \frac{1}{n} \sin\left(2\pi n f_o t\right) \qquad (6.19)$$

Comparing the results for $M = 51$ with and without the Lanczos sigma factor is shown in **Figure 6.8**.

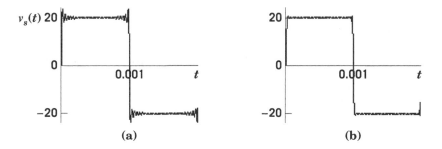

Figure 6.8 Step function produced by Fourier series: (a) series with 51 harmonics with noticeable Gibbs phenomenon; and (b) same series using the Lanczos sigma factor.

Even and odd functions

Quite often, periodic functions may be odd or even functions, which simplify matters as far as the Fourier coefficients are concerned. By definition, a function $f(t)$ is even when

$f(-t) = f(t)$, which is graphically symmetric with respect to the vertical axis. On the other hand, a function $f(t)$ is odd when $f(-t) = -f(t)$, which is graphically symmetric with respect to the origin. An example of odd and even functions is the step and triangular functions shown in **Figure 6.3** and **Figure 6.5**, respectively. In this case, you can take a shortcut when calculating Fourier coefficients, as shown in **Table 6.1**.

Fourier series	$f(t) = a_o + \sum\limits_{n=1}^{\infty}\left[a_n \cos\left(\dfrac{n\pi}{L}t\right) + b_n \sin\left(\dfrac{n\pi}{L}t\right) \right]$; $L = \dfrac{p}{2}$
Fourier coefficients for a general periodic function	$a_o = \dfrac{1}{2L}\int_{-L}^{L} f(t)\,dt$ $a_n = \dfrac{1}{L}\int_{-L}^{L} f(t)\cos\left(\dfrac{n\pi}{L}t\right)dt$; $b_n = \dfrac{1}{L}\int_{-L}^{L} f(t)\sin\left(\dfrac{n\pi}{L}t\right)dt$
Fourier coefficients for an even periodic function	$a_o = \dfrac{1}{L}\int_{0}^{L} f(t)\,dt$; $a_n = \dfrac{2}{L}\int_{0}^{L} f(t)\cos\left(\dfrac{n\pi}{L}t\right)dt$ $b_n = 0$
Fourier coefficients for an odd periodic function	$a_o = 0$; $a_n = 0$ $b_n = \dfrac{2}{L}\int_{0}^{L} f(t)\sin\left(\dfrac{n\pi}{L}t\right)dt$

Table 6.1 Fourier coefficients summary.

Example 6.3

Consider the current periodic function, $i(t)$, as shown in **Figure 6.5**, over the interval $-\infty < t < \infty$.

1. Find the Fourier series of $i(t)$.
2. Show graphically how well the addition of successive harmonics represents $i(t)$ over one period in the interval $-\dfrac{p}{2} \le t \le \dfrac{p}{2}$.

Solution:

1. Find the Fourier series of $i(t)$.

The triangular periodic function shown in **Figure 6.5** is symmetric with respect to the vertical axis. Thus, it is an even function. From **Table 6.1**, $b_n = 0$ so that

$$i(t) = a_o + \sum_{n=1}^{\infty} a_n \cos\left(\frac{n\pi}{L}t\right) \qquad (6.20)$$

with $a_o = \frac{1}{L}\int_0^L i(t)\,dt$ and $a_n = \frac{2}{L}\int_0^L i(t)\cos\left(\frac{n\pi}{L}t\right)dt$, where $L = \frac{p}{2} = 1s$. From Eq. (6.3)

$i(t) = 200t$ over $0 \le t < 1$. Thus, $a_o = \int_0^1 200t\,dt = 200\left(\frac{1}{2}t^2\right)\Big|_0^1 = 100$.

$a_n = \frac{2}{L}\int_0^L i(t)\cos\left(\frac{n\pi}{L}t\right)dt = 400\int_0^1 t\cos(n\pi t)\,dt$. Integrating by parts or using CAS yields

$a_n = 400\dfrac{\cos(n\pi) + n\pi\sin(n\pi) - 1}{n^2\pi^2}$. However, $\sin(n\pi) = 0$ for integer $n \ge 1$. Moreover,

$\cos(n\pi) - 1 = 0$ for even n, and $\cos(n\pi) - 1 = -2$ for odd n, so that $a_n = -\dfrac{800}{n^2\pi^2}$ Thus,

$$i(t) = 100 - \frac{800}{\pi^2}\sum_{n=1}^{M}\frac{1}{n^2}\cos(\pi nt) \qquad (6.21)$$

for an odd integer $M \ge 1$.

2. Show graphically how well the addition of successive harmonics represents $i(t)$ over one period in the interval $-\dfrac{p}{2} \le t \le \dfrac{p}{2}$.

Figure 6.9 shows the Fourier series representation of the triangular periodic function for one period over the interval $-1 \le t \le 1$, using Eq. (6.21). Here, the Fourier series converges quite well to the desired triangular function with six harmonics, as shown in the bottom left figure. Notice that there is no Gibbs phenomenon for this continuous function. Notice also the rapid decrease of amplitude that is proportional to $1/n^2$ as opposed to $1/n$ for the step function in Example 6.2.

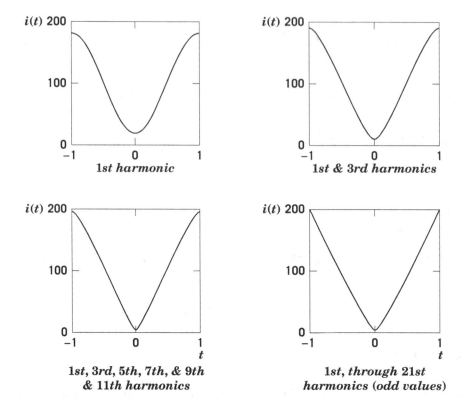

Figure 6.9 Fourier series representation of the triangular periodic function in **Figure 6.5**.

Spatial Fourier series

Before leaving this section, it should be noted that the foregoing Fourier analysis in the time domain is equally applicable to functions that are periodic in space. However, instead of the fundamental period p, we speak of the wavelength, λ; and instead of the fundamental circular frequency, ω_o, we speak of the fundamental wavenumber, κ_o, where κ_o is in $\dfrac{rad}{m}$ when x is in m. The analogies are summarized in **Table 6.2**. The Fourier coefficients are found in the same manner, with no need to repeat them in this table.

Temporal Fourier series	**Spatial Fourier series**
fundamental period: $p = 2L$	fundamental wavelength: $\lambda = 2L$
fundamental circular frequency, $\omega_o = \dfrac{2\pi}{p}$	fundamental wavenumber, $\kappa_o = \dfrac{2\pi}{\lambda}$
$f(t) = a_o + \displaystyle\sum_{n=1}^{\infty}\left[a_n \cos(n\omega_o t) + b_n \sin(n\omega_o t)\right]$, or $f(t) = a_o + \displaystyle\sum_{n=1}^{\infty}\left[a_n \cos\left(\dfrac{n\pi}{L}t\right) + b_n \sin\left(\dfrac{n\pi}{L}t\right)\right]$	$f(x) = a_o + \displaystyle\sum_{n=1}^{\infty}\left[a_n \cos(n\kappa_o x) + b_n \sin(n\kappa_o x)\right]$ or $f(x) = a_o + \displaystyle\sum_{n=1}^{\infty}\left[a_n \cos\left(\dfrac{n\pi}{L}x\right) + b_n \sin\left(\dfrac{n\pi}{L}x\right)\right]$

Table 6.2 Fourier series in time and space.

In addition, for a spatial wave described as $f(x)$, moving at speed v, as shown in **Figure 6.10**, it is possible to convert to a temporal function, $f(t)$, and vice versa, by noting the relationship between space to time,

$$x = vt \qquad (6.22)$$

so that over one period,

$$\lambda = vp \qquad (6.23)$$

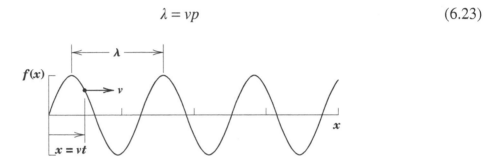

Figure 6.10 Sinusoidal motion of a wave in space at speed v.

Example 6.4

Consider the function $f(x) = \sin(\kappa_o x)$ moving at wave speed $v = 2\dfrac{m}{s}$, as shown in **Figure 6.10**. The wavelength is $\lambda = (40\pi)m$.

1. Find the fundamental frequency, ω_o.
2. Find $f(t)$.

Solution:

1. Find the fundamental frequency, ω_o.

Given the wavelength $\lambda = (40\pi)m$, then the fundamental wavenumber is $\kappa_o = \dfrac{2\pi}{40\pi} = 0.05\dfrac{rad}{m}$.

From Eq. (6.23) $p = \dfrac{\lambda}{v} = \dfrac{40\pi}{2} = (20\pi)s$ so that $\omega_o = \dfrac{2\pi}{p} = \dfrac{2\pi}{20\pi} = 0.1\dfrac{rad}{s}$.

2. Find $f(t)$.

Substituting $\kappa_o = 0.05\dfrac{rad}{m}$ and Eq. (6.22) in $f(x) = \sin(\kappa_o x)$ yields $f(t) = \sin(0.1t)$ that matches $f(t) = \sin(\omega_o t)$.

In the next section, you will be introduced to the derivation of the Fourier transform by applying a limiting process to Fourier series. The applications of Fourier transform are vast; however, we aim in this guide to introduce you to the concepts of frequency spectrum and bandwidth, and set the stage for the discrete Fourier transform that is covered in the following chapter.

6.4 Continuous Fourier Transform

Let us begin with Eq. (6.7), $f(t) = a_o + \sum_{n=1}^{\infty} \left[a_n \cos(n\omega_o t) + b_n \sin(n\omega_o t) \right]$ which can be represented compactly as a complex function. From Euler's formulas

$$e^{i\theta} = \cos(\theta) + i\sin(\theta) \tag{6.24}$$

and

$$e^{-i\theta} = \cos(\theta) - i\sin(\theta) \tag{6.25}$$

you can add Eqs. (6.24) and (6.25) to express $\cos(\theta)$ as

$$\cos(\theta) = \frac{e^{i\theta} + e^{-i\theta}}{2} \tag{6.26}$$

and subtract to express $\sin(\theta)$ as

$$\sin(\theta) = \frac{e^{i\theta} - e^{-i\theta}}{2i} \tag{6.27}$$

Notwithstanding the algebraic details, what you should know is that substituting Eqs. (6.26) and (6.27) into the Fourier series, Eq. (6.7), along with the Fourier coefficients (6.10) and (6.11), yields the complex form of the Fourier series as $f(t) = \sum_{n=-\infty}^{\infty} \left(\frac{1}{p} \int_{-p/2}^{p/2} f(t) e^{-in\omega t} dt \right) e^{in\omega t}$, or

$$f(t) = \sum_{n=-\infty}^{\infty} \left(\frac{\omega_o}{2\pi} \int_{-p/2}^{p/2} f(t) e^{-in\omega t} dt \right) e^{in\omega t} \tag{6.28}$$

Equation (6.28) may be considered the basis for defining the Fourier transform by setting the period, p, in the limit to infinity. Setting $p \to \infty$ is not true for a periodic function that has a repetitive pattern and a finite period. However, $p \to \infty$ is applicable to a nonperiodic function as follows. A nonperiodic, also known as aperiodic, is defined as a function that does not have a period. In that sense, defining the period of a nonperiodic function as infinity would satisfy the nonexistence of periodicity. This limiting process is the essence of the Fourier transform so that the summation in Eq.(6.28) is converted to an integral. Letting $p \to \infty$ sets $\omega_o \to 0$ based on Eq. (6.6). In order to transform Eq. (6.28) to an integral equation, define $\Delta\omega = \omega_o$ as the variable of integration so that in the limit $p \to \infty$, $\Delta\omega \to 0$. In this case, Eq. (6.28) reads

$$f(t) = \lim_{\substack{\Delta\omega \to 0 \\ n=-\infty}} \sum_{n=-\infty}^{\infty} \left(\frac{\Delta\omega}{2\pi} \int_{-\infty}^{\infty} f(t) e^{-in\omega t} dt \right) e^{in\omega t} \qquad (6.29)$$

In the limit, the summation is replaced with an integral so that Eq. (6.29) reads

$$f(t) = \frac{1}{2\pi} \int_{-\infty}^{\infty} \left(\int_{-\infty}^{\infty} f(t) e^{-i\omega t} dt \right) e^{i\omega t} d\omega \qquad (6.30)$$

The Fourier transform is defined from Eq. (6.30) using the inner integral between parentheses, which varies from textbook to textbook and software to software. The following are two versions:

Version 1: Replace $\frac{1}{2\pi}$ with the product $\frac{1}{\sqrt{2\pi}} \frac{1}{\sqrt{2\pi}}$ in Eq. (6.30) as

$$f(t) = \frac{1}{\sqrt{2\pi}} \int_{-\infty}^{\infty} \left(\frac{1}{\sqrt{2\pi}} \int_{-\infty}^{\infty} f(t) e^{-i\omega t} dt \right) e^{i\omega t} d\omega \qquad (6.31)$$

Define the term between parentheses as the Fourier transform,

$$\hat{f}(\omega) = \frac{1}{\sqrt{2\pi}} \int_{-\infty}^{\infty} f(t) e^{-i\omega t} dt \qquad (6.32)$$

so that Eq. (6.31) yields the inverse of $\hat{f}(\omega)$ as

$$f(t) = \frac{1}{\sqrt{2\pi}} \int_{-\infty}^{\infty} \hat{f}(\omega) e^{i\omega t} d\omega \qquad (6.33)$$

Version 2: Define the Fourier transform, without any constant factor, as

$$\hat{f}(\omega) = \int_{-\infty}^{\infty} f(t) e^{-i\omega t} dt \qquad (6.34)$$

so that Eq. (6.30) yields the inverse of $\hat{f}(\omega)$ as

$$f(t) = \frac{1}{2\pi} \int_{-\infty}^{\infty} \hat{f}(\omega) e^{i\omega t} d\omega \qquad (6.35)$$

Table 6.3 is an example of how some CAS adapt either version 1 or version 2.

Fourier transform of function $f(t)$	Inverse Fourier transform	Mathematica	Matlab	Maple
$\hat{f}(\omega) = \dfrac{1}{\sqrt{2\pi}} \int_{-\infty}^{\infty} f(t)e^{-i\omega t}\,dt$	$f(t) = \dfrac{1}{\sqrt{2\pi}} \int_{-\infty}^{\infty} \hat{f}(\omega)e^{i\omega t}\,d\omega$	✓		
$\hat{f}(\omega) = \int_{-\infty}^{\infty} f(t)e^{-i\omega t}\,dt$	$f(t) = \dfrac{1}{2\pi} \int_{-\infty}^{\infty} \hat{f}(\omega)e^{i\omega t}\,d\omega$		✓	✓

Table 6.3 Two versions of the Fourier transform and its inverse.

Definition of the Fourier transform

Undoubtedly, one can arrive at the same results using either version 1 or version 2 for the definition of the transform and its inverse. In this guide, the Fourier transform of a function $f(t)$ is defined as the integral operator

$$\hat{f}(\omega) = \mathscr{F}[f(t)] = \int_{-\infty}^{\infty} f(t)e^{-i\omega t}\,dt \tag{6.36}$$

over the interval $-\infty < \omega < \infty$.

provided the integral exists. The inverse Fourier transform of Eq. (6.36) is defined as

$$f(t) = \frac{1}{2\pi} \int_{-\infty}^{\infty} \hat{f}(\omega)e^{i\omega t}\,d\omega \tag{6.37}$$

over the interval $-\infty < t < \infty$.

Properties of the Fourier transform

The Fourier transform is a linear operator; that is, given two functions $f_1(t)$ and $f_2(t)$ whose Fourier transforms exist,

$$\mathscr{F}\left[a\,f_1(t) + b\,f_2(t)\right] = a\,\mathscr{F}\left[f_1(t)\right] + b\,\mathscr{F}\left[f_2(t)\right] \tag{6.38}$$

where a and b are constants.

Frequency spectrum

In general, $\hat{f}(\omega)$ can be written as a complex number as

$$\hat{f}(\omega) = \hat{f}_r(\omega) + i\,\hat{f}_i(\omega) \tag{6.39}$$

where $\hat{f}_r(\omega)$ and $\hat{f}_i(\omega)$ are the real and imaginary parts of the Fourier transform, respectively. The magnitude of the Fourier transform in Eq. (6.39), given by

$$\left|\hat{f}(\omega)\right| = \sqrt{\hat{f}_r(\omega)^2 + \hat{f}_i(\omega)^2} \tag{6.40}$$

is defined as the frequency spectrum. An example of the frequency spectrum, $\left|\hat{f}(\omega)\right|$, of a real function $f(t)$ is shown in **Figure 6.11a**, which is a continuous function over the interval $-\infty < \omega < \infty$. For a real function $f(t)$, $\left|\hat{f}(\omega)\right|$ is an even function and therefore symmetric about the vertical axis at $\omega = 0$. Thus, it suffices to plot $\left|\hat{f}(\omega)\right|$ over the interval $\omega \geq 0$, as shown in **Figure 6.11b**. The plot of $\left|\hat{f}(\omega)\right|$ versus frequency provides valuable information about frequency content in $f(t)$ and dominant frequencies in a compounded function.

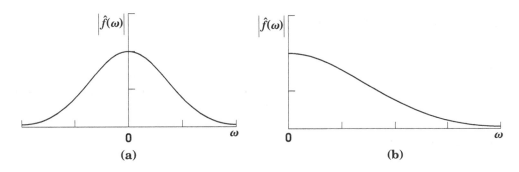

Figure 6.11 Example of a frequency spectrum: (a) $\left|\hat{f}(\omega)\right|$ over the interval $-\infty < \omega < \infty$; and (b) $\left|\hat{f}(\omega)\right|$ over the interval $\omega \geq 0$.

Example 6.5

Consider the nonperiodic function

$$f(t) = \begin{cases} 0 & if \quad t < 0 \\ e^{-0.1t}\sin(t) & if \quad t \geq 0 \end{cases} \tag{6.41}$$

as shown in **Figure 6.12**.

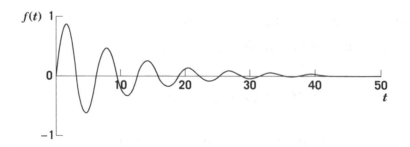

Figure 6.12 Nonperiodic function $f(t)$ as defined by Eq. (6.41).

1. Find the Fourier transform of $f(t)$ using Eq. (6.36).
2. Show that the frequency spectrum is an even function, and plot it over the intervals $-\infty < \omega < \infty$ and $\omega \geq 0$.

Solution:

1. Find the Fourier transform of $f(t)$ using Eq. (6.36).

Using Eq. (6.36),

$$\hat{f}(\omega) = \int_{-\infty}^{\infty} f(t)e^{-i\omega t}\,dt = \left[\int_{-\infty}^{0}(0)e^{-i\omega t}\,dt + \int_{0}^{\infty} e^{-0.1t}\sin(t)e^{-i\omega t}\,dt\right] = \int_{0}^{\infty} e^{-(0.1+i\omega)t}\sin(t)\,dt$$

Integrating by parts or using **CAS** yields

$$\hat{f}(\omega) = \frac{100}{101 + 20i\omega - 100\omega^2} \tag{6.42}$$

2. Show that the frequency spectrum is an even function, and plot it over the intervals $-\infty < \omega < \infty$ and $\omega \geq 0$.

From Eq. (6.40), the frequency spectrum is

$$\left|\hat{f}(\omega)\right| = \frac{100}{\left|101 + 20i\omega - 100\omega^2\right|} = \frac{100}{\sqrt{\left(101 - 100\omega^2\right)^2 + 400\omega^2}} \tag{6.43}$$

Replacing ω with $-\omega$ in Eq. (6.43) shows that $\left|\hat{f}(-\omega)\right| = \left|\hat{f}(\omega)\right|$ Thus, $\left|\hat{f}(\omega)\right|$ is an even function. Graphically, the function is depicted in **Figure 6.13**.

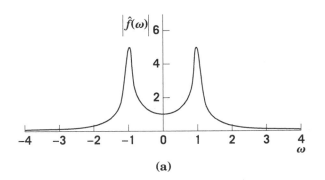 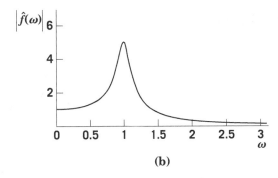

(a)   (b)

Figure 6.13 Frequency spectrum: (a) over the interval $-\infty < \omega < \infty$; and (b) over the interval $\omega \geq 0$.

Figure 6.13 indicates that $f(t)$ has a frequency content even though it is a nonperiodic function. The peak at $\omega = 1$ is indicative of a dominant harmonic function at that frequency. That is not accidental because $f(t) = e^{-0.1t}\sin(t)$ contains $\sin(t)$ that has a frequency $\omega = 1\dfrac{rad}{s}$.

Note also that if you proceed with this problem using the Fourier transform, Eq. (6.32) in version 1, you would obtain

$$\hat{f}(\omega) = \frac{1}{\sqrt{2\pi}} \frac{100}{101 + 20i\omega - 100\omega^2} \tag{6.44}$$

$$\left|\hat{f}(\omega)\right| = \frac{1}{\sqrt{2\pi}} \frac{100}{\sqrt{\left(101 - 100\omega^2\right)^2 + 400\omega^2}} \tag{6.45}$$

as shown in **Figure 6.14**. Thus, as far as frequency content is concerned, you would arrive at the same conclusion as in **Figure 6.13**. In fact, plotting Eq. (6.45) as $\sqrt{2\pi}\left|\hat{f}(\omega)\right|$ versus frequency yields the same plot as in **Figure 6.13**.

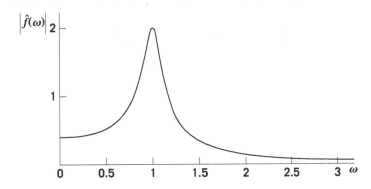

Figure 6.14 Frequency spectrum scaled by $\dfrac{1}{\sqrt{2\pi}}$ over the interval over the interval $\omega \geq 0$.

Bandwidth

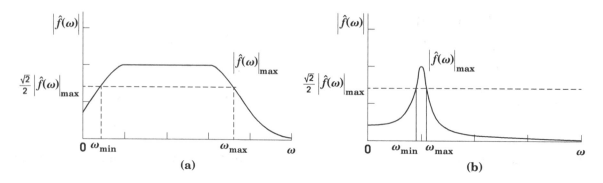

Figure 6.15 Bandwidth of typical frequency spectra: (a) wide bandwidth; and (b) narrow bandwidth.

Consider two typical frequency spectra, as shown in **Figure 6.15**. Each curve is characterized by a maximum amplitude, $\left|\hat{f}(\omega)\right|_{max}$. Consider also a frequency spectrum

level defined as $\dfrac{\sqrt{2}}{2}\left|\hat{f}(\omega)\right|_{max}$ that intersects a curve at two frequencies, ω_{min} and ω_{max}. The

bandwidth, ω_b is defined as $\omega_b = \omega_{max} - \omega_{min}$, which may be described as a wide bandwidth as in **Figure 6.15a**, or a narrow bandwidth as in **Figure 6.15b**. A wide bandwidth means that frequency content about $f(t)$ is carried fairly intact over a wide range of frequencies, whereas narrow bandwidth means that frequency content is carried fairly intact only over a small range.

Next you will learn how to find the frequency spectrum and bandwidth for functions commonly encountered in engineering.

Frequency spectrum and bandwidth of common functions in engineering

For all the functions described below, refer to **Table 6.4** for figures.

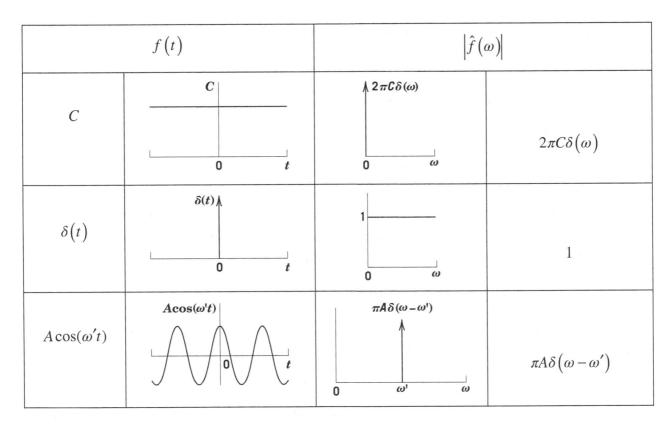

$f(t)$		$\left|\hat{f}(\omega)\right|$	
C	C, constant shown over t with 0	$2\pi C\delta(\omega)$ spike at 0	$2\pi C\delta(\omega)$
$\delta(t)$	$\delta(t)$ spike at 0	1, constant over ω	1
$A\cos(\omega't)$	$A\cos(\omega't)$ oscillation	$\pi A\delta(\omega-\omega')$ spike at ω'	$\pi A\delta\left(\omega-\omega'\right)$

Table 6.4 Summary of temporal functions and their frequency spectra commonly encountered in engineering.

Frequency spectrum and bandwidth of a constant

The Fourier transform of a constant function, $f(t) = C$, is found by Eq. (6.36) as $\hat{f}(\omega) = C\int_{-\infty}^{\infty} e^{-i\omega t}\,dt$. Using CAS yields $\hat{f}(\omega) = 2\pi C\delta(\omega)$. Thus, the frequency spectrum of a constant function is

$$\left|\hat{f}(\omega)\right| = 2\pi C\delta(\omega) \tag{6.46}$$

where $\delta(\omega) = \begin{cases} 0 & \omega \neq 0 \\ \infty & \omega = 0 \end{cases}$ is the Dirac delta function, which was defined in Eq. (5.5), Chapter 5.

The Dirac delta function causes the frequency spectrum of a constant to vanish everywhere except at $\omega = 0$, where there is a Dirac delta spike, as shown in **Table 6.4**. Based on the figure of $\left|\hat{f}(\omega)\right|$, the frequency spectrum of a constant has zero bandwidth when compared to **Figure 6.15**.

Frequency spectrum and bandwidth of the Dirac delta function

The Fourier transform of the Dirac delta function, $f(t) = \delta(t)$, is found by applying the Fourier transform given by Eq. (6.36) and using CAS, which yield $\hat{f}(\omega) = \int_{-\infty}^{\infty} \delta(t) e^{-i\omega t} dt = 1$. Therefore, the frequency spectrum is

$$\left| \hat{f}(\omega) \right| = 1 \tag{6.47}$$

as shown in **Table 6.4**. The frequency spectrum of $\delta(t)$ has an infinite bandwidth, which means that $\delta(t)$ is an ideal impulse as an input forcing function to systems in order to test responses over a wide range of frequencies as you will learn in Example 6.8.

Frequency spectrum and bandwidth of a sinusoid

The Fourier transform of a sinusoid function, for example, $f(t) = A\cos(\omega' t)$, with amplitude A and frequency ω', is found by applying Eq. (6.36) and using CAS, which yields

$$\hat{f}(\omega) = \int_{-\infty}^{\infty} A\cos(\omega' t) e^{-i\omega t} dt = \pi A \left[\delta(\omega + \omega') + \delta(\omega - \omega') \right] \tag{6.48}$$

Using the definition of a shifted Dirac delta function given by Eq. (5.5) in Chapter 5, $\delta(\omega + \omega')$ and $\delta(\omega - \omega')$ represent shifted Dirac delta functions at $-\omega'$ and ω', respectively. For a frequency spectrum limited to $\omega \geq 0$, $\delta(\omega + \omega')$, which is to the left of $\omega = 0$, is dropped from Eq. (6.48) so that

$$\hat{f}(\omega) = \int_{0}^{\infty} A\cos(\omega' t) e^{-i\omega t} dt = \pi A \delta(\omega - \omega') \tag{6.49}$$

Therefore, the frequency spectrum over the interval $\omega \geq 0$ is

$$\left| \hat{f}(\omega) \right| = \pi A \delta(\omega - \omega') \tag{6.50}$$

as shown in **Table 6.4**, which indicates that the bandwidth of a cosine function is zero. The cosine frequency spectrum is the same for a sine function, $f(t) = A\sin(\omega' t)$.

In general, a linear superposition of harmonics will have as many Dirac delta functions in the frequency spectrum graph at positions corresponding to the frequencies of the harmonics.

Example 6.6

Consider the periodic function,

$$f(t) = 2\sin(t) + 5\cos(2t) + 10\sin(3t) + 5\cos(4t) + 2\sin(5t) \tag{6.51}$$

as shown in **Figure 6.16**.

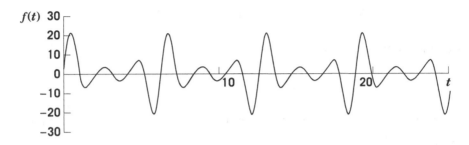

Figure 6.16 Plot of $f(t)$ defined in Eq. (6.51).

1. Find the frequency spectrum of $f(t)$.
2. Determine which harmonic is dominant.

Solution:

1. Find the frequency spectrum of $f(t)$.

Using the Fourier transform property, Eq. (6.38) yields

$$\hat{f}(\omega) = 2\,\mathscr{F}\!\left[\sin(t)\right] + 5\,\mathscr{F}\!\left[\cos(2t)\right] + 10\,\mathscr{F}\!\left[\sin(3t)\right] + 5\,\mathscr{F}\!\left[\cos(4t)\right] + 2\,\mathscr{F}\!\left[\sin(5t)\right]$$

The Fourier transform of each sinusoid is found from Eq. (6.49) as

$$\hat{f}(\omega) = 2\pi\delta(\omega-1) + 5\pi\delta(\omega-2) + 10\pi\delta(\omega-3) + 5\pi\delta(\omega-4) + 2\pi\delta(\omega-5)$$

For a given frequency ω in the set $(1,2,3,4,5)$, one Dirac delta function exists while the other four functions vanish, based on the definition of the Dirac delta function. For example, $\hat{f}(2) = 5\pi\delta(\omega-2)$ so that its frequency spectrum is $\left|\hat{f}(2)\right| = 5\pi\delta(\omega-2)$. On that basis,

$$\left|\hat{f}(\omega)\right| = 2\pi\delta(\omega-1) + 5\pi\delta(\omega-2) + 10\pi\delta(\omega-3) + 5\pi\delta(\omega-4) + 2\pi\delta(\omega-5)$$

as shown in **Figure 6.17**.

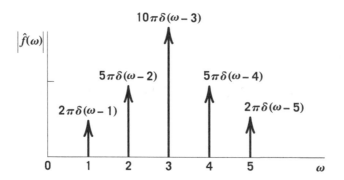

Figure 6.17 Frequency spectrum spikes corresponding to $f(t)$ as defined in Eq. (6.51).

2. Determine which harmonic is dominant.

From **Figure 6.17** the largest Dirac delta function is at $\omega = 3$. Thus the sinusoid $10\sin(3t)$ is the dominant harmonic in $f(t)$.

Example 6.7

Consider the frequency spectrum $\left|\hat{f}(\omega)\right| = 2$.

1. Find the function $f(t)$ that corresponds to $\left|\hat{f}(\omega)\right|$.

Solution:

From **Table 6.4**, $\left|\hat{f}(\omega)\right| = 1$ corresponds to $f(t) = \delta(t)$. Thus, for $\left|\hat{f}(\omega)\right| = 2$, the corresponding function is $f(t) = 2\delta(t)$.

Reconsider the simulation of the impact force described in Example 5.4, Chapter 5. In that case, we analyzed an impulse forcing function, $f(t)$, that simulated an electronic hammer striking a pier, as shown in **Figure 6.18**. The forcing function is what you control in an engineering system. If the impact forcing function can excite the natural frequency of the pier, information about the strength of the pier can be determined as you may learn in a complementary example in Chapter 7, Discrete Fourier Transform.

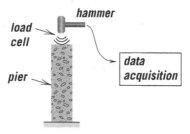

Figure 6.18 Impulsive test of concrete piers.

In order to excite piers of different sizes and shapes that have a wide range of natural frequencies, it would be effective if the forcing function $f(t)$ inherently contained as wide a bandwidth as possible as demonstrated in the next example.

Example 6.8 Frequency Spectrum and Bandwidth of an Impulse Hammer

For the concrete pier case study described above, consider a test carried out by striking the top of the pier with an impulse hammer connected to a load cell, as shown in **Figure 6.18**. The natural frequency of the pier is expected to be approximately $650 \frac{rad}{s}$. The impact of the hammer induces an impulse force on the pier, $F(t)$, over a very short time interval. For the purpose of identifying the desirable bandwidth, consider four such tests with different time intervals and amplitudes.

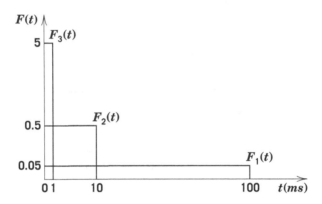

Figure 6.19 Impulse forces used in the pier impulse test.

Three tested forces, $F_1(t)$, $F_2(t)$, and $F_3(t)$, are rectangular force pulses, as shown in **Figure 6.19**, which are given by $F_1(t) = 0.05\left[u(t) - u(t - 0.1)\right]$, $F_2(t) = 0.5\left[u(t) - u(t - 0.01)\right]$, and $F_3(t) = 5\left[u(t) - u(t - 0.001)\right]$, all in *Newton* (*N*), where t is in s. The fourth force is a Dirac delta function, $F_4(t) = 0.005\delta(t)$, also in *N*. All forces have the same impulse, $I = F\Delta t = 0.005 N.s$.

1. Find the frequency spectra of the forcing functions and plot them in the frequency domain over $\omega \geq 0$.
2. Find the bandwidth of each frequency spectrum.

Solution:

1. Find the frequency spectra of the forcing functions and plot them in the frequency domain over $\omega \geq 0$.

Using Eq. (6.36), the Fourier transform of $F_1(t)$ is

$$\hat{f}_1(\omega) = \int_{-\infty}^{\infty} F_1(t) e^{-i\omega t} dt = 0.05 \int_0^{0.1} e^{-i\omega t} dt = -\frac{0.05}{i\omega} e^{-i\omega t} \Big|_0^{0.1}$$

which yields

$$\hat{f}_1(\omega) = -\frac{0.05}{i\omega} \left[e^{-0.1(i\omega)} - 1 \right] = \frac{0.05}{\omega} \left[i \left(\cos(0.1\omega) - 1 \right) + \sin(0.1\omega) \right] \tag{6.52}$$

Using Eq. (6.40) to find the frequency spectrum for $F_1(t)$ yields

$$\left| \hat{f}_1(\omega) \right| = \frac{0.05}{\omega} \sqrt{2 \left[1 - \cos(0.1\omega) \right]} \tag{6.53}$$

Following a similar procedure for $F_2(t)$, and $F_3(t)$ yields

$$\left| \hat{f}_2(\omega) \right| = \frac{0.5}{\omega} \sqrt{2 \left[1 - \cos(0.01\omega) \right]} \tag{6.54}$$

$$\left| \hat{f}_3(\omega) \right| = \frac{5}{\omega} \sqrt{2 \left[1 - \cos(0.001\omega) \right]} \tag{6.55}$$

For $F_4(t) = 0.005\delta(t)$, use the second row in **Table 6.4**,

$$\left| \hat{f}_4(\omega) \right| = 0.005 \tag{6.56}$$

The frequency spectra given by Eqs. (6.53)–(6.56) are shown **Figure 6.20**.

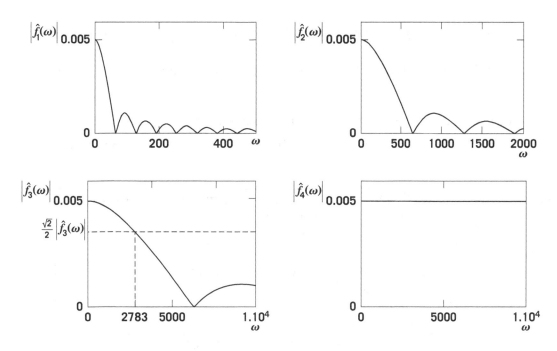

Figure 6.20 Frequency spectra for the tested forces.

2. Find the bandwidth of each frequency spectrum.

The bandwidth of the Dirac delta function is infinity, which is ideal. However, it is impractical to apply a hammer pulse over a zero time interval. Thus, by inspecting **Figure 6.20**, it appears that the rectangular force pulse, $F_3(t)$, is the best option insofar as the bandwidth is concerned. For numerical comparison, all frequency spectra have a maximum value $\left|\hat{f}(\omega)\right|_{max} = 0.005$ at $\omega = 0$, as shown in each figure, so that $\frac{\sqrt{2}}{2}\left|\hat{f}(\omega)\right|_{max} = 3.54\left(10^{-3}\right)$. Substituting this value in Eq. (6.55) yields a nonlinear equation for the bandwidth of $F_3(t)$,

$$3.54\left(10^{-3}\right) = \frac{5}{\omega}\sqrt{2\left[1-\cos(0.001\omega)\right]} \tag{6.57}$$

Using a root-finding algorithm in CAS to solve Eq. (6.57) yields the upper end of the bandwidth, $\omega_{max} = 2783\frac{rad}{s}$. Furthermore, $\omega_{min} = 0$ so that the bandwidth is

$\omega_b = \omega_{max} = 2783\frac{rad}{s}$.

Alternatively, you may write Eq. (6.57) as $3.54\left(10^{-3}\right)\omega = 5\sqrt{2\left[1-\cos(0.001\omega)\right]}$, define two functions $r_1(t) = 3.54\left(10^{-3}\right)\omega$ for the left hand term of the equation and

$r_2(t) = 5\sqrt{2[1-\cos(0.001\omega)]}$ for the right hand term, and plot $r_1(t)$ and $r_2(t)$ versus ω. The solution $\omega_{max} = 2783 \dfrac{rad}{s}$ would be where $r_1(t) = r_2(t)$, that is where the two functions intersect.

Repeating the same procedure for the rectangular force pulses, $F_1(t)$ and $F_2(t)$ using Eqs. (6.53) and (6.55), respectively, yields their respective bandwidths, as shown in **Table 6.5**.

Force	Amplitude (N)	Time interval (ms)	Impulse N.s	Bandwidth (rad/s)	Bandwidth (Hz)
$F_4(t)$	∞ such that the area under the Dirac delta is equal to 0.005	0	0.005	∞	∞
$F_3(t)$	5	1	0.005	2783	443
$F_2(t)$	0.5	10	0.005	278.3	44.3
$F_1(t)$	0.05	100	0.005	27.83	4.43

Table 6.5 Summary of tested rectangular pulses and Dirac delta function along with their bandwidths.

The pier expected natural frequency of $650 \dfrac{rad}{s}$ is outside the narrow bandwidths of $F_1(t)$ and $F_2(t)$. Therefore, $F_3(t)$ is best suited as a practical test function for this experiment.

In this applied engineering example, you learned how the time interval, over which a rectangular pulse function is applied, affects the frequency spectrum. In order to have the same impulse magnitude for different forcing functions, you must decrease the time interval of the forcing function as you increase its magnitude, and as a result the bandwidth of the frequency spectrum increases.

HOMEWORK PROBLEMS FOR CHAPTER SIX

Problems for Section 6.2

For Problems 1-3,

a. Find the fundamental circular frequency of $g(t)$, ω_o.

b. Express $g(t)$ in terms of ω_o, and identify its harmonics.

c. Define $f(t) = \sin(\omega_o t)$ as a reference function associated with the fundamental frequency, and demonstrate graphically that $g(t)$ and $f(t)$ have the same frequency.

1. $g(t) = \sin(2\pi t) + \sin(5\pi t) + \sin(7\pi t)$.

2. $g(t) = \sin(4\pi t) + \sin(8\pi t)$.

3. $g(t) = \sin(12\pi t) + \sin(15\pi t)$.

Problems for Section 6.3

For Problems 4–6, consider the piecewise continuous, and periodic function $f(t)$ over the time interval $-\infty < t < \infty$, where t is in s.

a. Find the fundamental period, p, and fundamental circular frequency, ω_o.

b. Express $f(t)$ as a piecewise continuous function over one period.

c. Find the Fourier series of $f(t)$. Whenever possible, take advantage of the odd or even property of a function.

d. Show graphically how well the addition of successive harmonics represents $f(t)$ over one period.

4.

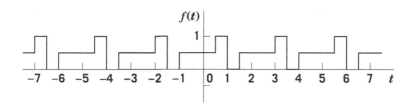

5.

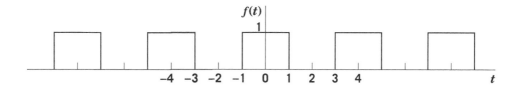

6.

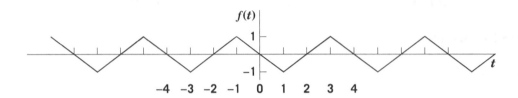

For Problems 7–9, consider the continuous, and periodic function $f(t)$ over the time interval $-\infty < t < \infty$, where t is in s.

a. If a particle propagates over the function $f(t)$ in time with a horizontal speed of $v = 4 m/s$, determine the equivalent function that describes the path of the particle in space, $f(x)$.

b. Determine the fundamental wavelength, λ, and the fundamental wavenumber, κ_o.

c. Plot $f(x)$ over one wavelength.

7. Problem 4.
8. Problem 5.
9. Problem 6.

10. Consider the analysis of the valve motion in an engine assembly as part of a study on engine efficiency. A typical cross section of the assembly consists of an overhead cam, rocker arm, spring, and valve, as shown in **Figure 6.21a**. The cam is designed to cause the valve to oscillate as it rotates. As a first look at the analysis of the oscillating problem, the assembly is modeled as a mass, spring, and damper system, as shown in **Figure 6.21b**. As the valve opens and closes, the oscillating cam exerts a force $F(t)$ on the rocker arm, which is a periodic function in time, as shown in **Figure 6.22**.

a. Express $F(t)$ as a Fourier series. The necessary data is shown in the figure. Take advantage of the odd or even property of a function.

b. Plot $F(t)$ over the interval $0 \le t \le p$, and report the number of terms required in the series in order to limit the % error between $F(0.02)$ and the true value to less than 1%.

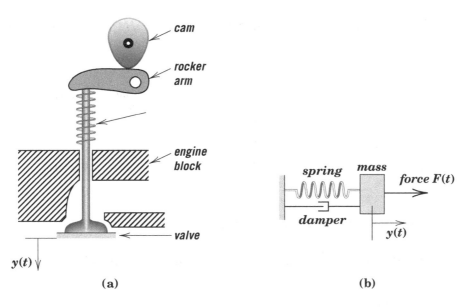

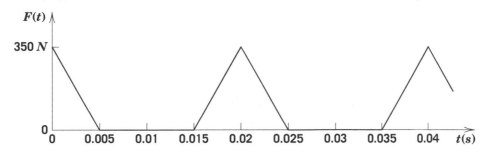

Figure 6.21 Part of an automobile engine: (a) the engine components; and (b) the engine mass, spring, and damper model.

Figure 6.22 Force as a function of time exerted by the cam.

11. In a full-wave silicon-diode rectifier circuit, the instantaneous current $i(t)$, in *ampere*, is described by $i(t) = 10|\sin(\omega t)|$, where t is in s, and ω is the circular frequency that corresponds to 60 Hz. as shown in **Figure 6.23**.

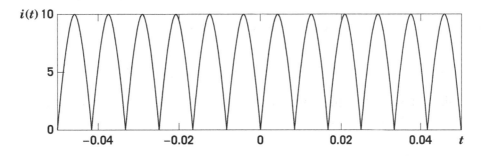

Figure 6.23 Rectifier current.

a. Find the circular frequency in *rad/s*.
b. Express $i(t)$ as a Fourier series. Take advantage of the odd or even property of a function.

12. Consider the height function of the railroad tracks cross section, $H(x)$, as shown in **Figure 6.24**, with the specified geometry.

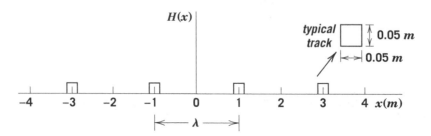

Figure 6.24 Spatial periodic function for railroad tracks.

a. Find the Fourier series of $H(x)$, where the wavelength $\lambda = 2m$ is defined from center to center of two tracks.

b. Plot $H(x)$ over one wavelength.

13. In a manufacturing company, aluminum cans are fabricated from a sheet metal strip, as shown in **Figure 6.25**.

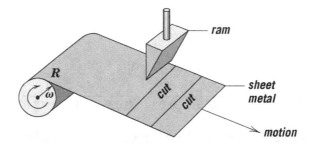

Figure 6.25 Process of sheet metals.

In this application, the process is to be controlled by a motor via a computer algorithm. Therefore, an engineer is charged with programming the sheet acceleration with a periodic function in time. The process begins with the sheet metal accelerating from rest with a value $a = \dfrac{1}{3}\dfrac{m}{s^2}$ for 1 *sec*. Then, the sheet moves at a uniform speed, that is, zero acceleration for 0.5 *sec*.

Next, the sheet decelerates with $a = -\dfrac{1}{3}\dfrac{m}{s^2}$ for 1 *sec*. Finally, the system dwells for 0.5 *sec*; that is, no motion takes place during this time interval, while a hydraulic ram applies an impulse force of $225N$ to cut the sheet metal before the cycle is repeated.

a. Plot the acceleration $a(t)$ as a piecewise continuous function over one period, $p = 3s$.

b. Find the Fourier series of $a(t)$.

c. Plot the Fourier series of $a(t)$ over the interval $0 \le t \le p$.

14. In a pharmaceutical application, vials are filled with a chemical compound by positioning them on a rotary table driven by a motor, as shown in **Figure 6.26**. The process cycle begins with the motor turning uniformly at an angular acceleration $\alpha = 1$ *revolution* $/ s^2$ for 0.5 *sec*. Then, the motor decelerates to zero speed at an acceleration $\alpha = -1$ *revolution* $/ s^2$ for 0.5 *sec*. Finally, the system dwells for 1 *sec*; that is, no motion takes place during this time interval, while a vial is being filled with the chemical compound, before the cycle is repeated.

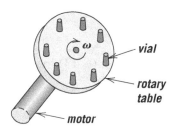

Figure 6.26 Vials positioned on a rotary table driven by a motor.

a. Plot the angular acceleration $\alpha(t)$ as a piecewise continuous function over one period, $p = 2s$.

b. Find the Fourier series of $\alpha(t)$.

c. Plot the Fourier series of $\alpha(t)$ over the interval $0 \le t \le p$.

Problems for Sections 6.4

For Problems 15 –18,

a. Find the Fourier transform of $f(t)$ using Eq. (6.36).

b. Find the frequency spectrum and show it is an even function.

c. Plot the frequency spectrum $F(\omega)$ over the interval $\omega \ge 0$.

15. $f(t) = u(t)\sin(t)$, where $u(t)$ is the unit step function.

16. $f(t) = 10\delta(t-1)$.

17. $f(t) = 10\sin(10t)$.

18. $f(t) = e^{-t^2}$.

Given the frequency spectra in Problems 19–21:

a. Plot the frequency spectrum for the given function.

b. Find the function $f(t)$ in the time domain that corresponds to the frequency spectrum using cosine functions.

c. Plot $f(t)$ over one period.

d. Identify the dominant harmonic in $f(t)$.

19. $\left|\hat{f}(\omega)\right| = \delta(\omega - \pi) + 5\delta(\omega - 2\pi) + 4\pi\delta(\omega - 3\pi) + 3\delta(\omega - 4\pi) + 2\delta(\omega - 5\pi)$.

20. $\left|\hat{f}(\omega)\right| = \delta(\omega) + \delta(\omega - \pi) + \pi\delta(\omega - 3\pi) + \delta(\omega - 4) + 10\delta(\omega - 5)$.

21. $\left|\hat{f}(\omega)\right| = \delta(\omega) + 4\pi\delta(\omega - 3\pi) + 3\delta(\omega - 4)$.

22. Revisit Example 6.5.

a. Find $\left|\hat{f}(\omega)\right|_{max}$.

b. Find $\dfrac{\sqrt{2}}{2}\left|\hat{f}(\omega)\right|_{max}$ and use its value in Eq. (6.43) in order to find the bandwidth limits, ω_{min} and ω_{max} in $\dfrac{rad}{s}$.

c. Report the bandwidth in $\dfrac{rad}{s}$.

For Problems 23–26,

a. Plot $\left|\hat{f}(\omega)\right|$ over the interval $\omega \geq 0$.

b. Find the bandwidth.

23. $\left|\hat{f}(\omega)\right| = \dfrac{1}{1 + \omega}$.

24. $\left|\hat{f}(\omega)\right| = \dfrac{1}{\sqrt{1 + \omega^2}}$.

25. $\left|\hat{f}(\omega)\right| = \dfrac{1}{\sqrt{(1 - \omega^2) + (0.01\omega)^2}}$.

26. $\left|\hat{f}(\omega)\right| = 10$.

27. Revisit Example 6.8: Frequency Spectrum and Bandwidth of an Impulse Hammer. Consider the case where a rectangular pulse of 1-N is applied for 10^{-4} s.

a. Determine the Fourier transform of the pulse using Eq. (6.36).
b. Determine the frequency spectrum and plot it over an adequate frequency interval in order to identify several lobes, for example, four lobes along the ω axis.
c. Find the bandwidth.

CHAPTER SEVEN

DISCRETE FOURIER TRANSFORM

7.1 Introduction

In the early chapters on ODEs, the solution $y(t)$ was viewed as a system response to a forcing function, $r(t)$, as shown in **Figure 7.1a**. For a given function $r(t)$, the ODE predicts the behavior of a first-order or second-order system via the response, $y(t)$.

(a)   (b)

Figure 7.1 Depiction of an ODE as a system with input forcing function and response: (a) t-domain; and (b) s-domain.

Similarly, a system response was also predicted by the Laplace transform, as shown in **Figure 7.1b**, where $R(s)$, the Laplace transform of $r(t)$, was combined with the Laplace transform of the ODE, which yielded an algebraic equation in $Y(s)$. The system response $y(t)$ was found from the inverse Laplace transform of $Y(s)$.

Analogously, the Fourier transform, Eq. (6.36), and its inverse, Eq. (6.37), introduced in Chapter 6, could also be used to solve ODEs. However, in Chapter 6, we used the Fourier transform for a different purpose, namely, to find the frequency spectrum and identify the dominant frequency of a given function $f(t)$ and frequency content in a bandwidth. For example, in testing various rectangular force pulses in Example 6.8, Frequency Spectrum and Bandwidth of an Impulse Hammer, the frequency spectrum and bandwidth guided us to select the most suitable force to apply to a concrete pier. However, we expended no effort to model the concrete pier by an ODE and find its response, $y(t)$. In fact, that is exactly our goal for the following reason. In testing a pier as part of a maintenance schedule, a forcing function is applied to the pier, for example, a rectangular pulse, and the response $y(t)$ is measured directly, as shown in **Figure 7.2**, in the form of an output voltage.

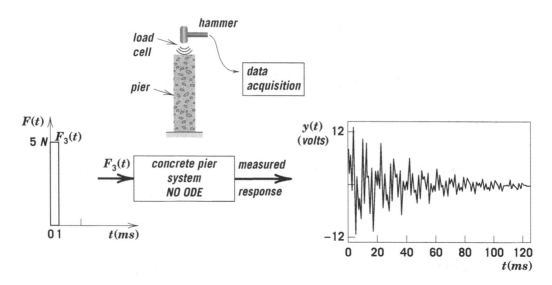

Figure 7.2 Pier system subject to a controlled input force with measured output.

The question that you may ask is, how is the Fourier transform relevant to **Figure 7.2**? The answer to that question has two parts. First, the Fourier transform applied to several test forcing functions as illustrated in Example 6.8, which were well defined, provided frequency spectra and bandwidth information. On the basis of the expected natural frequency of the pier, the choice fell on the 5-N rectangular pulse, $F_3(t)$, over a 1-ms time interval. At this point, we have no more utilization of function. Second, with the recommended force, $F_3(t)$, the electronic hammer is applied to the pier, which resulted in a pier response, $y(t)$, as shown by voltage signal over a time interval in **Figure 7.2**. Although we have an idea about the range of the pier natural frequency, for example,

$400 < \omega_n < 700 \dfrac{rad}{s}$, we shall determine it exactly by applying the Fourier transform to the

output signal, which is the theme of this chapter. That said, the Fourier transform shall not be applied to the signal as a continuous function, but rather as a discrete signal—hence, the name discrete Fourier transform (DFT).

In this chapter, you will learn about the discrete Fourier transform to find the frequency spectrum of data by a method called sampling.

Engineering case studies

Concrete pier test case

In the first case study, the DFT is used to evaluate the strength of concrete piers supporting a bridge, as shown in **Figure 7.3a**. Piers are tested periodically to determine their strength because concrete deteriorates due to environmental conditions, for example, acid rain. One of the tests that can be conducted for that purpose is to strike the pier top impulsively with an electronic hammer connected to a load cell, as shown in **Figure 7.2**. The impact of the hammer causes a standing wave in the vibrating pier, as shown in **Figure 7.3b**. The pier response is defined by its surface velocity, which is measured as a function of time in a form proportional to voltage, using an electronic transducer. A typical pier voltage signal is shown as an output response, $y(t)$, in **Figure 7.2**.

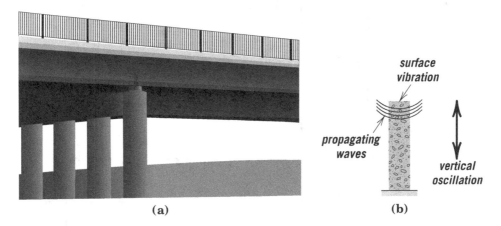

Figure 7.3 Concrete piers supporting a bridge: (a) pier inspection; and (b) propagating waves in the pier struck with an electronic hammer.

The seemingly random voltage response is due to two phenomena: the several natural frequencies induced in the pier by the impulse force, and the distortion of the natural frequencies due to imperfections in the concrete. Hidden in the response function is a particular harmonic that is a dominant natural frequency. By sampling the response at discrete points in time and then taking the DFT of the sampled points, it is possible to determine this natural frequency. Normally, the Department of Transportation (DOT) would already have established standard relationships between concrete strength and natural frequencies that are categorized for weak concrete and strong concrete. Thus, in reference to the DOT standard, the natural frequency found in the discrete frequency spectrum (DFS) would indicate whether the pier has reached its life-cycle limit for either maintenance or replacement.

Elevator safety case

Another interesting safety case study concerns the misalignment diagnosis of elevator guide rail joints to determine if they induce excessive vibration during a maintenance inspection. Consider the elevator cab and guide rails as shown in **Figure 7.4a**. An elevator cab is restrained from lateral movement by the guide rails and spring-loaded roller guides, as shown in the top section of **Figure 7.4b**. These components are designed to produce a smooth ride in the elevator. The guide rails in tall buildings are very long, built in sections stacked end-to-end, and bolted to a frame. After many cycles of elevator use, the joints where the guide rail ends meet become loose owing to vibration, thus causing misalignment that becomes a bump, as shown in the bottom section of **Figure 7.4b**. As the elevator cab passes over the bump, it oscillates and a measurable lateral vibration appears, which may pose safety problems.

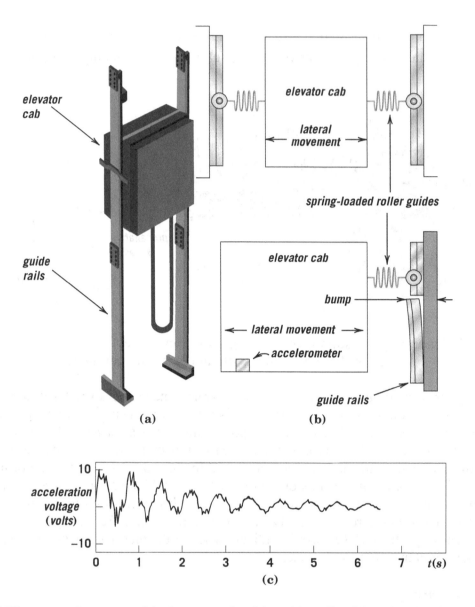

Figure 7.4 Elevator cab system: (a) elevator cab with guide rails; (b) spring-loaded roller guides and guide rail misalignment; and (c) a typical acceleration response in the form of voltage.

In order to monitor the condition of the guide rails, tests are run using an accelerometer, which is an electronic instrument for acceleration measurement as a function of time, as shown in **Figure 7.4b**. A typical acceleration response, proportional to voltage, is shown in **Figure 7.4c**. Notice that the acceleration looks erratic. This seemingly random function is due to two phenomena: the complex interaction between the elevator roller guide and the guide rail bump; and the distortion of the natural frequencies caused by external disturbances that transmit vibration to the rails, for example induced vibrations caused by passing vehicles. Sampling the acceleration function at discrete points in time and taking its DFT may spot the existence of a bump by detecting a dominant harmonic frequency in the DFS.

In the next section, the DFT and its corresponding DFS are developed, starting with an introduction to sampling. In this chapter, unless otherwise specified, t stands for time, and the units are s for time, Hz for frequency f and $\dfrac{rad}{s}$ for circular frequency ω.

7.2 Discrete functions

Consider a known function $f(t)$ over an interval $0 \le t \le t_{max}$, as shown in **Figure 7.5a**.

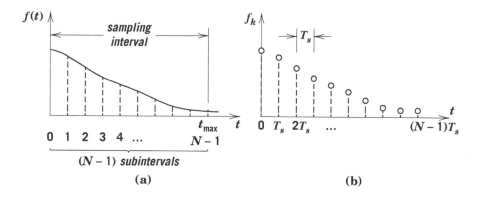

Figure 7.5 Discretized continuous function: (a) selection of N points along the t-axis; and (b) values of the function at N discrete points.

Divide the interval $0 \le t \le t_{max}$ into N discrete points so that there are $N-1$ subintervals with equal width, T_s, known as the sampling period, which is given by

$$T_s = \frac{t_{max}}{N-1} \tag{7.1}$$

For a large value of N, $N-1 \approx N$, so that

$$T_s = \frac{t_{max}}{N} \tag{7.2}$$

Equation (7.2) shall be used as part of the DFT formulation. Furthermore, we shall use the same equation in examples even when N is small, simply to clarify concepts with manageable mathematics. In real applications, N is quite large.

The corresponding sampling frequency is given by

$$f_s = \frac{1}{T_s} \tag{7.3}$$

The time at each discrete point is given by

$$t_k = kT_s \qquad (7.4)$$

for $k = 0, 1, 2, ..., N-1$.

When $f(t)$ is evaluated at $t_k = kT_s$, we write $f(t_k)$ or simply f_k and say that $f(t)$ is sampled at the discrete points. The discrete point function f_k is shown in **Figure 7.5b**.

Example 7.1

Consider the following function in the interval $t \geq 0$,

$$f(t) = e^{-0.1t} \sin(t) \qquad (7.5)$$

1. Take $N = 20$ and discretize Eq. (7.5) over the interval $0 \leq t \leq t_{max}$ using Eq. (7.1) where $t_{max} = 19s$.
2. Reconstruct the function from the discrete points. Plot on a single graph f_k using discrete points and $f(t)$ as a continuous curve for comparison.
3. Repeat questions 1 and 2 using Eq. (7.2).

Solution:

1. Take $N = 20$ and discretize Eq. (7.5) over the interval $0 \leq t \leq t_{max}$ using Eq. (7.1) where $t_{max} = 19s$.

From Eq. (7.1), $T_s = \dfrac{t_{max}}{N-1} = \dfrac{19}{19} = 1s$. The discrete function f_k is evaluated at each discrete point using Eq. (7.4) for $k = 0, 1, 2, ..., 19$, so that $f_k = e^{-0.1(kT_s)} \sin(kT_s) = e^{-0.1k} \sin(k)$.

2. Reconstruct the function from the discrete points. Plot on a single graph f_k using discrete points and $f(t)$ as a continuous curve for comparison.

The results are listed in **Table 7.1a** and plotted along with $f(t)$ given by Eq. (7.5) in **Figure 7.6a**.

k	t_k	f_k	k	t_k	f_k
0	0	0	10	10	−0.200
1	1	0.761	11	11	−0.333
2	2	0.744	12	12	−0.162
3	3	0.105	13	13	0.115
4	4	−0.507	14	14	0.244
5	5	−0.582	15	15	0.1451
6	6	−0.153	16	16	−0.0580
7	7	0.326	17	17	−0.176
8	8	0.445	18	18	−0.124
9	9	0.168	19	19	0.0224

k	t_k	f_k	k	t_k	f_k
0	0	0	10	9.5	−0.029
1	0.95	0.740	11	10.45	−0.301
2	1.9	0.783	12	11.4	−0.294
3	2.85	0.216	13	12.35	−0.062
4	3.8	−0.418	14	13.3	0.177
5	4.75	−0.621	15	14.25	0.239
6	5.7	−0.311	16	15.2	0.106
7	6.65	0.184	17	16.15	−0.085
8	7.6	0.453	18	17.1	−0.178
9	8.55	0.326	19	18.05	−0.118

(a)   **(b)**

Table 7.1 Discrete points f_k of Eq. (7.5): (a) $T_s = 1s$; and (b) $T_s = 0.95s$.

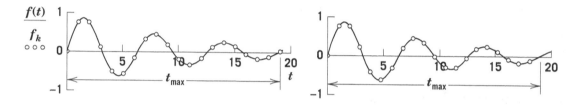

Figure 7.6 The function $f(t)$ and its discrete function f_k: (a) $T_s = 1s$; and (b) $T_s = 0.95s$.

3. Repeat questions 1 and 2 using Eq. (7.2).

From Eq. (7.2), $T_s = \dfrac{t_{max}}{N} = \dfrac{19}{20} = 0.95s$. The discrete function f_k is evaluated at each discrete point using Eq. (7.4) for $k = 0, 1, 2..., 19$, so that $f_k = e^{-0.1(kT_s)} \sin(kT_s) = e^{-0.095k} \sin(0.95k)$. The results are listed in **Table 7.1b** and plotted along with $f(t)$ given by Eq. (7.5) in **Figure 7.6b**. If you study **Table 7.1b**, you can notice that the discrete points are short by one sample point using Eq. (7.2). This is acceptable as long as N is large enough to reproduce the function from discrete points without loss of the function's essential feature over the sampling interval, for example, its sinusoidal decay in this case, as will be demonstrated in the next example. Furthermore, it also should be obvious that the sampled data differ with T_s, as shown in **Figure 7.7**.

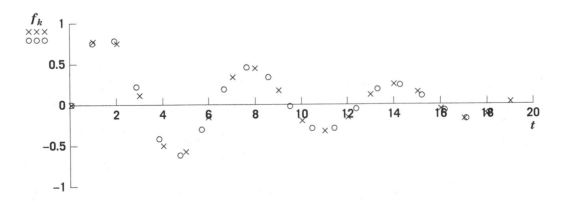

Figure 7.7 The function $f(t)$ reconstructed from its discrete points, f_k: (a) x symbol for $T_s = 1s$; and (b) o symbol for $T_s = 0.95s$.

What if the function is sampled with fewer points? Let us explore this question in the next example.

Example 7.2

Consider the following function in the interval $t \geq 0$,

$$f(t) = e^{-0.1t} \sin(t) \tag{7.6}$$

1. Take $N = 6$ and discretize Eq. (7.6) over the interval $0 \leq t \leq t_{max}$ using Eq. (7.2) where $t_{max} = 6\pi$.
2. Reconstruct the function from the discrete points.

Solution:

1. Take $N = 6$ and discretize Eq. (7.6) over the interval $0 \leq t \leq t_{max}$ using Eq. (7.2) where $t_{max} = 6\pi$.

From Eq. (7.2), $T_s = \dfrac{t_{max}}{N} = \dfrac{6\pi}{6} = \pi$. The discrete function f_k is evaluated at each discrete point using Eq.(7.4) for $k = 0, 1, 2, 3, 4, 5$, so that

$$f_k = e^{-0.1(kT_s)} \sin(kT_s) = e^{-0.1k\pi} \sin(\pi k) \tag{7.7}$$

Equation (7.7) clearly shows that $f_k = 0$ because $\sin(\pi k) = 0$ for $k = 0, 1, 2, 3, 4, 5$.

2. Reconstruct the function from the discrete points.

As you can see from **Figure 7.8**, the zero discrete points are meaningless. The given sampling interval and number of points yielded a sampling period that caused failure to reconstruct the essential feature of the function from discrete points.

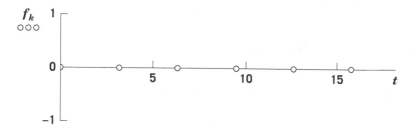

Figure 7.8 Failure to reconstruction the function $f(t)$ from its discrete function f_k.

The selection of the sampling period can be remedied by invoking a theorem known as the Nyquist rate.

The Nyquist sampling theorem

The Nyquist sampling theorem states that the sampling frequency, f_s, given by Eq. (7.3), must satisfy the inequality

$$f_s > 2f_{exp} \tag{7.8}$$

where f_{exp} is an expected frequency component in $f(t)$. One way to establish f_{exp} is to have an idea about the highest frequency in the signal. Another way to establish f_{exp} is based on physical grounds; for example, in the elevator misalignment problem case study presented in the introduction, f_{exp} can be predicted to a good degree of reliability from a simple mass-spring model, as discussed later on in Example 7.5. If the sampling frequency does not satisfy Eq. (7.8), then you may obtain erroneous results from the DFS.

Example 7.3

Consider the following function in the interval $t \geq 0$,

$$f(t) = e^{-0.1t} \sin(t) \tag{7.9}$$

1. Discretize Eq. (7.9) over the interval $0 \leq t \leq t_{max}$ using Eq. (7.2) where $t_{max} = 6\pi$ so that the sampling period satisfies the Nyquist sampling theorem.
2. Reconstruct the function from the discrete points.

Solution:

1. Discretize Eq. (7.9) over the interval $0 \le t \le t_{max}$ using Eq. (7.2) where $t_{max} = 6\pi$ so that the sampling period satisfies the Nyquist sampling theorem.

The function $\sin(t)$ in Eq. (7.9) has the form $\sin(\omega t)$ so that $\omega = 1\dfrac{rad}{s}$, which is the highest frequency in this particular function. Thus, the expected frequency is $f_{exp} = \dfrac{\omega}{2\pi} = \dfrac{1}{2\pi} Hz$. By the Nyquist sampling theorem, Eq. (7.8), $f_s > 2\left(\dfrac{1}{2\pi}\right) Hz$ or $f_s > 0.315 Hz$. For definiteness, choose $f_s = 0.5 Hz$, so that the sampling period is $T_s = 2s$. From Eq. (7.2), $N = \dfrac{t_{max}}{T_s} = \dfrac{6\pi}{2} = 9.4$, which should be an integer. Obviously, setting $N = 10$ is the proper choice, which results in a sampling period slightly smaller than $T_s = 2s$. Using $N = 10$ yields $T_s = \dfrac{t_{max}}{N} = \dfrac{6\pi}{10} = 1.885s$.

The discrete function f_k is evaluated at each discrete point using Eq. (7.9) for $k = 0, 1, 2, \ldots, 9$, with $T_s = 1.885s$,

$$f_k = e^{-0.1(kT_s)} \sin(kT_s) = e^{-0.1885k} \sin(1.885k) \tag{7.10}$$

2. Reconstruct the function from the discrete points.

The discrete points, as shown in **Figure 7.9**, reconstructed the essential feature of $f(t)$ successfully, that is, a sinusoid decaying exponentially, in comparison to the continuous function shown as a solid curve.

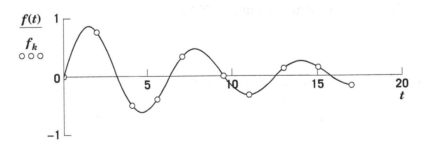

Figure 7.9 Successful reconstruction of function $f(t)$ from its discrete function f_k.

With this background in discretization, you shall learn next about the DFT.

7.3 Discrete Fourier Transform and Discrete Frequency Spectrum

Discrete Fourier transform

The DFT is derived from the continuous Fourier transform,

$$\hat{f}(\omega) = \int_{-\infty}^{\infty} f(t) e^{-i\omega t} dt \qquad (7.11)$$

over the interval $-\infty < \omega < \infty$, which was defined in Eq. (6.36), Chapter 6.

In Chapter 6, the frequency spectrum was shown to be symmetric for real functions. Thus, Eq. (7.11) can be evaluated over the interval $0 \le \omega < \infty$,

$$\hat{f}(\omega) = \int_{0}^{\infty} f(t) e^{-i\omega t} dt \qquad (7.12)$$

From a discrete point of view, Eq. (7.12) necessitates the discretization of two variables, namely, time t and frequency ω, both of which appear in the equation. We already demonstrated how to discretize the time domain. The frequency domain is discretized using

$$\omega_m = m\omega_{max} \qquad (7.13)$$

for $m = 0, 1, 2, ..., N-1$, where $\omega_{max} = \dfrac{2\pi}{t_{max}} = \dfrac{2\pi}{NT_s}$ is defined as the fundamental frequency.

With the discrete points in the time domain and frequency domain well defined, we now are in a position to convert the integral to a Riemann sum by replacing the integral in Eq. (7.12), $\int_{0}^{\infty}$

with $\sum_{k=0}^{N-1}$, $f(t)$ with f_k, ω with Eq. (7.13), t with Eq. (7.4), and dt with T_s, so that the DFT is

$$\hat{f}(\omega_m) \simeq \sum_{k=0}^{N-1} f_k e^{-i\omega_m t_k} T_s = \sum_{k=0}^{N-1} f_k e^{-i\left(\frac{2\pi m}{NT_s}\right)(kT_s)} T_s, \text{ which simplifies to}$$

$$\hat{f}(\omega_m) = T_s \sum_{k=0}^{N-1} f_k e^{\frac{-2\pi imk}{N}} \qquad (7.14)$$

Replacing T_s in Eq. (7.14) with its definition in Eq. (7.2) yields

$$\hat{f}(\omega_m) = t_{max} \frac{1}{N} \sum_{k=0}^{N-1} f_k e^{\frac{-2\pi imk}{N}} \qquad (7.15)$$

The definition of the DFT is based on the right-hand summation in Eq. (7.15), with and without N. The following are several versions:

Version 1:
$$\hat{f}_m = \frac{1}{\sqrt{N}} \sum_{k=0}^{N-1} f_k e^{\frac{-2\pi imk}{N}} \qquad (7.16)$$

Version 2:
$$\hat{f}_m = \sum_{k=0}^{N-1} f_k e^{\frac{-2\pi imk}{N}} \qquad (7.17)$$

These versions vary from book to book and software to software. **Table 7.2** is an example of how some CAS adapt one of the versions.

Version	DFT	Mathematica	Matlab	Maple
Version 1	$\hat{f}_m = \frac{1}{\sqrt{N}} \sum_{k=0}^{N-1} f_k e^{\frac{-2\pi imk}{N}}$	default		default
Version 2	$\hat{f}_m = \sum_{k=0}^{N-1} f_k e^{\frac{-2\pi imk}{N}}$		default	

Table 7.2 Two versions of the discrete Fourier transform.

Insofar as our objective is concerned—that is, finding the dominant frequency in a signal—the particular version used does not matter because version 1 differs from version 2 by $\frac{1}{\sqrt{N}}$ scale on the vertical axis of the DFS. Thus, the dominant frequency will be detected uniquely as a peak in the DFS in version 1 or version 2. In this chapter, version 2 is used.

Matrix form of the DFT

Define

$$a = e^{\frac{-2\pi i}{N}} \qquad (7.18)$$

so that Eq. (7.17) reads

$$\hat{f}_m = \sum_{k=0}^{N-1} a^{mk} f_k \qquad (7.19)$$

for $m = 0, 1, 2, \ldots, N-1$ and $k = 0, 1, 2, \ldots, N-1$. Thus, the $N \times N$ matrix form of Eq. (7.19) is given by

$$\begin{bmatrix} \hat{f}_0 \\ \hat{f}_1 \\ \hat{f}_2 \\ \vdots \\ \hat{f}_{N-1} \end{bmatrix} = \begin{bmatrix} a^{(0)(0)} & a^{(0)(1)} & a^{(0)(2)} & \cdots & a^{(0)(N-1)} \\ a^{(1)(0)} & a^{(1)(1)} & a^{(1)(2)} & \cdots & a^{(1)(N-1)} \\ a^{(2)(0)} & a^{(2)(1)} & a^{(2)(2)} & \cdots & a^{(2)(N-1)} \\ \cdots & \cdots & \cdots & \cdots & \cdots \\ a^{(N-1)(0)} & a^{(N-1)(1)} & a^{(N-1)(2)} & \cdots & a^{(N-1)(N-1)} \end{bmatrix} \begin{bmatrix} f_0 \\ f_1 \\ f_2 \\ \vdots \\ f_{N-1} \end{bmatrix} \qquad (7.20)$$

Note that the elements raised to power zero in the first row and first column are equal to one. Thus, Eq. (7.20) is simplified to

$$
\begin{bmatrix} \hat{f}_0 \\ \hat{f}_1 \\ \hat{f}_2 \\ \vdots \\ \hat{f}_{N-1} \end{bmatrix} = \begin{bmatrix} 1 & 1 & 1 & \cdots & 1 \\ 1 & a^{(1)(1)} & a^{(1)(2)} & \cdots & a^{(1)(N-1)} \\ 1 & a^{(2)(1)} & a^{(2)(2)} & \cdots & a^{(2)(N-1)} \\ \cdots & \cdots & \cdots & \cdots & \cdots \\ 1 & a^{(N-1)(1)} & a^{(N-1)(2)} & \cdots & a^{(N-1)(N-1)} \end{bmatrix} \begin{bmatrix} f_0 \\ f_1 \\ f_2 \\ \vdots \\ f_{N-1} \end{bmatrix} \tag{7.21}
$$

Discrete frequency spectrum

Analogous to the frequency spectrum as the magnitude of the Fourier transform in Chapter 6, the DFS is the magnitude of $\hat{f}_m$, $\left| \hat{f}_m \right|$ for $m = 0, 1, 2, \ldots, N-1$.

Steps to generate the DFS

1. Using the Nyquist sampling theorem given by Eq. (7.8), determine the sampling period T_s based on an expected frequency.
2. Choose the sampling time interval $0 \le t \le t_{max}$.
3. Find N from Eq. (7.2) and round it up to an even integer.
4. Recalculate T_s.
5. Find f_k by sampling function $f(t)$.
6. Calculate $\hat{f}_m$ from Eq. (7.21).
7. Plot $\left| \hat{f}_m \right|$ over the interval $0 \le m \le \dfrac{N}{2}$. The complex exponential function in Eq. (7.17) possesses periodicity so that $\left| \hat{f}_m \right|$ is portrayed with symmetry. The interval $0 \le m \le \dfrac{N}{2}$ should contain the necessary information about the dominant frequency for engineering systems.

Example 7.4

Consider the following function in the interval $t \ge 0$,

$$
f(t) = e^{-0.1t} \sin(t) \tag{7.22}
$$

In Example 7.3, the sampling frequency was chosen as $f_s = 0.5 Hz$ based on the Nyquist sampling theorem. Thus, $T_s = \dfrac{1}{f_s} = 2s$. In order to demonstrate the algebraic steps in Eq. (7.21) manageably, the sampling interval is limited to $t_{\max} = 2\pi$, which is one period of $\sin(t)$. From Eq. (7.2), $N = \dfrac{t_{\max}}{T_s} = \dfrac{2\pi}{2} = 3.142$. Thus, set $N = 4$.

1. Calculate the sampling period using $N = 4$.
2. Find the discrete points, f_k, for $k = 0, 1, 2, \ldots, N-1$.
3. Solve Eq. (7.21) for $m = 0, 1, 2, \ldots, N-1$.
4. Evaluate $\left|\hat{f}_m\right|$ for $m = 0, 1, 2, \ldots, N-1$.
5. Reevaluate $\left|\hat{f}_m\right|$ for $m = 0, 1, 2, \ldots, \dfrac{N}{2}$.
6. Identify the dominant frequency by plotting $\left|\hat{f}_m\right|$ over the interval $0 \le m \le \dfrac{N}{2}$.

Solution:

1. Calculate the sampling period using $N = 4$.

From Eq. (7.2), $T_s = \dfrac{t_{\max}}{N} = \dfrac{2\pi}{4} = 1.571s$.

2. Find the discrete points, f_k, for $k = 0, 1, 2, 3$.

The discrete function f_k is evaluated at each discrete point using Eq. (7.22) with $T_s = 1.571s$,

$$f_k = e^{-0.1(kT_s)} \sin(kT_s) = e^{-0.1571k} \sin(1.571k) \tag{7.23}$$

For $k = 0, 1, 2, 3$, Eq. (7.23) yields

$$\begin{bmatrix} f_0 \\ f_1 \\ f_2 \\ f_3 \end{bmatrix} = \begin{bmatrix} 0 \\ 0.885 \\ 0 \\ -0.624 \end{bmatrix} \tag{7.24}$$

3. Solve Eq. (7.21) for $m = 0, 1, 2, 3$.

$$\begin{bmatrix} \hat{f}_0 \\ \hat{f}_1 \\ \hat{f}_2 \\ \hat{f}_3 \end{bmatrix} = \begin{bmatrix} 1 & 1 & 1 & 1 \\ 1 & a^{(1)(1)} & a^{(1)(2)} & a^{(1)(3)} \\ 1 & a^{(2)(1)} & a^{(2)(2)} & a^{(2)(3)} \\ 1 & a^{(3)(1)} & a^{(3)(2)} & a^{(3)(3)} \end{bmatrix} \begin{bmatrix} f_0 \\ f_1 \\ f_2 \\ f_3 \end{bmatrix} \tag{7.25}$$

Substituting Eq. (7.18) for each element in the 4 x 4 matrix in Eq. (7.25) yields

$$
\begin{bmatrix} \hat{f}_0 \\ \hat{f}_1 \\ \hat{f}_2 \\ \hat{f}_3 \end{bmatrix} = \begin{bmatrix} 1 & 1 & 1 & 1 \\ 1 & e^{\frac{-\pi i}{2}} & e^{-\pi i} & e^{\frac{-3\pi i}{2}} \\ 1 & e^{-\pi i} & e^{-2\pi i} & e^{-3\pi i} \\ 1 & e^{\frac{-3\pi i}{2}} & e^{-3\pi i} & e^{\frac{-9\pi i}{2}} \end{bmatrix} \begin{bmatrix} f_0 \\ f_1 \\ f_2 \\ f_3 \end{bmatrix}
$$

(7.26)

Using Euler's formula $e^{i\theta} = \cos(\theta) + i\sin(\theta)$ for each element in the 4 x 4 matrix in Eq. (7.26) yields

$$
\begin{bmatrix} \hat{f}_0 \\ \hat{f}_1 \\ \hat{f}_2 \\ \hat{f}_3 \end{bmatrix} = \begin{bmatrix} 1 & 1 & 1 & 1 \\ 1 & -i & -1 & i \\ 1 & -1 & 1 & -1 \\ 1 & i & -1 & -i \end{bmatrix} \begin{bmatrix} f_0 \\ f_1 \\ f_2 \\ f_3 \end{bmatrix}
$$

(7.27)

Reading the matrix in rows yields

$$
\begin{aligned}
\hat{f}_0 &= f_0 + f_1 + f_2 + f_3 \\
\hat{f}_1 &= f_0 - if_1 - f_2 + if_3 \\
\hat{f}_2 &= f_0 - f_1 + f_2 - f_3 \\
\hat{f}_3 &= f_0 + if_1 - f_2 - if_3
\end{aligned}
$$

(7.28)

Substituting the discrete values, Eq. (7.24) in Eq. (7.28) yields

$$
\begin{aligned}
\hat{f}_0 &= f_1 + f_3 = 0.23 \\
\hat{f}_1 &= -if_1 + if_3 = -1.479i \\
\hat{f}_2 &= -f_1 - f_3 = -0.23 \\
\hat{f}_3 &= if_1 - if_3 = 1.479i
\end{aligned}
$$

(7.29)

4. Evaluate $\left|\hat{f}_m\right|$ for $m = 0, 1, 2, 3$.
From Eq. (7.29),

$$\left|\hat{f}_0\right| = 0.23$$
$$\left|\hat{f}_1\right| = 1.479$$
$$\left|\hat{f}_2\right| = 0.23 \qquad\qquad (7.30)$$
$$\left|\hat{f}_3\right| = 1.479$$

Note that $\left|\hat{f}_0\right| = \left|\hat{f}_2\right|$ and $\left|\hat{f}_1\right| = \left|\hat{f}_3\right|$ so that a shortcut can be taken as demonstrated next.

5. Reevaluate $\left|\hat{f}_m\right|$ for $m = 0, 1, 2, \ldots, \dfrac{N}{2}$.

For $\dfrac{N}{2} = 2$, $m = 0, 1$ so that Eq. (7.27) is rewritten as $\begin{bmatrix} \hat{f}_0 \\ \hat{f}_1 \end{bmatrix} = \begin{bmatrix} 1 & 1 & 1 & 1 \\ 1 & -i & -1 & i \\ 1 & -1 & 1 & -1 \\ 1 & i & -1 & -i \end{bmatrix} \begin{bmatrix} f_0 \\ f_1 \\ f_2 \\ f_3 \end{bmatrix}$ that yields

$$\hat{f}_0 = f_0 + f_1 + f_2 + f_3 = 0.23$$
$$\hat{f}_1 = f_0 - if_1 - f_2 + if_3 = -1.479i$$

Thus, $\begin{aligned} \left|\hat{f}_0\right| &= 0.23 \\ \left|\hat{f}_1\right| &= 1.479 \end{aligned}$.

6. Identify the dominant frequency by plotting $\left|\hat{f}_m\right|$ over the interval $0 \le m \le \dfrac{N}{2}$.

For $N = 4$, the DFS, $\left|\hat{f}_m\right|$ is shown in **Figure 7.10** over the interval $0 \le m \le 2$. **Figure 7.10** shows that the dominant frequency takes place at $m = 1$. From Eq. (7.13), $\omega_m = m\omega_{\max}$, where

$\omega_{\max} = \dfrac{2\pi}{NT_s}$, yields $\omega_1 = \dfrac{2\pi}{4(1.571)} = 1\dfrac{rad}{s}$, which is in fact the frequency of $\sin(t)$ in Eq. (7.22).

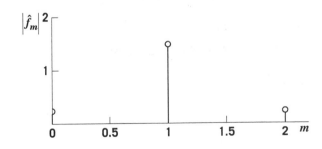

Figure 7.10 Discrete Fourier spectrum (DFS) for Eq. (7.22).

Example 7.5 Elevator Oscillation Due to Rail Misalignment

Revisit the case study of the elevator described in the introduction, as shown in **Figure 7.4**, where the lateral acceleration of the elevator is measured in *volts* in order to determine any rail misalignment. A sample test using an accelerometer is shown in **Figure 7.11** over the time interval $0 \le t \le 6.5s$.

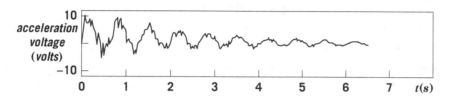

Figure 7.11 Lateral acceleration signal in *volts* over a 6.5-*s* time interval.

When the elevator cab encounters a bump, it oscillates between the two springs, as shown in **Figure 7.4b**. Thus, it is possible to model the cab of mass m connected to two springs, each with stiffness k, as a second-order system, as shown in **Figure 7.12**, which is given by $\dfrac{d^2x}{dt^2} + \omega_n^2 x = 0$,

where the natural frequency of the elevator cab is given by $\omega_n = \sqrt{\dfrac{2k}{m}}$ with a corresponding

frequency in *Hz*, $f_n = \dfrac{\omega_n}{2\pi} = \dfrac{1}{\pi}\sqrt{\dfrac{k}{2m}}$. The frequency f_n would be considered the expected frequency $f_{\exp}$.

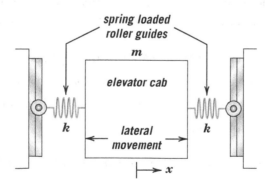

Figure 7.12 Modeling the cab between two spring-loaded roller guides as a second-order system.

k	Acceleration f_k	k	Acceleration f_k	k	Acceleration f_k	k	Acceleration f_k
0	0	8	−3.12	16	0.958	24	0.001866
1	−0.946	9	1.954	17	−0.102	25	−0.0807
2	1.153	10	−0.433	18	−0.551	26	0.0617
3	−0.494	11	−1.035	19	0.9	27	0.00819
4	−0.784	12	2.11	20	−0.944	28	−0.0755
5	2.21	13	−2.58	21	0.762	29	0.1
6	−3.27	14	2.44	22	−0.474	30	−0.0664
7	3.61	15	−1.82	23	0.1925	31	−0.01526

Table 7.3 Discrete acceleration data in *volts* with a sampling period $T_S = 0.2s$.

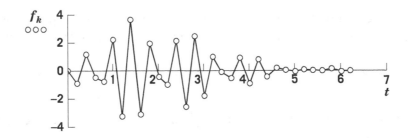

Figure 7.13 Discrete acceleration data in *volts* with a sampling period $T_S = 0.2s$.

The acceleration signal, $f(t)$ in **Figure 7.11** is sampled so as to produce 32 evenly spaced data points at 0.2 s intervals, as shown in **Table 7.3** and **Figure 7.13**. The lines connecting the discrete points, as shown in the figure are used simply to add clarity about data trend.

1. Show that the sampling period, $T_s = 0.2s$, satisfies the Nyquist sampling theorem.
2. Identify the dominant frequency in the DFS, and compare it with f_n to determine if the rails are misaligned. The criterion for misalignment is when f_n differs by more than 3% when compared to the expected frequency.

Data: Mass of the elevator: $m = 350kg$. Elevator spring stiffness: $k = 35000\dfrac{N}{m}$.

Solution:

1. Show that the sampling period, $T_s = 0.2s$, satisfies the Nyquist sampling theorem.

Using the given data of mass and stiffness, $f_{exp} = \dfrac{1}{\pi}\sqrt{\dfrac{k}{2m}} = \dfrac{1}{\pi}\sqrt{\dfrac{35000}{2(350)}} = 2.251Hz$. The Nyquist

sampling theorem requires that $f_s > 2f_{exp}$; that is, $f_s > 4.5Hz$, which implies $T_s < 0.222s$. Thus, $T_s = 0.2s$ satisfies the Nyquist criterion.

2. Identify the dominant frequency in the DFS, and compare it with f_n to determine if the rails are misaligned. The criterion for misalignment is when f_n differs by more than 3% when compared to the expected frequency.

The discrete Fourier transform is given by Eq. (7.21),

$$\begin{bmatrix} \hat{f}_0 \\ \hat{f}_1 \\ \hat{f}_2 \\ \vdots \\ \hat{f}_{31} \end{bmatrix} = \begin{bmatrix} 1 & 1 & 1 & \cdots & 1 \\ 1 & a^{(1)(1)} & a^{(1)(2)} & \cdots & a^{(1)(31)} \\ 1 & a^{(2)(1)} & a^{(2)(2)} & \cdots & a^{(2)(31)} \\ \cdots & \cdots & \cdots & \cdots & \cdots \\ 1 & a^{(31)(1)} & a^{(31)(2)} & \cdots & a^{(31)(31)} \end{bmatrix} \begin{bmatrix} f_0 \\ f_1 \\ f_2 \\ \vdots \\ f_{31} \end{bmatrix} \qquad (7.31)$$

Substituting Eq. (7.18) for each element in the 32 x 32 matrix in Eq. (7.31) yields

$$\begin{bmatrix} \hat{f}_0 \\ \hat{f}_1 \\ \hat{f}_2 \\ \vdots \\ \hat{f}_{31} \end{bmatrix} = \begin{bmatrix} 1 & 1 & 1 & \cdots & 1 \\ 1 & e^{\frac{-\pi i}{16}} & e^{\frac{-\pi i}{8}} & \cdots & e^{\frac{-31\pi i}{16}} \\ 1 & e^{\frac{-\pi i}{8}} & e^{\frac{-\pi i}{4}} & \cdots & e^{\frac{-31\pi i}{8}} \\ \cdots & \cdots & \cdots & \cdots & \cdots \\ 1 & e^{\frac{-31\pi i}{16}} & e^{\frac{-31\pi i}{8}} & \cdots & e^{\frac{-961\pi i}{16}} \end{bmatrix} \begin{bmatrix} f_0 \\ f_1 \\ f_2 \\ \vdots \\ f_{31} \end{bmatrix} \qquad (7.32)$$

For $N = 32$, the DFS, $\left|\hat{f}_m\right|$ is shown in **Figure 7.14** over the interval $0 \le m \le 16$.

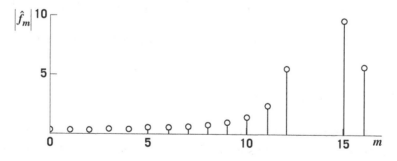

Figure 7.14 Discrete Fourier spectrum (DFS) for the acceleration voltage.

Figure 7.14 shows that the dominant frequency takes place at $m = 15$. From Eq. (7.13), $\omega_m = m\omega_{max}$, where $\omega_{max} = \dfrac{2\pi}{NT_s}$, yields $\omega_{15} = 15\dfrac{2\pi}{32(0.2)} = 14.73\dfrac{rad}{s}$. Thus, the frequency in

Hz is $f_{15} = \dfrac{\omega_{15}}{2\pi} = 2.34 Hz$. This frequency differs by approximately 4% from the theoretical natural frequency of the elevator, $f_n = 2.25 Hz$. Hence, the rails are misaligned and need adjustment.

In this applied engineering example, you have used the DFS found by the DFT of the acceleration as a diagnostic tool to determine if there is excessive vibration in the elevator due to rail misalignment.

You should note that the algebra looks overwhelming as the matrix, Eq. (7.21), gets very large. However, the problem is straightforward in CAS because all it takes is the array generation of the discrete points, f_k, followed by invoking Eq. (7.17) directly, and computing $\left| \hat{f}_m \right|$, with no need to setup the problem as a matrix.

As the number of samples increases, the computation of Eq. (7.17) becomes computationally expensive. The computation time can be reduced drastically by invoking an algorithm known as the fast Fourier transform, which is introduced next.

7.4 Fast Fourier Transform

The number of computations required to calculate the DFT using Eq. (7.17), or its equivalent $N \times N$ matrix, Eq. (7.21), is on the order of N^2. If you study the coefficient matrix Eq. (7.21), you can see that its elements are symmetric with respect to the diagonal; for example, element $a^{(1)(2)}$ has the same exponents as element $a^{(2)(1)}$. In that regard, the computation that is done twice can be cut by half. That's where the fast Fourier transform (FFT) algorithm comes in, which is devised to reduce the number of computations to a magnitude on the order of $N \log_2(N)$. The relative growth of the DFT with respect to the FFT is defined by the ratio

$\dfrac{N^2}{N \log_2(N)}$ or $\dfrac{N}{\log_2(N)}$. For example, when $N = 1000$, it takes the DFT 10^6 calculations, and the FFT 9966 calculations with a relative growth of 100, that is 100 times more calculations for the DFT than the FFT.

Example 7.6 Bridge Pier Failure Analysis

Revisit the case study of the concrete pier described in the introduction. A graph supplied by the DOT, as shown in **Figure 7.15**, describes a standing wave for this class of concrete piers in terms of the wave velocity, defined as

$$v = 4 f_n L \qquad (7.33)$$

versus concrete compressive strength, σ, where $L = 10\ m$ is the length of the pier and f_n is the natural frequency at which the surface vibrates. The regions of weak and strong concrete are shown in the graph for $\sigma \leq 25\ MPa$ and $\sigma > 25\ MPa$, respectively. The pier in this bridge had a compressive strength of 30 MPa when it was originally built.

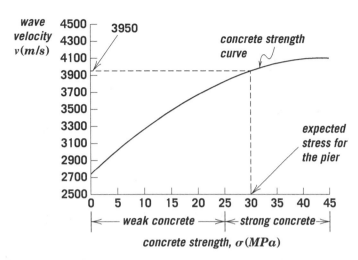

Figure 7.15 DOT specification of wave velocity v versus compressive strength of concrete σ.

Figure 7.16 shows the velocity response signal from a test carried out on a 40-year-old pier over the time interval $0 \le t \le 0.155s$.

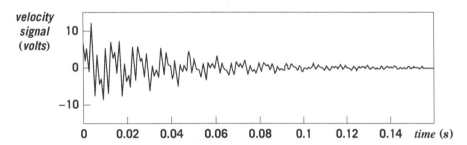

Figure 7.16 Velocity signal as a function of time for a 40-year-old pier.

The velocity signal, $f(t)$, in **Figure 7.16** is sampled so as to produce 32 evenly spaced data points at 5-*ms* intervals, as shown in **Table 7.4** and **Figure 7.17**. The lines connecting the discrete points, as shown in the figure are used simply to add clarity about data trend.

k	Velocity f_k	k	Velocity f_k	k	Velocity f_k	k	Velocity f_k
0	5	8	−0.347	16	0.381	24	0.158
1	−4.07	9	−0.772	17	−0.09	25	−0.381
2	1.408	10	1.278	18	−0.269	26	0.441
3	1.58	11	−1.302	19	0.595	27	−0.351
4	−3.66	12	1.096	20	−0.778	28	0.177
5	4.24	13	−0.87	21	0.749	29	0.0002
6	−3.46	14	0.708	22	−0.52	30	−0.118
7	1.923	15	−0.572	23	0.177	31	0.149

Table 7.4 Discrete velocity data in *volts* with a sampling period $T_s = 5ms$.

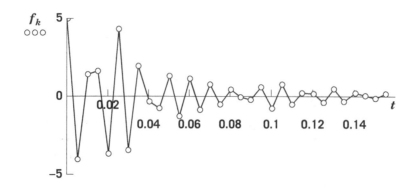

Figure 7.17 Discrete velocity data in *volts* with a sampling period $T_s = 5ms$.

1. Show that the sampling period, $T_s = 0.005s$, satisfies the Nyquist sampling theorem.
2. Identify the dominant frequency in the DFS, and compare it with f_n to determine if the concrete has deteriorated.
3. What is the relative growth of the number of computations of the DFT with respect to the FFT?

Solution:

1. Show that the sampling period, $T_s = 0.005s$, satisfies the Nyquist sampling theorem.

Since the pier in this bridge had a compressive strength of 30 *MPa* when it was originally built, the expected wave speed is $v = 3950 \dfrac{m}{s}$, as shown in **Figure 7.15**. Substituting the given data into Eq. (7.33), $3950 = 4 f_n (10)$ yields the expected natural frequency, $f_{exp} = 98.75 Hz$. The Nyquist sampling theorem requires that $f_s > 2 f_{exp}$; that is, $f_s > 197.5 Hz$, which implies a sampling period $T_s < 5.063 \times 10^{-3} s$. Thus, $T_s = 0.005s$ satisfies the Nyquist theorem.

2. Identify the dominant frequency in the DFS, and compare it with f_n to determine if the concrete has deteriorated.

The frequency spectrum found by CAS is plotted with discrete points in **Figure 7.18**.

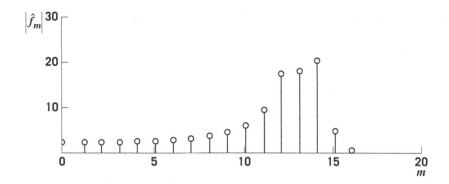

Figure 7.18 DFS by the FFT, $T_s = 0.005 s$ and $N = 32$.

Figure 7.18 shows that the dominant frequency takes place at $m = 14$. From Eq. (7.13), $\omega_m = m\omega_{max}$,

where $\omega_{max} = \dfrac{2\pi}{NT_s}$, yields $\omega_{14} = 14 \dfrac{2\pi}{32(0.005)} = 549.8 \dfrac{rad}{s}$. Thus, the frequency in Hz is

$f_{14} = \dfrac{\omega_{14}}{2\pi} = 87.5 Hz$ that yields from Eq. (7.33) a wave velocity $v = 4(87.5)(10) = 3500 \dfrac{m}{s}$. The

concrete compressive strength that corresponds to this wave speed is $15\ MPa$, as shown in **Figure 7.19**, which is in the weak concrete area. Therefore, the pier has poor strength and does need maintenance.

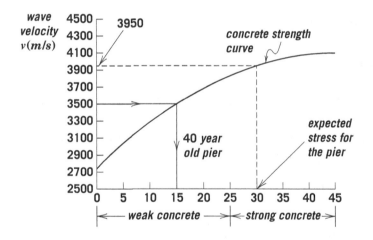

Figure 7.19 40-year-old pier compressive strength and corresponding wave speed.

3. What is the relative growth of the number of computations of the DFT with respect to the FFT?

For $N = 32$, the relative growth is $\dfrac{N}{\log_2(N)} = 6.4$, that is the FFT is approximately 6 times faster than the DFT even with this small number of data points.

In this applied engineering example, you have used the DFS found by the FFT as a tool to determine that the concrete pier has poor strength and needs maintenance.

HOMEWORK PROBLEMS FOR CHAPTER SEVEN

Problems for Section 7.2

1. Consider the following function in the interval $t \geq 0$,

$$f(t) = e^{-\sqrt{t}} \sin(2t) \tag{7.34}$$

a. Take $N = 20$ and discretize Eq. (7.34) over the interval $0 \leq t \leq 10s$ using Eq. (7.2).

b. Reconstruct the function from the discrete points. Plot on a single graph f_k using discrete points and $f(t)$ as a continuous curve for comparison.

In Problems 2–5, consider the function $f(t)$ in the interval $t \geq 0$.

a. Find the sampling period T_s used to sample $f(t)$ based on the Nyquist sampling theorem.

2. $f(t) = e^{-t} \cos(10t)$

3. $f(t) = \cos(t) + \cos(2t) + \sin(3t)$

4. $f(t) = \cos(t)$

5. $f(t) = e^{-t} \left[\sin(2t) + \sin(3t) \right]$

Problems for Section 7.3

In Problems 6–8, consider the function over the interval $0 \leq t \leq 5$ with 10 sample points.

a. Show that the sampling period will satisfy the Nyquist sampling theorem.

b. Find the discrete points, f_k, for $k = 0, 1, 2, \ldots, N-1$.

c. Solve Eq. (7.21) for $m = 0, 1, 2, \ldots, \dfrac{N}{2}$.

d. Evaluate and plot $\left| \hat{f}_m \right|$ for $m = 0, 1, 2, \ldots, \dfrac{N}{2}$.

6. $f(t) = \cos(4t)$.

7. $f(t) = e^{-0.5t} \sin(\pi t)$.

8. $f(t) = \cos(\pi t) + \sin(2\pi t)$.

9. Resolve Example 7.4 using $T_s = 2$ and $N = 10$.

a. Find the discrete points, f_k, for $k = 0, 1, 2, \ldots, N-1$.

b. Solve Eq. (7.21) for $m = 0, 1, 2, \ldots, \dfrac{N}{2}$.

c. Evaluate and plot $\left|\hat{f}_m\right|$ for $m = 0, 1, 2, \ldots, \dfrac{N}{2}$.

10. A pump-motor assembly, as shown in **Figure 7.20**, is routinely monitored for bearing misalignment or wear. Under normal conditions, the pump and motor run at 200 *rpm* $\left(20.94\dfrac{rad}{s}\right)$, and 1000 *rpm* $\left(104.7\dfrac{rad}{s}\right)$, respectively. The condition of the pump and motor is monitored by attaching accelerometers to their housings. A typical sampling record for the pump is shown in **Table 7.5**. The sampling period is $T_s = 0.03s$. If no maintenance is required, a peak appears on the DFS plot at $20.94\dfrac{rad}{s}$, which is the expected frequency, $f_{\exp}$ of the pump. If the pump requires maintenance, there will be an additional peak on the DFS plot, 3 to 4 times higher than the expected frequency.

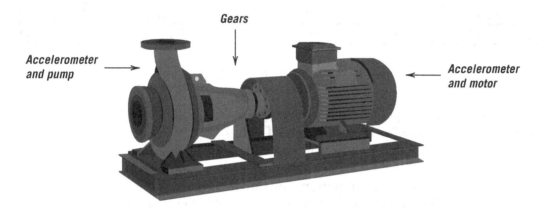

Figure 7.20 Motor and pump with acceleration instrumentation.

a. Show that the sampling period satisfies the Nyquist sampling theorem.
b. Use the DFT to find the DFS for the data in **Table 7.5**.
c. Determine from the DFS if the pump requires maintenance.

K	Acceleration f_k	k	Acceleration f_k	k	Acceleration f_k	k	Acceleration f_k
0	0	8	−4.09	16	−18.5	24	18.37
1	18.47	9	−18.4	17	−4.94	25	1.58
2	4.426	10	−0.63	18	−3.77	26	−18.45
3	4.259	11	18.48	19	−18.4	27	−5.295
4	18.46	12	4.768	20	−1.26	28	−3.459
5	0.316	13	3.931	21	18.46	29	−18.33
6	−18.47	14	18.42	22	5.117	30	−1.895
7	−4.596	15	0.949	23	3.614	31	18.43

Table 7.5 Pump acceleration data, $T_s = 0.03s$.

11. Revisit Problem 10 and resolve it for the motor using the sampled record as shown in **Table 7.6**. The sampling period is $T_s = 0.01s$. The normal angular speed of the motor is $\omega = 1000rpm = 104.7\dfrac{rad}{s}$. If the motor requires maintenance, the DFS plot will show an additional peak 2 to 3 times higher than the expected frequency.

k	Acceleration f_k	k	Acceleration f_k	k	Acceleration f_k	k	Acceleration f_k
0	0	8	0.0285	16	0.057	24	−0.171
1	20.79	9	−0.021	17	−20.8	25	20.81
2	0.007	10	0.0355	18	-0.13	26	0.093
3	−0.007	11	−20.77	19	20.81	27	−0.064
4	0.014	12	−0.085	20	0.071	28	0.099
5	−20.78	13	20.8	21	−0.05	29	−20.75
6	−0.043	14	0.0499	22	0.078	30	−0.213
7	20.79	15	−0.036	23	−20.8	31	20.82

Table 7.6 Motor acceleration data, $T_s = 0.01s$.

   a. Show that the sampling period satisfies the Nyquist sampling theorem.
   b. Use the DFT to find the DFS for the data in **Table 7.6**.
   c. Determine from the DFS if the motor requires maintenance.

12. Helicopter inspection is very important for safety. One inspection method for crack growth testing is based on attaching an accelerometer to the tail section, as shown in **Figure 7.21**. A typical acceleration record is shown in **Table 7.7**, where the sampling period is $T_s = 0.01s$.

Accelerometer

Figure 7.21 Helicopter tail section and attached accelerometer.

The specification for a crack-free tail is based on an expected frequency, $f_{\exp}$, given by the helicopter blades' speed, $\omega = 1860rpm = 194.8\dfrac{rad}{s}$. This frequency would appear as a peak on the DFS plot. If there are destructive cracks in the tail section, there will be an additional peak in the DFS, with a magnitude of at least 50% of the expected peak.

a. Show that $T_s = 0.01s$ satisfies the Nyquist sampling theorem.
b. Use the DFT to find the DFS for the data in **Table 7.7**.
c. Determine from the DFS of the tail section if there are major cracks.

k	Acceleration f_k	k	Acceleration f_k	k	Acceleration f_k	k	Acceleration f_k
0	0	8	0.108	16	−0.422	24	0.447
1	0.848	9	−0.787	17	1.041	25	−1.1
2	−0.767	10	0.464	18	−0.452	26	0.332
3	−0.319	11	0.534	19	−0.522	27	0.671
4	1.036	12	−1.112	20	0.871	28	−0.736
5	−0.258	13	0.369	21	−0.148	29	−0.092
6	−0.944	14	0.828	22	−0.91	30	0.967
7	0.918	15	−0.765	23	0.815	31	−0.825

Table 7.7 Tail section acceleration data, $T_s = 0.01s$.

13. Revisit Example 7.5: Elevator Oscillation Due to Rail Misalignment. In this problem, the elevator mass is $m = 600kg$, and the spring stiffness is $k = 50000 \dfrac{N}{m}$. The acceleration sampled data are given in **Table 7.8**, where the sampling period is $T_s = 0.2s$. The natural frequency of the elevator in Hz is $f_n = \dfrac{\omega_n}{2\pi} = \dfrac{1}{\pi}\sqrt{\dfrac{k}{2m}}$, which is considered the expected frequency, f_{exp}.

k	Acceleration f_k	k	Acceleration f_k	k	Acceleration f_k	k	Acceleration f_k
0	0	8	−0.1462	16	−0.4734	24	0.2390
1	−0.1290	9	−0.1889	17	0.5139	25	−0.0590
2	0.1207	10	0.4712	18	−0.3991	26	−0.1171
3	0.0389	11	−0.6061	19	0.1768	27	0.2380
4	−0.2772	12	0.5517	20	0.0767	28	−0.2751
5	0.4785	13	−0.3311	21	−0.2827	29	0.2280
6	−0.5407	14	0.0209	22	0.3848	30	−0.1209
7	0.4198	15	0.2777	23	−0.3637	31	−0.0071

Table 7.8 Discrete acceleration data (*volts*) for $T_S = 0.2s$.

a. Show that $T_S = 0.2s$ satisfies the Nyquist sampling theorem.
b. Use the DFT to find the DFS for the data in **Table 7.8**.

c. Identify the dominant frequency component in the DFS, and compare it with f_n to determine if the rails are misaligned. The criterion for misalignment in this elevator is when f_n differs by more than 3% when compared to the expected frequency.

14. Revisit Example 7.6: Bridge Pier Failure Analysis. Resolve the problem for another identical pier on the other end of the bridge. Both piers had the same strength when they were built. The sampled velocity data are given in **Table 7.9**, where the sampling period is $T_s = 4ms$.

k	Velocity f_k	k	Velocity f_k	k	Velocity f_k	k	Velocity f_k
0	0	8	0.760	16	0.854	24	-0.082
1	0.274	9	–0.509	17	–1.03	25	–0.424
2	–0.480	10	–0.016	18	0.754	26	0.755
3	0.546	11	0.589	19	–0.155	27	–0.829
4	–0.418	12	–0.935	20	–0.486	28	0.645
5	0.098	13	0.864	21	0.896	29	–0.262
6	0.321	14	–0.383	22	–0.926	30	–0.222
7	–0.667	15	–0.295	23	0.601	31	0.675

Table 7.9 Discrete velocity data for $T_s = 4ms$.

a. Use the DFT to find the DFS for the data in **Table 7.9**.
b. Identify the dominant frequency component in the DFS, and convert it to f_n; determine if the pier under consideration has deteriorated.

Problems for Section 7.4

15. Resolve Problem 10 using the FFT.
16. Resolve Problem 11 using the FFT.
17. Resolve Problem 12 using the FFT.
18. Resolve Problem 13 using the FFT.
19. Resolve Problem 14 using the FFT.

CHAPTER EIGHT

INTRODUCTION TO COMPUTATIONAL TECHNIQUES

8.1 Introduction

In Chapters 1–5, you modeled engineering problems by ODEs that had analytical solutions. However, for nonlinear ODEs, an analytical solution may not exist. For example, in Section 1.3, Chapter 1, the initial value first-order ODE, Eq. (1.47), $A\dfrac{dh}{dt}+k_o\sqrt{h}=\dot{Q}_i(t)$, which describes the fluid level in a tank, $h(t)$, is nonlinear by virtue of the quadratic term $\sqrt{h}$. This equation does not have an exact solution. However, it is possible to obtain a numerical solution of the ODE over the domain $t>t_0$ starting with a given initial condition (IC) $h(t_0)=h_o$. In Section 3.4, Chapter 3, the boundary value problem described by the second-order ODE, Eq. (3.30), $\dfrac{d^2T}{dx^2}+a_1(x)\dfrac{dT}{dx}+a_0(x)T=f(x)$, where $a_1(x)=\dfrac{1}{k}\dfrac{dk}{dx}$, $a_0(x)=-\dfrac{hP_e}{kA}$, and $f(x)=-\dfrac{hP_e}{kA}T_f$, models the spatial distribution of temperature, $T(x)$, over the domain $a<x<b$ with given Dirichlet, Neumann, or Robin boundary conditions (BCs) (see Table 3.3). The ODE is linear but may not have an analytical solution if conductivity varies spatially by a prescribed function, $k(x)$. In this case, it is possible to obtain a numerical solution of the ODE.

In this chapter, you are introduced to the finite difference method (FDM) and the finite element method (FEM) to solve engineering problems numerically by various discretization schemes. In the FDM, the physical domain is divided into discrete points called nodes, as shown in **Figure 8.1a**. In the FEM, the domain is divided into shapes that could be rectangular, triangular, tetrahedral, and so on, called elements, as shown in **Figure 8.1b**.

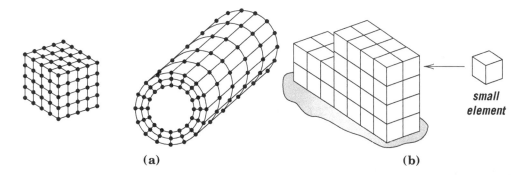

(a)   **(b)**

Figure 8.1 Numerical discretization: (a) finite difference discretization; and (b) finite element discretization.

You will learn that the FDM discretization is relatively straightforward in the time domain and in the space domain for simple geometries, as shown in **Figure 8.1a**. However, the FDM may not handle complex geometries as effectively as the FEM. In the FEM, as shown in **Figure 8.1b**, the elements are assembled into a matrix that describes the whole structure. Furthermore, the matrix includes information about all the elements, such as material properties. The advantage of the FEM lies in its ability to handle complex geometries with nonuniform grids. For unsteady problems that lead to partial differential equations, it is possible to combine the FEM with FDM; the FEM provides spatial solutions at a given time, whereas the FDM advances the solution in time.

In Section 8.2, the FDM is introduced to solve initial value problems using two techniques, the Euler method and the Runge–Kutta method. In Section 8.3, the FDM is introduced for boundary value problems using the direct method. In Section 8.4 the FEM is introduced.

8.2 The Finite Difference Method

At the introductory level, implementation of the Euler method is restricted to first-order initial value problems. The Runge–Kutta method will be applied to higher-order ODEs in the form of a system of first-order ODEs. The FDM will be applied to linear and nonlinear ODEs.

First-order ODEs

In Chapter 1, you learned about linear first-order ODEs as $\dfrac{dy}{dt} + a_0(t)y = r(t)$. Whenever the coefficient $a_0(t)$ is also a function of y, for example, $a_0(t,y)$, the ODE is nonlinear,

$$\frac{dy}{dt} + a_0(t,y)y = r(t) \tag{8.1}$$

Example 8.1

Consider the first-order nonlinear ODE $A\dfrac{dh}{dt} + k_o\sqrt{h} = \dot{Q}_i(t)$ in $h(t)$.

1. Express the ODE in the form of Eq. (8.1) and identify $a_o(t,h)$.

Solution:

Rewrite $A\dfrac{dh}{dt} + k_o\sqrt{h} = \dot{Q}_i(t)$ as $A\dfrac{dh}{dt} + k_o h^{\frac{1}{2}} = \dot{Q}_i(t)$, or $\dfrac{dh}{dt} + \dfrac{k_o}{A}h^{-\frac{1}{2}}h = \dfrac{\dot{Q}_i(t)}{A}$. Thus, in comparison with Eq. (8.1), $a_o(t,h) = \dfrac{k_o}{A}h^{-\frac{1}{2}}$. In this case, it is $a_o(h) = \dfrac{k_o}{A}h^{-\frac{1}{2}}$.

■

Suppose for a moment you know the analytical solution of Eq. (8.1), $y(t)$, for a given IC $y(t_0) = y_o$ over a domain of interest, $t_0 \le t \le t_N$, which is depicted graphically in **Figure 8.2a**. In this case, you know not only the solution, but also the local slope $y' = \dfrac{dy}{dt}$. What if you do not know the analytical solution of Eq. (8.1), which is the theme of this chapter? In this case, you resort to a numerical solution by discretizing the domain $t_0 \leq t \leq t_N$ into $N+1$ nodes, t_0, t_1, ..., t_n, t_{n+1}, ..., t_N, as shown in **Figure 8.2b**. Next, rewrite the ODE as $\dfrac{dy}{dt} = r(t) - a_o(t,y)y$, or

$$y' = f(t,y) \tag{8.2}$$

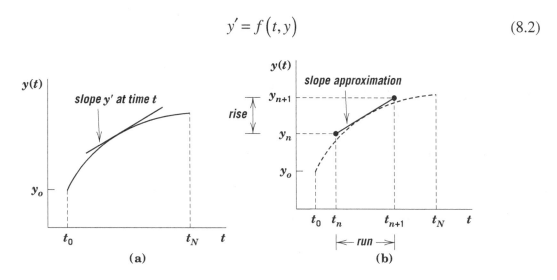

Figure 8.2 Solution of the ODE: (a) presumed solution with a local slope; and (b) unknown solution with slope approximation.

where $f(t,y) = r(t) - a_o(t,y)y$, and imagine the existence of a slope, $y' = \dfrac{dy}{dt}$, between two intermediate nodes, t_n and t_{n+1} with corresponding unknown values of the function, y_n and y_{n+1}. Approximate this imaginary slope as rise over run,

$$\frac{\Delta y}{\Delta t} = \frac{y_{n+1} - y_n}{t_{n+1} - t_n} \tag{8.3}$$

so that the end result is a discrete algebraic equation in $y(t)$. Replacing y' in Eq. (8.2) with Eq. (8.3) allows you to predict y_1 at t_1 from the IC y_0 at t_0; y_2 at t_2 from y_1 at t_1, and so on and so forth up to y_N at t_N over the domain of interest, $t_0 \leq t \leq t_N$, as long as you know where to evaluate the nodal value of $f(t,y)$ on the right-hand side of Eq.(8.2). The nodal value of $f(t,y)$ is handled by several schemes as follows.

Forward Euler method or explicit Euler method

The FDM is known as the forward Euler method or explicit Euler method when $f(t, y)$ in Eq. (8.2) is evaluated at node (t_n, y_n) so that Eqs. (8.2) and (8.3) yield $\frac{y_{n+1} - y_n}{t_{n+1} - t_n} = f(t_n, y_n)$ or

$$y_{n+1} = y_n + \Delta t f(t_n, y_n) \tag{8.4}$$

where $\Delta t = t_{n+1} - t_n$. The method is explicit because y_n is known at t_n to yield the right-hand side of Eq. (8.4), and hence the next value, y_{n+1} directly.

Backward Euler method or implicit Euler method

The FDM is known as the backward Euler method or the implicit Euler method when $f(t, y)$ in Eq. (8.2) is evaluated at (t_{n+1}, y_{n+1}) so that Eqs. (8.2) and (8.3) yield

$$y_{n+1} = y_n + \Delta t f(t_{n+1}, y_{n+1}) \tag{8.5}$$

The method is implicit because y_{n+1} may be embedded in nonlinear terms on the right-hand side, which does not permit you to calculate its value directly. Quite often, y_{n+1} is found indirectly by a root-finding method, which is a standard technique in CAS.

Improved Euler method

Another version of the Euler method is known as the improved Euler method and consists of two steps defined as the predictor and corrector steps. In the predictor step, an intermediate value, y_{n+1}^* is predicted as

$$y_{n+1}^* = y_n + \Delta t f(t_n, y_n) \tag{8.6}$$

The corrector step is given by

$$y_{n+1} = y_n + \frac{1}{2} \Delta t \left[f(t_n, y_n) + f(t_{n+1}, y_{n+1}^*) \right] \tag{8.7}$$

Example 8.2 Tank Flow Governed by a Linear ODE: All Euler Methods

Revisit the fluid level in a tank, $h(t)$, in Example 1.7, Chapter 1, where t is in s and h is in m.

The ODE is governed by $\tau \frac{dh}{dt} + h = k\theta$ with IC $h(0) = 0$, where $\tau = 60s$, $k = 0.197 \frac{m}{degree}$, and

$$\theta(t) = \begin{cases} \dfrac{t}{6} & 0 \leq t < 60 \\ 10 & t \geq 60 \end{cases}.$$

1. Take $\Delta t = 0.1\tau$ and find h_1 at t_1 and h_2 at t_2 by the forward Euler method, backward Euler method, and improved Euler method.
2. Compare all methods with the exact solution using absolute percent error.
3. Plot the solutions and the error over the range $0 \le t \le 30s$ for clear visualization.

Solution:

For all methods, cast the ODE in the form of Eq. (8.2) as $h' = \dfrac{1}{\tau}\left(k\theta - h\right)$.

For $0 \le t < 60$, $\theta(t) = \dfrac{t}{6}$ so that $h' = f(t,h)$, where $f(t,h) = \dfrac{1}{\tau}\left(\dfrac{k}{6}t - h\right)$.

For $t \ge 60$, $\theta(t) = 10$ so that $h' = f(t,h)$, where $f(t,h) = \dfrac{1}{\tau}\left(10k - h\right)$, or simply

$f(h) = \dfrac{1}{\tau}\left(10k - h\right)$.

$t_0 = 0$ and $\Delta t = 0.1(60) = 6s$. Thus, $t_1 = t_0 + \Delta t = 6s$, and $t_2 = t_1 + \Delta t = 12s$, which are in the

range $0 \le t < 60$. Thus, use $h' = f(t,h)$, where $f(t,h) = \dfrac{1}{\tau}\left(\dfrac{k}{6}t - h\right)$.

1. Take $\Delta t = 0.1\tau$ and find h_1 at t_1 and h_2 at t_2 by the forward Euler method, backward Euler method, and improved Euler method.

Forward Euler method:

Based on Eq. (8.4), $h_{n+1} = h_n + \Delta t f\left(\theta_n, h_n\right) = h_n + \dfrac{\Delta t}{\tau}\left(\dfrac{k}{6}t_n - h_n\right)$. For simplicity, shift indices, for

example,

$$h_n = h_{n-1} + \dfrac{\Delta t}{\tau}\left(\dfrac{k}{6}t_{n-1} - h_{n-1}\right)$$

so that you start counting at $n = 1$.

$n = 1$: $t_1 = 6s$, $h_1 = h_0 + \dfrac{\Delta t}{\tau}\left(\dfrac{k}{6}t_0 - h_0\right) = 0$

$n = 2$: $t_2 = 12s$, $h_2 = h_1 + \dfrac{\Delta t}{\tau}\left(\dfrac{k}{6}t_1 - h_1\right) = 0 + 0.1\left(6\dfrac{0.197}{6} - 0\right) = 0.0197$

Backward Euler method:

Based on Eq. (8.5), $h_{n+1} = h_n + \frac{\Delta t}{\tau}\left(\frac{k}{6}t_{n+1} - h_{n+1}\right)$. Here, by the nature of the linear ODE, h_{n+1}

terms are collected from both sides of the equation, which yield $h_{n+1}\left(1 + \frac{\Delta t}{\tau}\right) = h_n + \frac{\Delta t}{\tau}\frac{k}{6}t_{n+1}$, or

$h_{n+1} = \left(1 + \frac{\Delta t}{\tau}\right)^{-1}\left(h_n + \frac{\Delta t}{\tau}\frac{k}{6}t_{n+1}\right)$. Shifting indices yields

$$h_n = \left(1 + \frac{\Delta t}{\tau}\right)^{-1}\left(h_{n-1} + \frac{\Delta t}{\tau}\frac{k}{6}t_n\right)$$

$n = 1: t_1 = 6s$, $h_1 = \left(1 + \frac{\Delta t}{\tau}\right)^{-1}\left(h_0 + \frac{\Delta t}{\tau}\frac{k}{6}t_1\right) = (1.1)^{-1}\left[0 + 0.1\left(\frac{0.197}{6}\right)6\right] = 0.01971$

$n = 2: t_2 = 12s$, $h_2 = \left(1 + \frac{\Delta t}{\tau}\right)^{-1}\left(h_1 + \frac{\Delta t}{\tau}\frac{k}{6}t_2\right) = (1.1)^{-1}\left[0.01971 + 0.1\left(\frac{0.197}{6}\right)12\right] = 0.0521$

Improved Euler method:

From Eq. (8.6), $h^*_{n+1} = h_n + \frac{\Delta t}{\tau}\left(\frac{k}{6}t_n - h_n\right)$. From Eq. (8.7),

$h_{n+1} = h_n + \frac{1}{2}\frac{\Delta t}{\tau}\left(\frac{k}{6}t_n - h_n + \frac{k}{6}t_{n+1} - h^*_{n+1}\right)$, or $h_{n+1} = h_n + \frac{1}{2}\frac{\Delta t}{\tau}\left[\frac{k}{6}(t_n + t_{n+1}) - h_n - h^*_{n+1}\right]$

Shifting indices yields

$$h^*_n = h_{n-1} + \frac{\Delta t}{\tau}\left(\frac{k}{6}t_{n-1} - h_{n-1}\right)$$

$$h_n = h_{n-1} + \frac{1}{2}\frac{\Delta t}{\tau}\left[\frac{k}{6}(t_{n-1} + t_n) - h_{n-1} - h^*_n\right]$$

$n = 1: t_1 = 6s$, $h^*_1 = h_0 + \frac{\Delta t}{\tau}\left(\frac{k}{6}t_0 - h_0\right) = 0$, and

$h_1 = h_0 + \frac{1}{2}\frac{\Delta t}{\tau}\left[\frac{k}{6}(t_0 + t_1) - h_0 - h^*_1\right] = 0 + \frac{0.1}{2}\left(\frac{0.197}{6}\right)(0 + 6) = 0.00985$

$n = 2: t_2 = 12s$, $h^*_2 = h_1 + \frac{\Delta t}{\tau}\left(\frac{k}{6}t_1 - h_1\right) = 0.00985 + 0.1\left(6\frac{0.197}{6} - 0.00985\right) = 0.0286$

$h_2 = h_1 + \frac{1}{2}\frac{\Delta t}{\tau}\left[\frac{k}{6}(t_1 + t_2) - h_1 - h^*_2\right] = 0.00985 + \frac{0.1}{2}\left(18\frac{0.197}{6} - 0.00985 - 0.0286\right) = 0.0375$

2. Compare all methods with the exact solution using absolute percent error.

The exact solution of the ODE is given in Example 1.7 as

$$
h(t) = \begin{cases} 1.97 \left(e^{-\frac{t}{60}} - 1 \right) + 0.0328t & 0 \le t < 60 \\ -3.38e^{-\frac{t}{60}} + 1.97 & t \ge 60 \end{cases}
\tag{8.8}
$$

The numerical values of the three schemes and exact solution are shown in **Table 8.1**.

Time (s)	Exact $h(t)$ Eq. (8.8)	Forward Euler method		Backward Euler method		Improved Euler method	
	values	values	% error	values	% error	values	% error
6	0.00933	0	100	0.0179	92	0.00985	6
12	0.0365	0.0197	46	0.0521	43	0.0375	3

Table 8.1 A tabular comparison with the exact solution of the numerical values obtained by three Euler methods.

3. Plot the solutions and the error over the range $0 \le t \le 30s$ for clear visualization.

The numerical solutions of all three schemes can be obtained relatively easily in a spreadsheet over the time range of interest. For clear visualization of the numerical schemes and their errors, the range is limited to $0 \le t \le 30s$, as shown in **Figure 8.3a** for $h(t)$ and **Figure 8.3b** for the percent error. **Table 8.1** and the figure show that the forward and backward Euler methods have the same order-of-magnitude absolute error, which is on the order of 100% at 6s and 15% at 30s. On the other hand, the improved Euler method starts with 6% at 6s and decays to 1% at 30s. Thus, one can conclude that this method is more accurate than the forward and backward Euler methods. After 78s, the absolute error of the forward and backward Euler methods is less than 1% with an average of 0.4%, whereas the improved Euler method has an average error of 0.03% over $78 \le t \le 360s$. Recall from Figure 1.15 in Example 1.7 that $h(t)$ reaches a steady state at approximately 360s. Notice also that the solution is underestimated by the forward Euler method and overestimated by the backward Euler method on the concave-up portion of the solution, as shown in **Figure 8.3a**. The reverse takes place on the concave-down portion of the solution, as shown in **Figure 8.3c**.

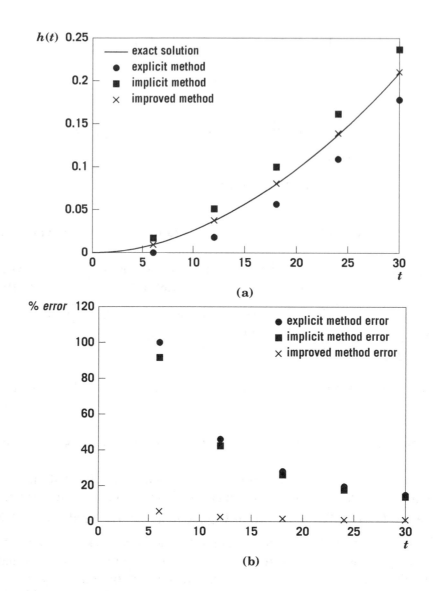

(a)

(b)

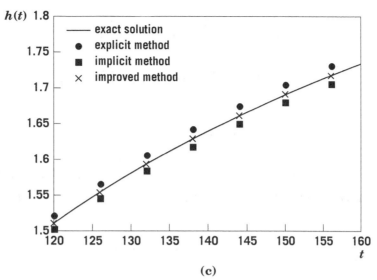

(c)

Figure 8.3 A graphical comparison with the exact solution of the numerical values obtained by three Euler methods: (a) early stages of the solution; (b) early stages of the absolute error; and (c) late stages of the solution.

■

Example 8.3 Tank Flow Governed by a Nonlinear ODE: Backward Euler Method.

Consider the level of fluid in a tank, governed by the nonlinear ODE, Eq. (1.47) in Chapter 1,
$A\dfrac{dh}{dt}+k_o\sqrt{h}=k_v\theta$, where t is in s, h is in m, and θ is in degrees. The tank area is $A=3m^2$. The

flow coefficient is $k_o=0.08\dfrac{m^3}{s\sqrt{m}}$, and the valve coefficient is $k_v=0.01\dfrac{m^3}{s\left(degree\right)}$. The

unusual units of k_o and k_v balance $k_o\sqrt{h}$ and $k_v\theta$, respectively, to m^3/s as required by the

leading term in the ODE, $A\dfrac{dh}{dt}$. The IC is $h\left(0\right)=0$. At $t=0$, the valve is suddenly opened from

a closed position to $5°$ (step function).

1. Take $\Delta t=6s$ and find h_1 at t_1 by the backward Euler method.

Solution:

Cast the ODE in the form of Eq. (8.2) as $\dfrac{dh}{dt}=\dfrac{1}{A}\left(k_v\theta-k_o\sqrt{h}\right)$, or $h'=f\left(\theta,h\right)$, where

$f\left(\theta,h\right)=\dfrac{1}{A}\left(k_v\theta-k_o\sqrt{h}\right)$. Based on Eq. (8.5), $h_{n+1}=h_n+\dfrac{\Delta t}{A}\left(k_v\theta_{n+1}-k_o\sqrt{h_{n+1}}\right)$. Here, no

attempt will be made to collect h_{n+1} terms, notwithstanding its possibility, so that the scheme remains effectively implicit in h_{n+1}. Shifting indices yields

$$h_n = h_{n-1} + \frac{\Delta t}{A}\left(k_v \theta_n - k_o \sqrt{h_n}\right)$$

$$n = 1:\ t_1 = 6s,\ h_1 = h_o + \frac{\Delta t}{A}\left(k_v \theta_1 - k_o \sqrt{h_1}\right) = 0 + \frac{6}{3}\left(0.01(5) - 0.08\sqrt{h_1}\right),\ \text{or}$$

$$h_1 = 0.1 - 0.16\sqrt{h_1} \tag{8.9}$$

Solving Eq. (8.9) for h_1 in that form requires a root finding algorithm that is a standard technique in CAS. However, you can guess the solution graphically by defining from Eq. (8.9) a function $r(h_1)$ as

$$r(h_1) = -h_1 + 0.1 - 0.16\sqrt{h_1} \tag{8.10}$$

Plotting Eq. (8.10) versus h_1, as shown in **Figure 8.4**, indicates that the root is approximately 0.06. Invoking a root finder in CAS yields to double-precision $h_1 = 0.060609579423040$ and $r(h_1) = 0$.

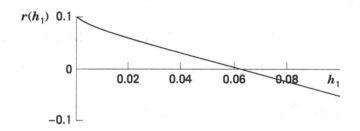

Figure 8.4 Root of Eq. (8.10).

In the event that you attempt to collect terms in h_1 in Eq. (8.9), for example, by rearranging Eq. (8.9) and squaring terms, $(h_1 - 0.1)^2 = 0.0256 h_1$, you will have $h_1^2 - 0.2256 h_1 + 0.01 = 0$. This quadratic equation has one double-precision root equal to 0.060609579423040 that satisfies both sides of Eq. (8.9). The second double-precision root is equal to 0.16499042057696, which is rejected because it does not satisfy Eq. (8.9). You should note that this special case of collecting terms in the implicit method does not work all the time, hence the need for a root-finding algorithm.

The accuracy of any scheme in the FDM is an essential feature of the numerical method. Another equally important feature is scheme stability, that is, whether or not the numerical process advances the solution with large errors that lead to unbounded results for a supposedly bounded solution. As far as accuracy is concerned, the scheme is normally described as first-order accurate, second-order accurate, and so on and so forth, based on the Taylor series expansion. For example, the forward Euler method is first-order accurate as follows. Consider the Taylor series expansion of y_{n+1} near t_n,

$$y_{n+1} = y_n + \Delta t\, y'_{t_n} + \frac{(\Delta t)^2}{2!} y''_{t_n} + \frac{(\Delta t)^3}{3!} y'''_{t_n} + \frac{(\Delta t)^4}{4!} y''''_{t_n} + \cdots \tag{8.11}$$

where the primes on the right-hand side are derivatives evaluated at t_n as required by the Taylor series. Substituting the first derivative, y'_{t_n}, with Eq. (8.2) yields

$$y_{n+1} = y_n + \Delta t\, f(t_n, y_n) + \frac{(\Delta t)^2}{2!} y''_{t_n} + \frac{(\Delta t)^3}{3!} y'''_{t_n} + \frac{(\Delta t)^4}{4!} y''''_{t_n} + \cdots$$

or

$$\frac{y_{n+1} - y_n}{\Delta t} = f(t_n, y_n) + \frac{\Delta t}{2!} y''_{t_n} + \frac{(\Delta t)^2}{3!} y'''_{t_n} \frac{(\Delta t)^3}{4!} y''''_{t_n} + \cdots \tag{8.12}$$

As $\Delta t \to 0$, $(\Delta t)^3 \ll (\Delta t)^2 \ll \Delta t$, which suggests keeping the first-order term, $\frac{\Delta t}{2!} y''_{t_n}$, and dropping all higher-order terms in Eq. (8.12), so that the latter is truncated approximately to

$$\frac{y_{n+1} - y_n}{\Delta t} \simeq f(t_n, y_n) + \frac{\Delta t}{2!} y''_{t_n} \tag{8.13}$$

where the equal sign in Eq. (8.12) is replaced with the approximation sign in Eq. (8.13). The term $\frac{\Delta t}{2!} y''_{t_n}$ is proportional to Δt, which is denoted $O(\Delta t)$ as $\Delta t \to 0$, where O is known as "on the order of," so that Eq. (8.13) reads

$$\frac{y_{n+1} - y_n}{\Delta t} \simeq f(t_n, y_n) + O(\Delta t) \tag{8.14}$$

We can interpret Eq. (8.4) as $\frac{y_{n+1} - y_n}{\Delta t} = f(t_n, y_n)$ in comparison with Eq. (8.14) by saying that the forward Euler method is a first-order accurate scheme, which suggests that halving Δt may cause the error to reduce linearly by a factor of two. However, a tendency to decrease Δt may not yield desirable results because of the cumulative increase in computer round-off error associated with the increased numerical steps.

If the Taylor series analysis is repeated by expanding y_{n+1} near t_{n+1}, it can be shown that the analysis yields the backward Euler method to $O(\Delta t)$, which offers no accuracy advantage over the forward Euler method. Its advantage, however, lies in stability, which is discussed shortly. As for the improved Euler method, the Taylor series expansion can also show that it is second-order accurate, or $O(\Delta t)^2$, so that halving Δt may cause an error reduction by a factor of four.

Recall from **Figure 8.2b** the slope y' where $\Delta y = y_{n+1} - y_n$ is the rise and $\Delta t = t_{n+1} - t_n$ is the run. Thus, multiplying the slope $y' = f(t, y)$ in Eq. (8.2) by the run yields a rise equal to $\Delta t f(t, y)$. Using this concept, define the rise $k_1 = \Delta t f(t_n, y_n)$ at (t_n, y_n) so that the improved Euler method predictor step, Eq. (8.6), is written as $y_{n+1}^* = y_n + k_1$. Also define a second rise $k_2 = \Delta t f(t_{n+1}, y_{n+1}^*) = \Delta t f(t_{n+1}, y_n + k_1)$, so that the corrector step, Eq. (8.7) is rewritten as $y_{n+1} = y_n + \frac{1}{2}(k_1 + k_2)$. Simply put, the improved Euler method can be viewed as a two-rise average scheme formulated as

$$y_{n+1} = y_n + \frac{1}{2}(k_1 + k_2) \tag{8.15}$$

where the first rise

$$k_1 = \Delta t f(t_n, y_n) \tag{8.16}$$

is evaluated at (t_n, y_n), and the second rise

$$k_2 = \Delta t f(t_{n+1}, y_n + k_1) \tag{8.17}$$

is evaluated at $(t_{n+1}, y_n + k_1)$.

This approach led to another high-order accurate scheme known as the Runge–Kutta method, which has several versions. The scheme that is ubiquitous in engineering is known as the fourth-order Runge–Kutta method, which is fourth-order accurate, or $O(\Delta t)^4$.

Fourth-order Runge–Kutta method for single first-order ODEs

In the fourth-order Runge–Kutta method, or RK4 for short, y_{n+1} is obtained from y_n via four intermediate rises, k_1, k_2, k_3, and k_4, given by

$$y_{n+1} = y_n + \frac{1}{6}(k_1 + 2k_2 + 2k_3 + k_4) \tag{8.18}$$

where

$$
\begin{aligned}
k_1 &= \Delta t f(t_n, y_n) \\
k_2 &= \Delta t f\left(t_n + \frac{\Delta t}{2}, y_n + \frac{k_1}{2}\right) \\
k_3 &= \Delta t f\left(t_n + \frac{\Delta t}{2}, y_n + \frac{k_2}{2}\right) \\
k_4 &= \Delta t f(t_n + \Delta t, y_n + k_3)
\end{aligned}
\tag{8.19}
$$

Considering the straightforward implementation of this explicit method for a first-order ODE, we shall set it up for a system of ODEs as explained next.

RK4 method for higher-order ODEs

In Chapter 4, you learned how to cast a second-order ODE into a system of first-order ODEs. For example, consider

$$\frac{d^2 y}{dt^2} + a_1 \frac{dy}{dt} + a_0 y = r(t) \tag{8.20}$$

If you define $\frac{dy}{dt} = z$, then Eq. (8.20) becomes $\frac{dz}{dt} + a_1 z + a_0 y = r(t)$. The two foregoing equations can be cast in the form of Eq. (8.2) as

$$\begin{aligned} y' &= z = f_1(t, y, z) \\ z' &= -a_1 z - a_0 y + r(t) = f_2(t, y, z) \end{aligned} \tag{8.21}$$

The numerical scheme of Eq. (8.21) by the RK4 method is summarized in **Table 8.2**, where a tilde is used with the rises of the z function for clarity. Notice that $f_1(t, y, z) = z$ reduces to $f_1(z) = z$, so that overloading the function with all three arguments is unnecessary. You should also note that k_1 through k_4, and $\tilde{k}_1$ through $\tilde{k}_4$ are intertwined pairwise; that is, to find k_2 you need $\tilde{k}_1$, and to find $\tilde{k}_2$ you need k_1 and $\tilde{k}_1$, and so on and so forth. Thus, it is recommended that you calculate at each step the pair k_1 and $\tilde{k}_1$, then k_2 and $\tilde{k}_2$, followed by k_3 and $\tilde{k}_3$, and finally k_4 and $\tilde{k}_4$. In total, you will have 10 calculations per iteration.

$y_{n+1} = y_n + \dfrac{1}{6}(k_1 + 2k_2 + 2k_3 + k_4);$ $\quad f_1(z) = z$	$z_{n+1} = z_n + \dfrac{1}{6}(\tilde{k}_1 + 2\tilde{k}_2 + 2\tilde{k}_3 + \tilde{k}_4);$ $\quad f_2(t, y, z) = -a_1 z - a_0 y + r(t)$
$k_1 = \Delta t f_1(z_n) \qquad = (\Delta t) z_n$	$\tilde{k}_1 = \Delta t f_2(t_n, y_n, z_n)$
$k_2 = \Delta t f_1\!\left(z_n + \dfrac{\tilde{k}_1}{2}\right) = (\Delta t)\left(z_n + \dfrac{\tilde{k}_1}{2}\right)$	$\tilde{k}_2 = \Delta t f_2\!\left(t_n + \dfrac{\Delta t}{2}, y_n + \dfrac{k_1}{2}, z_n + \dfrac{\tilde{k}_1}{2}\right)$
$k_3 = \Delta t f_1\!\left(z_n + \dfrac{\tilde{k}_2}{2}\right) = (\Delta t)\left(z_n + \dfrac{\tilde{k}_2}{2}\right)$	$\tilde{k}_3 = \Delta t f_2\!\left(t_n + \dfrac{\Delta t}{2}, y_n + \dfrac{k_2}{2}, z_n + \dfrac{\tilde{k}_2}{2}\right)$
$k_4 = \Delta t f_1(z_n + \tilde{k}_3) = (\Delta t)(z_n + \tilde{k}_3)$	$\tilde{k}_4 = \Delta t f_2(t_n + \Delta t, y_n + k_3, z_n + \tilde{k}_3)$

Table 8.2 The RK4 scheme for a second-order ODE reduced to a system of two first-order ODEs.

Example 8.4 Positioning Table Governed by Second-Order ODE: RK4 Method

Consider a positioning table used to fill vials in a pharmaceutical application. The table is driven by a DC motor, as shown in **Figure 8.5**. In this application, the angular velocity of the motor, ω, is governed by the second-order ODE

$$\frac{d^2\omega}{dt^2} + 2\zeta\omega_n\frac{d\omega}{dt} + \left(\omega_n^2\right)\omega = k_m\,\omega_n^2\,V \tag{8.22}$$

where $V = 1volt$ is the input voltage to the motor, $k_m = 150\dfrac{rad}{volt \cdot s}$ is the gain of the power amplifier,

$\omega_n = 23.94\dfrac{rad}{s}$ is the motor natural frequency, and $\zeta = 0.04$ is the motor damping ratio. The motor

ICs are $\omega_0 = \omega(0) = 0$ for the angular velocity and $\alpha_0 = \dfrac{d\omega(0)}{dt} = 10\dfrac{rad}{s^2}$ for the angular acceleration.

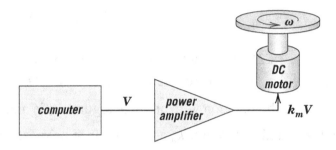

Figure 8.5 Positioning table.

1. Find $\omega(t)$ numerically using the RK4 method over the interval $0 \le t \le 0.7s$ by converting Eq. (8.22) to a system of two first-order ODEs. Use $\Delta t = 0.02s$.
2. Compare the numerical result with the exact solution.

Solution:

1. Find $\omega(t)$ numerically using the RK4 method over the interval $0 \le t \le 0.7s$ by converting Eq. (8.22) to a system of two first-order ODEs. Use $\Delta t = 0.02s$.

Define $\dfrac{d\omega}{dt} = \alpha$ so that Eq. (8.22) becomes $\dfrac{d\alpha}{dt} + 2\zeta\omega_n\alpha + \left(\omega_n^2\right)\omega = k_m\,\omega_n^2\,V$. Cast the two foregoing ODEs in the form of Eq. (8.2),

$$\begin{aligned}\omega' &= \alpha \\ \alpha' &= -\left(\omega_n^2\right)\omega - 2\zeta\omega_n\alpha + k_m\,\omega_n^2\,V\end{aligned} \tag{8.23}$$

The RK4 method applied to Eq. (8.23) is summarized in **Table 8.3**, where $f_1(\alpha) = \alpha$ and

$$f_2(\omega,\alpha) = -(\omega_n^2)\omega - 2\zeta\omega_n\alpha + k_m\,\omega_n^2\,V$$

$\omega_{i+1} = \omega_i + \dfrac{1}{6}(k_1 + 2k_2 + 2k_3 + k_4)$	$\alpha_{i+1} = \alpha_i + \dfrac{1}{6}(\tilde{k}_1 + 2\tilde{k}_2 + 2\tilde{k}_3 + \tilde{k}_4)$
$k_1 = \Delta t f_1(\alpha_i) \qquad = (\Delta t)\alpha_i$ $k_2 = \Delta t f_1\!\left(\alpha_i + \dfrac{\tilde{k}_1}{2}\right) = (\Delta t)\left(\alpha_i + \dfrac{\tilde{k}_1}{2}\right)$ $k_3 = \Delta t f_1\!\left(\alpha_i + \dfrac{\tilde{k}_2}{2}\right) = (\Delta t)\left(\alpha_i + \dfrac{\tilde{k}_2}{2}\right)$ $k_4 = \Delta t f_1(\alpha_i + \tilde{k}_3) = (\Delta t)(\alpha_i + \tilde{k}_3)$	$\tilde{k}_1 = \Delta t f_2(\omega_i, \alpha_i)$ $\tilde{k}_2 = \Delta t f_2\!\left(\omega_i + \dfrac{k_1}{2}, \alpha_i + \dfrac{\tilde{k}_1}{2}\right)$ $\tilde{k}_3 = \Delta t f_2\!\left(\omega_i + \dfrac{k_2}{2}, \alpha_i + \dfrac{\tilde{k}_2}{2}\right)$ $\tilde{k}_4 = \Delta t f_2(\omega_i + k_3, \alpha_i + \tilde{k}_3)$

Table 8.3 The RK4 scheme for a second-order ODE reduced to a system of two first-order ODEs.

The dummy index i is used instead of n to eliminate any confusion with the natural frequency subscript. Using the given data, $f_2(\omega,\alpha) = -573.12\omega - 1.9151\alpha + 85968.54$. It should be clear by now that shifting indices, as done earlier, is straightforward in order to start with $i = 1$, as shown in **Table 8.4**.

$\omega_i = \omega_{i-1} + \dfrac{1}{6}(k_1 + 2k_2 + 2k_3 + k_4)$	$\alpha_i = \alpha_{i-1} + \dfrac{1}{6}(\tilde{k}_1 + 2\tilde{k}_2 + 2\tilde{k}_3 + \tilde{k}_4)$
$k_1 = (\Delta t)\alpha_{i-1}$ $k_2 = (\Delta t)\left(\alpha_{i-1} + \dfrac{\tilde{k}_1}{2}\right)$ $k_3 = (\Delta t)\left(\alpha_{i-1} + \dfrac{\tilde{k}_2}{2}\right)$ $k_4 = (\Delta t)(\alpha_{i-1} + \tilde{k}_3)$	$\tilde{k}_1 = \Delta t f_2(\omega_{i-1}, \alpha_{i-1})$ $\tilde{k}_2 = \Delta t f_2\!\left(\omega_{i-1} + \dfrac{k_1}{2}, \alpha_{i-1} + \dfrac{\tilde{k}_1}{2}\right)$ $\tilde{k}_3 = \Delta t f_2\!\left(\omega_{i-1} + \dfrac{k_2}{2}, \alpha_{i-1} + \dfrac{\tilde{k}_2}{2}\right)$ $\tilde{k}_4 = \Delta t f_2(\omega_{i-1} + k_3, \alpha_{i-1} + \tilde{k}_3)$

Table 8.4 The RK4 scheme for a second-order ODE reduced to a system of two first-order ODEs with shifted indices.

For the pair k_1 and $\tilde{k}_1$:

$k_1 = \Delta t f_1(\alpha_0) = (\Delta t)\alpha_0 = 0.02(10) = 0.2$, and

$\tilde{k}_1 = \Delta t f_2(0,10) = 0.02(-19.151 + 85968.54) = 1719$.

For the pair k_2 and $\tilde{k}_2$:

$$k_2 = (\Delta t)\left(\alpha_0 + \frac{\tilde{k}_1}{2}\right) = 0.02\left(10 + \frac{1719}{2}\right) = 17.29$$

$$\tilde{k}_2 = \Delta t f_2 \left(\omega_0 + \frac{k_1}{2}, \alpha_0 + \frac{\tilde{k}_1}{2} \right) = \Delta t f_2 \left(0.1, 10 + \frac{1719}{2} \right) = \Delta t f_2 \left(0.1, 869.5 \right); \text{ thus}$$

$$\tilde{k}_2 = 0.02 \left[-573.12(0.1) - 1.9151(869.5) + 85968.54 \right] = 1685.$$

For the pair k_3 and $\tilde{k}_3$:

$$k_3 = (\Delta t) \left(\alpha_0 + \frac{\tilde{k}_2}{2} \right) = 0.02 \left(10 + \frac{1685}{2} \right) = 17.05$$

$$\tilde{k}_3 = \Delta t f_2 \left(\omega_0 + \frac{k_2}{2}, \alpha_0 + \frac{\tilde{k}_2}{2} \right) = \Delta t f_2 \left(\frac{17.29}{2}, 10 + \frac{1685}{2} \right) = \Delta t f_2 \left(8.645, 852.2 \right); \text{ thus}$$

$$\tilde{k}_3 = 0.02 \left[-573.12(8.645) - 1.9151(852.5) + 85968.54 \right] = 1588.$$

For the pair k_4 and $\tilde{k}_4$:

$$k_4 = (\Delta t)(\alpha_0 + \tilde{k}_3) = 0.02(10 + 1588) = 31.95$$

$$\tilde{k}_4 = \Delta t f_2 \left(\omega_0 + k_3, \alpha_0 + \tilde{k}_3 \right) = \Delta t f_2 \left(17.05, 10 + 1588 \right) == \Delta t f_2 \left(17.05, 1598 \right); \text{ thus}$$

$$\tilde{k}_4 = 0.02 \left[-573.12(17.05) - 1.9151(1598) + 85968.54 \right] = 1463.$$

Thus, $\omega_1 = \omega_0 + \frac{1}{6}(k_1 + 2k_2 + 2k_3 + k_4) = 0 + \frac{1}{6}[0.2 + 2(17.29) + 2(17.05) + 31.95] = 16.81 \frac{rad}{s}$,

and $\alpha_1 = \alpha_0 + \frac{1}{6}(\tilde{k}_1 + 2\tilde{k}_2 + 2\tilde{k}_3 + \tilde{k}_4) = 10 + \frac{1}{6}(1719 + 2(1685) + 2(1588) + 1463) = 1631 \frac{rad}{s^2}$.

The values of ω_1 and α_1 are used in the next iteration for ω_2 and α_2, and so on and so forth. Obviously, you should rely on CAS in order to carry out a complete simulation over $0 \le t \le 0.7s$, as shown in the second column of **Table 8.5**.

Time (s)	Angular velocity $\omega(t)$ in rad/s		Absolute error (%)
	RK4	Exact solution	
0	0.00	0.00	0
0.02	16.81	16.66	0.9000
0.04	62.40	62.11	0.4669
0.06	125.6	125.24	0.2874
⋮	⋮	⋮	⋮
0.66	229.6	229.73	0.05659
0.68	218.1	218.02	0.03669
0.7	192.0	191.75	0.1304

Table 8.5 Angular speed prediction using the RK4 method for $\Delta t = 0.02s$.

2. Compare the numerical result with the exact solution.

For an underdamped second-order ODE, that is, $0 < \zeta < 1$, the exact solution of Eq. (8.22) is given by

$$\omega(t) = -e^{-0.9576t}\left[150\cos(23.92\,t) + 6.005\sin(23.92\,t)\right] + 150 \qquad (8.24)$$

based on the tools you learned in Chapter 2, which is tabulated in the third column. The maximum error is less than 1%. The numerical results are also shown in **Figure 8.6** as points that fall right on the analytical solution. Note that the exact solution given by Eq. (8.24) has a damped natural frequency $\omega_d = 23.92\,\dfrac{rad}{s}$; therefore, the corresponding period is $p = \dfrac{2\pi}{\omega_d} = 0.26s$. If you study **Figure 8.6**, you can see that the numerical shape of $\omega(t)$ is well described by a good number of discrete points because of the choice of $\Delta t = 0.02s$. Thus, a rule of thumb for time increment choice is $\Delta t \le \dfrac{p}{10}$ in order to capture the sinusoid accurately. Otherwise, you may predict a lower frequency than that of the actual system if the numerical points have a big spread on the time axis. In this example $\Delta t \le \dfrac{0.26}{10} = 0.026$ would work.

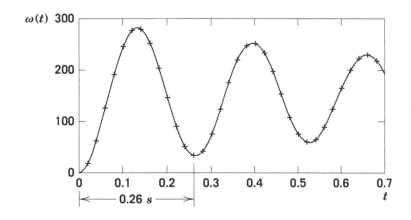

Figure 8.6 Angular speed of the positioning table.

In this engineering example, you learned how to carry out a numerical solution using the RK4 method for a second-order ODE, and estimate the error of the method in comparison to a known solution.

The stability of a numerical scheme is another feature equally important to accuracy. If the numerical scheme yields unbounded values for a supposedly bounded solution, the scheme is classified as unstable. In a homework problem, you may explore numerical stability. However, it suffices in this guide to state that explicit methods such as the forward Euler method, improved Euler method, and RK4 method have limited stability. That is, the foregoing methods are stable only for a range of Δt; otherwise, the method is unstable. Hence, we speak of the limited stability or conditionally stable scheme whenever there are conditions on Δt. If there are no conditions on Δt, then the scheme is classified as unconditionally stable. The backward Euler method is one such scheme; that is, the solution is stable no matter how large or small Δt. However, with large Δt accuracy may be lost. **Table 8.6** is a summary of the entire FDM schemes presented above.

Method	Accuracy order	Stability	Advantages	Disadvantages
Forward Euler	first-order	limited	simple to implement	accuracy
Backward Euler	first-order	unconditionally stable	stable	root finder for nonlinear equations
Improved Euler	second-order	limited	accuracy	limited stability
Runge–Kutta	fourth-order	limited	accuracy	multiple steps

Table 8.6 Summary of the FDM characteristics.

In the next section, you will apply the FDM to boundary value problems.

8.3 Boundary Value Problems

Second-order boundary value problems (BVPs) were covered in Section 3.2, Chapter 3, in the form of

$$\frac{d^2 y}{dx^2} + a_1(x)\frac{dy}{dx} + a_0(x)y = f(x) \qquad (8.25)$$

over the interval $a < x < b$, with Dirichlet, Neumann, and Robin BCs (Table 3.1), which are summarized in **Table 8.7**.

Type of boundary condition	Specified boundary conditions
Dirichlet	$y(a) = y_a$; $y(b) = y_b$
Neumann	$\dfrac{dy(a)}{dx} = y'_a$; $\dfrac{dy(b)}{dx} = y'_b$
Robin	$k_1 y(a) + k_2\, y'(a) = r_a$; $l_1 y(b) + l_2\, y'(b) = r_b$

Table 8.7 Boundary conditions for the BVP.

Finite difference formulation for BVPs: The direct method

As in the time domain, consider a spatial domain over $a < x < b$, as shown in **Figure 8.7**, which consists of $N+1$ discretized nodes, with two nodes at the boundaries and $N-1$ interior nodes. The boundary nodes are labeled x_0 at the left boundary and x_N at the right boundary. The interior nodes are labeled x_1, x_2, ..., x_{n-1}, x_n, x_{n+1}, ..., x_{N-1}. The spatial increment between two typical nodes x_n and x_{n+1} is $\Delta x = x_{n+1} - x_n$.

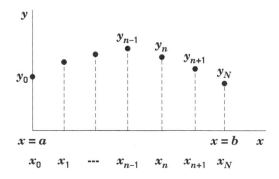

Figure 8.7 Spatial discretization with $N+1$ points including the boundaries.

Imagine a solution y_n at node x_n. This solution is obtained by converting Eq. (8.25) and two BCs from **Table 8.7** to a finite difference scheme, which leads to a system of algebraic equations solved simultaneously to generate the numerical solutions at the nodes of interest. There are several schemes to approximate $\dfrac{dy}{dx}$ and $\dfrac{d^2y}{dx^2}$, which differ by the degree of accuracy. At this introductory level, approximating $\dfrac{dy}{dx}$ and $\dfrac{d^2y}{dx^2}$ to $O(\Delta x)^2$ is quite robust in engineering problems. Guided by the Taylor series, the second-order accurate finite difference of $\dfrac{dy}{dx}$ at node x_n is given by

$$\left.\frac{dy}{dx}\right|_{x=x_n} = \frac{y_{n+1} - y_{n-1}}{2\Delta x} \tag{8.26}$$

Equation (8.26) is known as a central finite difference because y_n is centered between y_{n+1} and y_{n-1} The second-order accurate finite difference of $\dfrac{d^2y}{dx^2}$ at node x_n is given by

$$\left.\frac{d^2y}{dx^2}\right|_{x=x_n} = \frac{y_{n+1} - 2y_n + y_{n-1}}{\Delta x^2} \tag{8.27}$$

Discretization of a BVP governed by a second-order ODE

Substituting Eqs. (8.26) and (8.27) in Eq. (8.25) yields the finite difference equation at node x_n,

$$\frac{y_{n+1} - 2y_n + y_{n-1}}{\Delta x^2} + a_1(x_n)\frac{y_{n+1} - y_{n-1}}{2\Delta x} + a_0(x_n)y_n = f(x_n) \tag{8.28}$$

Equation (8.28) is evaluated at the interior nodes, $1 \leq n \leq N-1$, which will result in $N-1$ simultaneous equations to be solved along with two specified BCs as summarized in **Table 8.8**.

Type of boundary condition	Left BC $x = a$, or x_0	Right BC $x = b$, or x_N
Dirichlet	$y(a) = y_0$	$y(b) = y_N$
Neumann	$y'(a) = \dfrac{y_1 - y_0}{\Delta x}$; $\Delta x = x_1 - x_0$	$y'(b) = \dfrac{y_N - y_{N-1}}{\Delta x}$; $\Delta x = x_N - x_{N-1}$
Robin	$k_1 y_0 + k_2 \dfrac{y_1 - y_0}{\Delta x} = r_a$; $\Delta x = x_1 - x_0$	$l_1 y_N + l_2 \dfrac{y_N - y_{N-1}}{\Delta x} = r_b$; $\Delta x = x_N - x_{N-1}$

Table 8.8 Discretized boundary conditions for the BVP.

Equation (8.28) leads to a system of equations in matrix form, $AX = B$, with a direct solution as $X = A^{-1}B$, where X is the vector whose elements are the unknown y_1, y_2, ..., y_{N-1} at the interior nodes for Dirichlet BCs. Vector X may also contain y_0 or y_N whenever Neumann or Robin BCs are imposed. Note that the derivatives in the Neumann and Robin BCs are first-order accurate, or $O(\Delta x)$ for simplicity. However, it is possible to have higher-order accurate derivatives at the boundaries, which require additional nodes in the formulation.

Example 8.5 Cooling Rod: Mixed Boundary Condition

Revisit the cooling device for a computer model in Example 3.2, Chapter 3. The governing ODE is given by

$$\frac{d^2T}{dx^2} - 2(10^3)T = -5(10^4) \tag{8.29}$$

over the interval $0 \leq x \leq 0.0254m$. Consider the case of Dirichlet BC at $x = 0$,

$$T(0) = 80 \tag{8.30}$$

and Neumann BC at $x = 0.0254$,

$$\frac{dT(0.0254)}{dx} = 0 \tag{8.31}$$

1. Discretize the rod with five nodes and solve Eq. (8.29) with the given BCs by the direct method.
2. Compare the numerical results with the exact solution.
3. Are the proposed five nodes sufficient for an error less than 1% between the numerical solution and the exact solution?

Solution:

1. Discretize the rod with five nodes and solve Eq. (8.29) with the given BCs by the direct method.

Discretizing the rod with five nodes is shown in **Figure 8.8**, where $\Delta x = \dfrac{0.0254}{4} = 0.00635$.

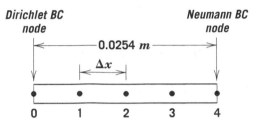

Figure 8.8 Discretized rod.

Comparing Eq. (8.29) to Eq. (8.25) yields $a_1 = 0$. Thus, the FDM scheme, Eq. (8.28), leads to

$$\frac{T_{n+1} - 2T_n + T_{n-1}}{\left(\Delta x\right)^2} - 2\left(10^3\right) T_n = -5\left(10^4\right) \tag{8.32}$$

Substituting $\Delta x = 0.00635$ in Eq. (8.32) and simplifying algebra yield

$$T_{n-1} - 2.0806 T_n + T_{n+1} = -2.02 \tag{8.33}$$

At the interior nodes, $1 \le n \le 3$, we have $T_0 - 2.0806 T_1 + T_2 = -2.02$ for $n = 1$, or

$$-2.0806 T_1 + T_2 = -2.02 - T_0 \tag{8.34}$$

For $n = 2$, $\qquad T_1 - 2.0806 T_2 + T_3 = -2.02 \tag{8.35}$

For $n = 3$, $\qquad T_2 - 2.0806 T_3 + T_4 = -2.02 \tag{8.36}$

Notice that the Dirichlet BC, Eq. (8.30), appears directly in Eq. (8.34) as T_0, which is moved to the right-hand side as a known value. However, T_4 at the right boundary node is an unknown because of the imposed Neumann BC at that node, Eq. (8.31). At that node, the derivative is approximated as $T'(b) = \dfrac{T_N - T_{N-1}}{x_N - x_{N-1}}$ based on **Table 8.8**. Thus, Eq. (8.31) is discretized as

$T'(0.0254) = \dfrac{T_4 - T_3}{x_4 - x_3} = 0$ that yields

$$-T_3 + T_4 = 0 \tag{8.37}$$

Casting Eqs. (8.34)–(8.37) in matrix form yields

$$\begin{bmatrix} -2.0806 & 1 & 0 & 0 \\ 1 & -2.0806 & 1 & 0 \\ 0 & 1 & -2.0806 & 1 \\ 0 & 0 & -1 & 1 \end{bmatrix} \begin{bmatrix} T_1 \\ T_2 \\ T_3 \\ T_4 \end{bmatrix} = \begin{bmatrix} -82.02 \\ -2.02 \\ -2.02 \\ 0 \end{bmatrix} \qquad (8.38)$$

Using CAS to find the nodal solutions for temperature by matrix inversion of Eq. (8.38) yields $T_1 = 70.28°C$, $T_2 = 64.21°C$, $T_3 = 61.29°C$, and $T_4 = 61.29°C$.

2. Compare the numerical results with the exact solution.

The exact solution of the differential equation that satisfies the Dirichlet and Neumann BCs is given by

$$T(x) = 5.142e^{20\sqrt{5}x} + 49.86e^{-20\sqrt{5}x} + 25 \qquad (8.39)$$

Table 8.9 is a summary of the numerical solution in comparison with the exact solution, Eq. (8.39).

Node index, n	Nodal coordinate (m)	FDM temperature ($°C$)	Exact temperature ($°C$)	% Error
1	0.00635	70.28	69.36	1.32
2	0.0127	64.21	62.33	3.01
3	0.0191	61.29	58.32	5.08
4	0.0254	61.29	57.02	7.47

Table 8.9 Comparison of FDM results with exact solution of temperature distribution in a rod.

3. Are the proposed five nodes sufficient for an error less than 1% between the numerical solution and the exact solution?

As you can see from Table 8.9, the percent error is greater than 1%; therefore, it is recommended to increase the number of nodes until the 1% upper bound for error is met.

In this engineering example, you learned how to carry out a numerical solution using the direct method for a BVP with mixed boundary conditions, discretized by the FDM, and to estimate the error of the method in comparison to a known solution.

In the next section, you will learn about the finite element method (FEM) applied to different engineering problems.

8.4 The Finite Element Method

The FEM is a powerful tool suitable for a host of problems with complex geometries in all areas of science and engineering. In this guide, the treatment is limited to 1D so that you learn basic concepts, terminology, and FEM formulation. Historically, the FEM has its origins in the force-displacement problem. So we will start with the force-displacement problem and then proceed to other types of engineering problems.

In 1D structure, the physical domain is modeled by N nodes with $N-1$ finite elements between the nodes, as shown in **Figure 8.9**.

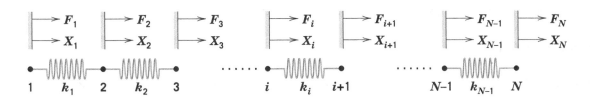

Figure 8.9 FEM model of 1D domain using springs for elements between nodal points.

Each element is modeled as a spring with stiffness k and translational displacement X under the action of forces in tension or compression. For example, with the first element between nodes 1 and 2 we associate stiffness k_1, displacement X_1 and force F_1 at node 1, and displacement X_2 and force F_2 at node 2, and so on and so forth. The displacements define the structure degrees of freedom (DOF). By convention, the displacement and force vectors are positive to the right, as shown in **Figure 8.9**. At any moving node, the force is external, for example, a pull, or push. In the absence of an external force at a moving node X_i we set $F_i = 0$. At a fixed node where the displacement is zero, there exists a force known as a reaction force.

The FEM approach can be used for actual springs, as you will learn in Example 8.8. However, you may also use spring elements to model the stiffness of structures, for example, in the design of laser support where the flange is modeled by springs. See, for example, **Figure 8.32** in homework Problem 28.

Global stiffness matrix

All forces, F_1, F_2, ..., F_N, as shown in **Figure 8.9** are called global forces. One essential feature of the FEM is to formulate a relationship between the global forces and the DOF X_1, X_2, ..., X_N as a matrix $F = KX$, or

$$
\begin{bmatrix} F_1 \\ F_2 \\ \vdots \\ F_i \\ F_{i+1} \\ \vdots \\ F_N \end{bmatrix} = \begin{bmatrix} - & - & - & - & - & - & - & - \\ - & - & - & - & - & - & - & - \\ - & - & - & - & - & - & - & - \\ - & - & - & - & - & - & - & - \\ - & - & - & - & - & - & - & - \\ - & - & - & - & - & - & - & - \\ - & - & - & - & - & - & - & - \end{bmatrix} \begin{bmatrix} X_1 \\ X_2 \\ \vdots \\ X_i \\ X_{i+1} \\ \vdots \\ X_N \end{bmatrix}
\tag{8.40}
$$

where the blank matrix is known as the global stiffness matrix, K. Matrix K is assembled from local stiffness matrices as explained next, which is another essential feature of the FEM.

Local element stiffness matrix

The local stiffness matrix is derived by considering the internal restoring forces in a typical spring element between two consecutive nodes, i and $i + 1$, with DOF X_i and X_{i+1}, respectively, as shown in **Figure 8.10**.

Figure 8.10 Internal restoring forces treated positive to the right per sign convention.

The internal restoring forces are labeled f_i and f_{i+1} to distinguish them from the global forces. f_i and f_{i+1} are shown to the right per sign convention in the FEM. Applying Hooke's law at the right node yields

$$
f_{i+1} = k_i \left(X_{i+1} - X_i \right)
\tag{8.41}
$$

for $X_{i+1} > X_i$. From $f_{i+1} + f_i = 0$, $f_i = -f_{i+1} = -k_i \left(X_{i+1} - X_i \right)$ or

$$
f_i = k_i \left(X_i - X_{i+1} \right)
\tag{8.42}
$$

Equations (8.41) and (8.42) can be written in matrix form as

$$
\begin{bmatrix} f_i \\ f_{i+1} \end{bmatrix} = \begin{bmatrix} k_i & -k_i \\ -k_i & k_i \end{bmatrix} \begin{bmatrix} X_i \\ X_{i+1} \end{bmatrix}
\tag{8.43}
$$

where $\begin{bmatrix} k_i & -k_i \\ -k_i & k_i \end{bmatrix}$ defines a local element stiffness matrix. Equation (8.43) yields for the first element,

$$\begin{bmatrix} f_1 \\ f_2 \end{bmatrix} = \begin{bmatrix} k_1 & -k_1 \\ -k_1 & k_1 \end{bmatrix} \begin{bmatrix} X_1 \\ X_2 \end{bmatrix} \tag{8.44}$$

and for the second element,

$$\begin{bmatrix} f_2 \\ f_3 \end{bmatrix} = \begin{bmatrix} k_2 & -k_2 \\ -k_2 & k_2 \end{bmatrix} \begin{bmatrix} X_2 \\ X_3 \end{bmatrix} \tag{8.45}$$

and so on and so forth. The local stiffness matrix is the building block for the stiffness matrix K in the global matrix assembly, Eq. (8.40).

Global stiffness matrix assembly

Matrix K is assembled from the local stiffness matrices of all elements as follows. The local stiffness of the first element between nodes 1 and 2, given by Eq. (8.44), $\begin{bmatrix} k_1 & -k_1 \\ -k_1 & k_1 \end{bmatrix}$, is ordered on a lattice consisting of columns 1 and 2, and rows 1 and 2, as shown in **Figure 8.11a**.

$$\begin{bmatrix} k_1 & -k_1 & - & - & - \\ -k_1 & k_1 & - & - & - \\ | & | & & & \\ | & | & & & \\ | & | & & & \end{bmatrix} \quad \begin{bmatrix} & | & | & & \\ - & k_2 & -k_2 & - & - \\ - & -k_2 & k_2 & - & - \\ & | & | & & \\ & | & | & & \end{bmatrix} \quad \begin{bmatrix} & & | & | & \\ & & | & | & \\ - & - & k_3 & -k_3 & - \\ - & - & -k_3 & k_3 & - \\ & & | & | & \end{bmatrix}$$

(a) **(b)** **(c)**

Figure 8.11 Local stiffness matrices for elements between nodes: (a) 1 and 2; (b) 2 and 3; and (c) 3 and 4.

The local stiffness of the second element between nodes 2 and 3, given by Eq. (8.45), $\begin{bmatrix} k_2 & -k_2 \\ -k_2 & k_2 \end{bmatrix}$, is ordered on a lattice consisting of columns 2 and 3, and rows 2 and 3, as shown in **Figure 8.11b**. Next, the local stiffness of the third element between nodes 3 and 4 is ordered on a lattice consisting of columns 3 and 4, and rows 3 and 4, as shown in **Figure 8.11c**, and so on and so forth. The addition of these local stiffness matrices, element by element, results in the global symmetric stiffness matrix K as shown in Eq. (8.46). Note that the elements on the main diagonal are positive, whereas the symmetric elements on the subdiagonal and superdiagonal are negative.

$$\begin{bmatrix} F_1 \\ F_2 \\ \vdots \\ F_i \\ \\ \vdots \\ F_N \end{bmatrix} = \begin{bmatrix} k_1 & -k_1 & 0 & 0 & 0 & 0 & 0 \\ -k_1 & k_1+k_2 & -k_2 & 0 & 0 & 0 & 0 \\ 0 & -k_2 & k_2+k_3 & -k_3 & 0 & 0 & 0 \\ 0 & 0 & -k_{i-1} & k_{i-1}+k_i & -k_i & 0 & 0 \\ 0 & 0 & 0 & -k_i & k_i+k_{i+1} & -k_{i+1} & 0 \\ 0 & 0 & 0 & 0 & -k_{N-2} & k_{N-2}+k_{N-1} & -k_{N-1} \\ 0 & 0 & 0 & 0 & 0 & -k_{N-1} & k_{N-1} \end{bmatrix} \begin{bmatrix} X_1 \\ X_2 \\ \vdots \\ X_i \\ X_{i+1} \\ \vdots \\ X_N \end{bmatrix} \quad (8.46)$$

Equation (8.46) is underdetermined; that is, there are more unknowns for the DOF and global forces than equations. Thus, in a given system certain displacements and/or forces are specified as part of the structure design so that Eq. (8.46) is determined.

Example 8.6

Consider a spring with stiffness $k = 1000\dfrac{N}{m}$. The spring is fixed at the left end. On the other end a 10-N force is applied to the right (tension), as shown in **Figure 8.12a**.

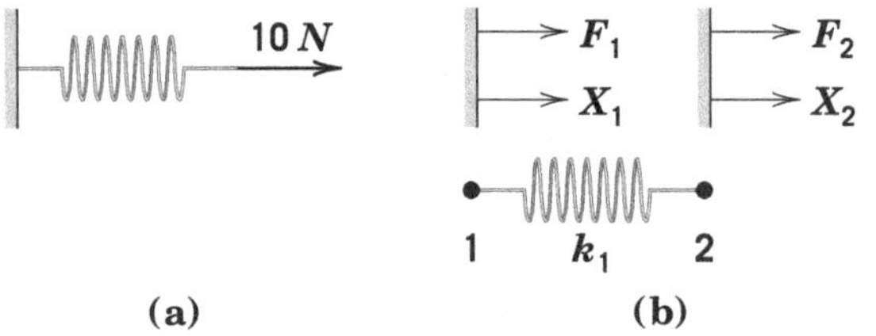

(a)   (b)

Figure 8.12 Spring-loaded structure: (a) one-element spring under tension; and (b) FEM model.

1. Discretize the structure as a FEM model.
2. Determine the displacement of the right end of the spring and the reaction force at the fixed end.

Solution:

1. Discretize the structure as a FEM model.

In FEM, the spring is modeled as one element, as shown in **Figure 8.12b**. Applying Eq. (8.46) to the element yields

$$\begin{bmatrix} F_1 \\ F_2 \end{bmatrix} = \begin{bmatrix} k_1 & -k_1 \\ -k_1 & k_1 \end{bmatrix} \begin{bmatrix} X_1 \\ X_2 \end{bmatrix} \quad (8.47)$$

2. Determine the displacement of the right end of the spring and the reaction force at the fixed end.

The spring is fixed at the left end; thus, $X_1 = 0$ so that F_1 is a reaction force. On the other end of the spring, a 10-N force is applied to the right; thus $F_2 = 10N$. Substituting $X_1 = 0$, $F_2 = 10N$, and the given stiffness $k_1 = 1000$ in Eq. (8.47) yields

$$\begin{bmatrix} F_1 \\ 10 \end{bmatrix} = \begin{bmatrix} 1000 & -1000 \\ -1000 & 1000 \end{bmatrix} \begin{bmatrix} 0 \\ X_2 \end{bmatrix} \tag{8.48}$$

Reading the second row in Eq. (8.48) as $10 = 1000X_2$ yields the displacement $X_2 = 0.01\, m$.

Reading the first row in Eq. (8.48) yields $F_1 = -1000X_2 = -1000(0.01) = -10N$. Thus, the reaction force acts to the left. The results are consistent with the sign convention that forces are positive to the right; that is, Newton's first law, $\sum F = 0$, or $F_2 - |F_1| = 0$, is satisfied as shown in **Figure 8.13**.

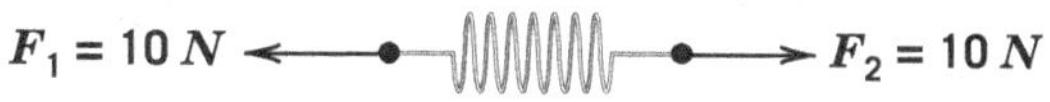

Figure 8.13 Free body diagram of the spring.

Example 8.7

Consider a three-element spring system, as shown in **Figure 8.14**, where a 10-N force is applied in compression at the right end. The left end is fixed. The stiffnesses are: $k_1 = 1000\dfrac{N}{m}$, $k_2 = 2000\dfrac{N}{m}$, and $k_3 = 3000\dfrac{N}{m}$.

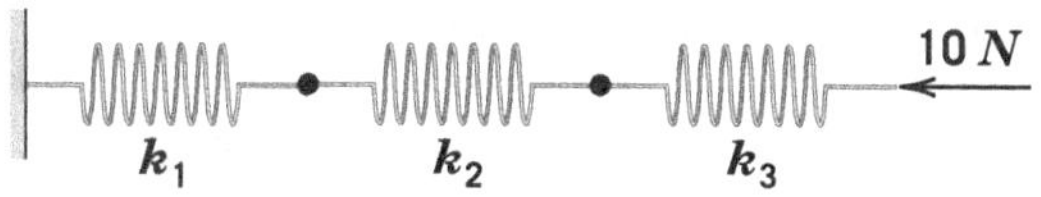

Figure 8.14 A three-element spring system in compression.

1. Discretize the structure as a FEM model using four nodes.
2. Determine the unknown displacements and forces.

Solution:

1. Discretize the structure as a FEM model using four nodes.

In FEM, the spring system is modeled with four nodes, as shown in **Figure 8.15**.

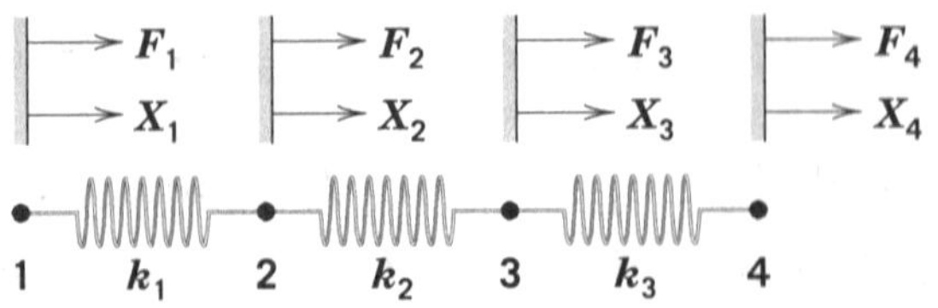

Figure 8.15 A three-element spring system modeled with four nodes.

Applying Eq. (8.46) to the spring system yields the global matrix equation

$$\begin{bmatrix} F_1 \\ F_2 \\ F_3 \\ F_4 \end{bmatrix} = \begin{bmatrix} k_1 & -k_1 & 0 & 0 \\ -k_1 & k_1+k_2 & -k_2 & 0 \\ 0 & -k_2 & k_2+k_3 & -k_3 \\ 0 & 0 & -k_3 & k_3 \end{bmatrix} \begin{bmatrix} X_1 \\ X_2 \\ X_3 \\ X_4 \end{bmatrix} \qquad (8.49)$$

2. Determine the unknown displacements and forces.

In Eq. (8.49), set $X_1 = 0$ for a fixed end and $F_4 = -10$ for a force in compression, which is opposite to our sign convention. Furthermore, there are no external forces at nodes 2 and 3; thus set $F_2 = F_3 = 0$. Substituting the foregoing information and given stiffnesses in Eq. (8.49) yields

$$\begin{bmatrix} F_1 \\ 0 \\ 0 \\ -10 \end{bmatrix} = \begin{bmatrix} 1000 & -1000 & 0 & 0 \\ -1000 & 3000 & -2000 & 0 \\ 0 & -2000 & 5000 & -3000 \\ 0 & 0 & -3000 & 3000 \end{bmatrix} \begin{bmatrix} 0 \\ X_2 \\ X_3 \\ X_4 \end{bmatrix} \qquad (8.50)$$

If you read the second, third, and fourth rows as three equations with three unknowns, you can write

$$\begin{bmatrix} 0 \\ 0 \\ -10 \end{bmatrix} = \begin{bmatrix} 3000 & -2000 & 0 \\ -2000 & 5000 & -3000 \\ 0 & -3000 & 3000 \end{bmatrix} \begin{bmatrix} X_2 \\ X_3 \\ X_4 \end{bmatrix}$$

which is solved for displacements X_2, X_3, and X_4 by matrix inversion in CAS as

$$\begin{bmatrix} 3000 & -2000 & 0 \\ -2000 & 5000 & -3000 \\ 0 & -3000 & 3000 \end{bmatrix}^{-1} \begin{bmatrix} 0 \\ 0 \\ -10 \end{bmatrix} = \begin{bmatrix} X_2 \\ X_3 \\ X_4 \end{bmatrix}$$

The results are $X_2 = -0.01m$, $X_3 = -0.015m$, and $X_4 = -0.018m$. These negative DOFs indicate that the spring system is displaced to the left as a result of the compressive force at node 4.

Next, substitute the found displacements in Eq. (8.50), and solve for the reaction force F_1 from the first row. The result is $F_1 = 10N$. According to our sign convention, F_1 acts to the right, which is consistent with Newton's first law for the overall system in static equilibrium, $\sum F = 0$.

Example 8.8 Spring Latch Model by FEM

Consider the design of a new spring latch for a cabinet drawer, as shown in **Figure 8.16a**. The mechanism consists of a handle; catch; thumb latch; and two springs, which are fixed to solid supports at their outer ends. To open the drawer, a force is exerted with the thumb on the latch and to the right, which causes the left spring to extend and the right spring to compress, as shown by the top view in **Figure 8.16b**. The spring model is shown in **Figure 8.16c**, where F_t represents the thumb force exerted on the latch.

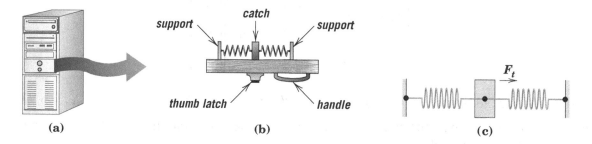

Figure 8.16 New cabinet latch design: (a) cabinet system; (b) thumb latch with catch supported between two springs; and (c) spring model of the latch.

The design requires the catch to displace at least 0.003 m. The stiffness of each spring is $350\frac{N}{m}$

1. Discretize the catch, springs, and supports as a FEM model.
2. Determine the displacement of the latch and the reaction forces at the fixed supports as a result of 4-N load exerted by the thumb. Will the catch clear the catch holder?

Solution:

1. Discretize the catch, springs, and supports as a FEM model.

By our FEM model, the spring system in **Figure 8.16c** is modeled as a three-node spring system, as shown in **Figure 8.17**.

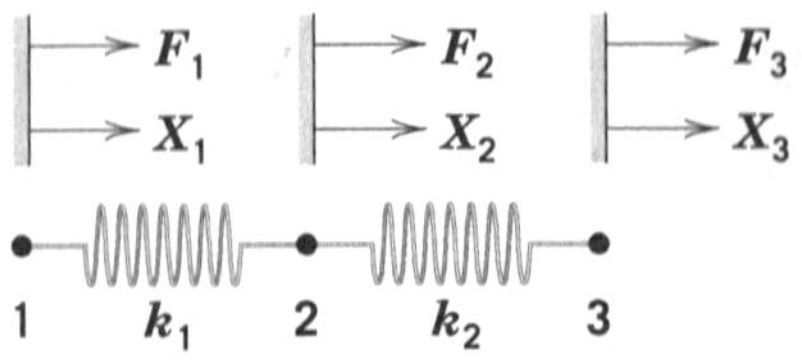

Figure 8.17 FEM model of the latch system.

Applying Eq. (8.46) to the spring system yields the global matrix equation

$$\begin{bmatrix} F_1 \\ F_2 \\ F_3 \end{bmatrix} = \begin{bmatrix} k_1 & -k_1 & 0 \\ -k_1 & k_1+k_2 & -k_2 \\ 0 & -k_2 & k_2 \end{bmatrix} \begin{bmatrix} X_1 \\ X_2 \\ X_3 \end{bmatrix} \qquad (8.51)$$

2. Determine the displacement of the latch and the reaction forces at the fixed supports as a result of a 4-N load exerted by the thumb. Will the catch clear the catch holder?

In this problem, $F_2 = F_t = 4N$. The displacements $X_1 = X_3 = 0$. Substituting the foregoing information and given stiffness in Eq. (8.51) yields

$$\begin{bmatrix} F_1 \\ 4 \\ F_3 \end{bmatrix} = \begin{bmatrix} 350 & -350 & 0 \\ -350 & 700 & -350 \\ 0 & -350 & 350 \end{bmatrix} \begin{bmatrix} 0 \\ X_2 \\ 0 \end{bmatrix} \qquad (8.52)$$

Reading the second row in Eq. (8.52) yields $4 = 700X_2$, so that the displacement $X_2 = 5.71mm$. Therefore, the catch clears the catch holder, which satisfies the 3 *mm* minimum displacement required in this design problem.

Substituting the value of X_2 into Eq. (8.52) and solving for the reaction forces yield, $F_1 = -2N$ and $F_3 = -2N$. Per our sign convention, the actual directions of the forces are shown in **Figure 8.18**, so that Newton's first law, $\sum F = 0$, or $4N - 2N - 2N = 0$, is satisfied.

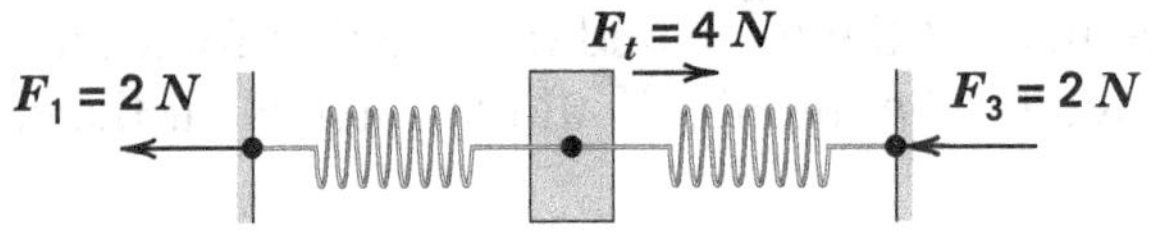

Figure 8.18 Actual force direction in the spring latch system.

If you think about the results intuitively, when the latch is pulled to the right, the left spring is stretched, which causes the reaction force F_1 to the left. Conversely, the right spring is compressed, which causes the reaction force F_3 to the left.

Next, you will learn how to apply the FEM to solve a heat transfer problem.

FEM for conduction heat transfer

The FEM is useful for solving heat transfer problems to predict heat rates and temperature distribution in a system. For simplicity, the focus is on steady-state systems with one mode of heat transfer only, namely, conduction; see Section 3.4, Chapter 3, and Eq. (3.25). Furthermore, the analysis is limited to 1D, for example, along the x axis, so that the fundamental equation is given by Fourier's law of conduction,

$$\dot{Q} = -kA \frac{dT}{dx} \qquad (8.53)$$

where $\dot{Q}$ is the rate of heat transfer in *Watts* (W), k is the thermal conductivity coefficient in $\frac{W}{m\,^{o}C}$ which depends on the material, A is the surface area normal to the direction of heat transfer in m^2, and T is temperature in ^{o}C. Equation (8.53) can be stated per unit area as

$$Q'' = -k \frac{dT}{dx} \qquad (8.54)$$

where $Q'' = \frac{\dot{Q}}{A}$ in $\frac{W}{m^2}$ is defined as the heat flux. The double prime is a reminder of flux.

As an example, consider a double-pane window, as shown in **Figure 8.19a**. The window has two large, thin glass panes separated by dead air space, as shown in the cross section in **Figure 8.19b**. Consider the case also in winter, where the glass interior surface temperature is at $20^{o}C$, and the glass exterior surface temperature is at $-10^{o}C$ so that heat transfer is from the room to the surrounding. The heat is conduction between nodes 1 and 2 in glass, nodes 2 and 3 in dead air space, and nodes 3 and 4 in glass, which causes a temperature distribution between nodes 1 and 4. This is an example of 1D heat conduction where the temperature distribution across the window will be uniform at any point regardless of the vertical position.

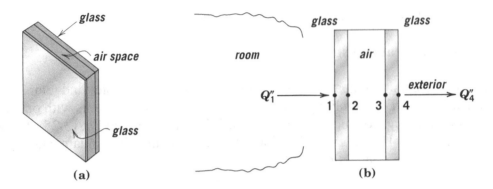

(a)   **(b)**

Figure 8.19 Conduction heat transfer: (a) double-pane window; and (b) window pane section.

Another example of 1D conduction heat transfer is the rear window of a car with a defroster, as shown in **Figure 8.20**. Here, a 1D model of the glass window includes a source term, Q_S'', due to the heater. This source term could be, for instance, the product of voltage and current per unit area.

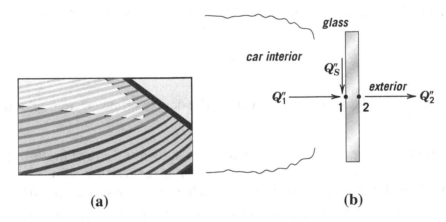

(a)   **(b)**

Figure 8.20 Rear window of a car with a defroster: (a) actual system; and (b) 1D model of glass.

In the finite element formulation, a system consisting of several components connected in series is modeled as conduction elements between nodal temperatures, as shown in **Figure 8.21**. For this element, the DOFs are the temperatures, T_1, T_2, and so on. The following restrictions apply at this introductory level:

- The temperature distribution is linear over each element.

- For N nodes, the heat fluxes at the boundaries, namely, Q_1'' at T_1 and Q_N'' at T_N, are conduction. Furthermore, Q_1'' and Q_N'' are nonzero whenever Dirichlet BCs are used at T_1 and T_N. However, $Q_1'' = 0$ if a Neumann BC is applied at node 1 as $\dfrac{dT(0)}{dx} = 0$ by virtue of Eq. (8.54). Similarly $Q_N'' = 0$ if the BC at node N is $\dfrac{dT(N)}{dx} = 0$.

- For intermediate nodes, $2 \leq i \leq N-1$, the heat flux, Q_i'' is an external source, for example, a heating element. When there is no external source, the corresponding heat flux is set to zero.
- By convention all heat fluxes are positive when they point toward the node, as shown in **Figure 8.21**.

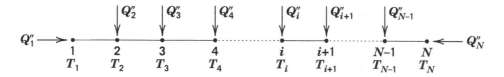

Figure 8.21 1D FEM conduction model.

For a linear temperature distribution model over the ith element of length $\Delta x = x_{i+1} - x_i = L_i$, between two nodes i and $i+1$, Eq. (8.54) yields a heat flux $Q_i'' = -k_i \dfrac{T_{i+1} - T_i}{x_{i+1} - x_i} = \dfrac{k_i}{L_i}\left(T_i - T_{i+1}\right)$ so that

the element conduction property is $\dfrac{k_i}{L_i}$. As in the force-displacement FEM, the relationship

between the global fluxes, Q_1'', Q_2'', …, Q_N'' and DOF, T_1, T_2, …, T_N is a global matrix formulated as $Q'' = KT$, where K is the conduction matrix. It can be shown by balancing energy on a local element in conjunction with Fourier's law of conduction that

$$
\begin{bmatrix} Q_1'' \\ Q_2'' \\ \vdots \\ Q_i'' \\ \vdots \\ Q_N'' \end{bmatrix}
=
\begin{bmatrix}
\dfrac{k_1}{L_1} & -\dfrac{k_1}{L_1} & - & - & - & - & - \\[2mm]
-\dfrac{k_1}{L_1} & \dfrac{k_1}{L_1}+\dfrac{k_2}{L_2} & -\dfrac{k_2}{L_2} & - & - & - & - \\[2mm]
- & -\dfrac{k_2}{L_2} & \dfrac{k_2}{L_2}+\dfrac{k_3}{L_3} & -\dfrac{k_3}{L_3} & - & - & - \\[2mm]
- & - & -\dfrac{k_{i-1}}{L_{i-1}} & \dfrac{k_{i-1}}{L_{i-1}}+\dfrac{k_i}{L_i} & -\dfrac{k_i}{L_i} & - & - \\[2mm]
- & - & - & -\dfrac{k_i}{L_i} & \dfrac{k_i}{L_i}+\dfrac{k_{i+1}}{L_{i+1}} & -\dfrac{k_{i+1}}{L_{i+1}} & - \\[2mm]
- & - & - & - & -\dfrac{k_{N-2}}{L_{N-2}} & \dfrac{k_{N-2}}{L_{N-2}}+\dfrac{k_{N-1}}{L_{N-1}} & -\dfrac{k_{N-1}}{L_{N-1}} \\[2mm]
- & - & - & - & - & -\dfrac{k_{N-1}}{L_{N-1}} & \dfrac{k_{N-1}}{L_{N-1}}
\end{bmatrix}
\begin{bmatrix} T_1 \\ T_2 \\ \vdots \\ T_i \\ \vdots \\ T_N \end{bmatrix}
\quad (8.55)
$$

which is analogous to the force-displacement formulation, Eq. (8.46), where the stiffness k is replaced with $\dfrac{k}{L}$. As in force-displacement formulation, specific DOF and/or fluxes are given so that Eq. (8.55) is a determined system of equations.

Example 8.9 Double-Pane Window by FEM

Consider the double-pane window model, as shown in **Figure 8.19**. Both glass panes are 4 *mm* thick, and the air space is 10 *mm* thick. The glass and air space of the window have thermal conductivity coefficients of $0.8\dfrac{W}{m^oC}$ and $0.3\dfrac{W}{m^oC}$, respectively. The exterior and interior surface temperatures of the window are -10^oC and 20^oC, respectively.

1. Discretize the window as a FEM conduction model.
2. Formulate the global conduction matrix.
3. Determine the temperatures at the glass and air space interface.
4. Determine the heat fluxes through the window and their directions.

Solution:

1. Discretize the window as a FEM conduction model.

The double-pane window model is shown in **Figure 8.22**, which conforms to the 1D model given in **Figure 8.21**, where all heat fluxes are shown into the nodes.

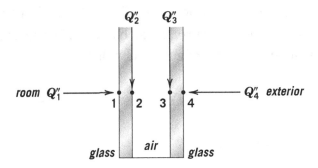

Figure 8.22 Double-pane window FEM discretization.

2. Formulate the global conduction matrix.

For the glass between nodes 1 and 2, $\dfrac{k_1}{L_1} = \dfrac{0.8}{0.004} = 200\dfrac{W}{m^2\,^oC}$. For the air space between nodes 2 and 3, $\dfrac{k_2}{L_2} = \dfrac{0.3}{0.01} = 30\dfrac{W}{m^2\,^oC}$, and for the glass between nodes 3 and 4, $\dfrac{k_3}{L_3} = \dfrac{0.8}{0.004} = 200\dfrac{W}{m^2\,^oC}$.

The global conduction matrix is given by Eq. (8.55) as

$$
\begin{bmatrix} Q_1'' \\ Q_2'' \\ Q_3'' \\ Q_4'' \end{bmatrix} = \begin{bmatrix} \dfrac{k_1}{L_1} & -\dfrac{k_1}{L_1} & 0 & 0 \\ -\dfrac{k_1}{L_1} & \dfrac{k_1}{L_1}+\dfrac{k_2}{L_2} & -\dfrac{k_2}{L_2} & 0 \\ 0 & -\dfrac{k_2}{L_2} & \dfrac{k_2}{L_2}+\dfrac{k_3}{L_3} & -\dfrac{k_3}{L_3} \\ 0 & 0 & -\dfrac{k_3}{L_3} & \dfrac{k_3}{L_3} \end{bmatrix} \begin{bmatrix} T_1 \\ T_2 \\ T_3 \\ T_4 \end{bmatrix}
\tag{8.56}
$$

There are no external sources such as heaters. Therefore, $Q_2'' = Q_3'' = 0$. Substituting the given temperatures and the $\dfrac{k}{L}$ values in Eq. (8.56) yields

$$
\begin{bmatrix} Q_1'' \\ 0 \\ 0 \\ Q_4'' \end{bmatrix} = \begin{bmatrix} 200 & -200 & 0 & 0 \\ -200 & 230 & -30 & 0 \\ 0 & -30 & 230 & -200 \\ 0 & 0 & -200 & 200 \end{bmatrix} \begin{bmatrix} 20 \\ T_2 \\ T_3 \\ -10 \end{bmatrix}
\tag{8.57}
$$

3. Determine the temperatures at the glass and air space interface.

Reading the second and third rows in Eq. (8.57) yields $\begin{aligned} 0 &= -4000 + 230T_2 - 30T_3 \\ 0 &= -30T_2 + 230T_3 + 2000 \end{aligned}$, or in matrix form

$$
\begin{bmatrix} 4000 \\ -2000 \end{bmatrix} = \begin{bmatrix} 230 & -30 \\ -30 & 230 \end{bmatrix} \begin{bmatrix} T_2 \\ T_3 \end{bmatrix} \quad \text{so that} \quad \begin{bmatrix} 230 & -30 \\ -30 & 230 \end{bmatrix}^{-1} \begin{bmatrix} 4000 \\ -2000 \end{bmatrix} = \begin{bmatrix} T_2 \\ T_3 \end{bmatrix}.
$$

The result is $T_2 = 16.54\ {}^{o}C$ and $T_3 = -6.54\ {}^{o}C$. The double-pane window temperature distribution is shown in **Figure 8.23** from inside toward the exterior, which varies as a piecewise linear function based on this FEM.

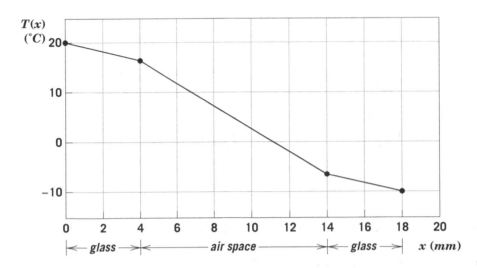

Figure 8.23 Double-pane window piecewise linear temperature distribution from the inside toward the exterior.

4. Determine the heat fluxes through the window and their directions.

The first row in Eq. (8.57) reads $Q_1'' = 4000 - 200(16.54) = 692.4 \frac{W}{m^2}$, whereas the last row reads

$Q_4'' = -200(-6.54) - 2000 = -692.4 \frac{W}{m^2}$. By our sign convention, the heat flux Q_1'' is correctly pointing into node 1, which means that heat is from the room to glass. However, $Q_4'' < 0$ is opposite to our sign convention, which means that the heat flux is from node 4 toward the exterior, as shown in **Figure 8.19b**. This is consistent with the physics of the problem; that is, one would expect the heat flux to be from high temperature to low temperature. Notice also that the magnitudes of both fluxes match as required by conservation of energy. In this engineering example, you learned how to apply the FEM to a heat conduction problem.

FEM for electrical resistor circuits

The FEM can be applied in electrical circuits to solve current and voltage distributions in a system. For simplicity, the focus is on resistive circuits. As an example, consider the Wheatstone Bridge powered by a battery with voltage E, as shown in **Figure 8.24a**. The circuit consists of one variable resistor R_1, as shown by the arrow across the resistor, and fixed resistors, R_2, R_3, R_4 and R_5. The voltage difference $V_4 - V_2$ across R_5 is defined as the bridge voltage output, written as V_b for simplicity. V_b varies with R_1. An application of such a circuit is in the display of automobile engine temperature by employing R_1 as a resistance temperature detector, as shown in **Figure 8.24b**. The display of temperature is via V_b on the dashboard, which is calibrated in units of temperature, for example, $V_b = 0\,volt$ when the engine is at rest and $V_b = 1\,volt$ when the engine reaches 140^oC. Thus, in the design of this circuit, it is necessary to determine the range of R_1 of the thermometer resistor, which will satisfy the bridge voltage output over the range $0 \le V_b \le 1$. For this purpose, the circuit is analyzed in order to develop a relationship between R_1 and V_b. This relationship is established by the FEM.

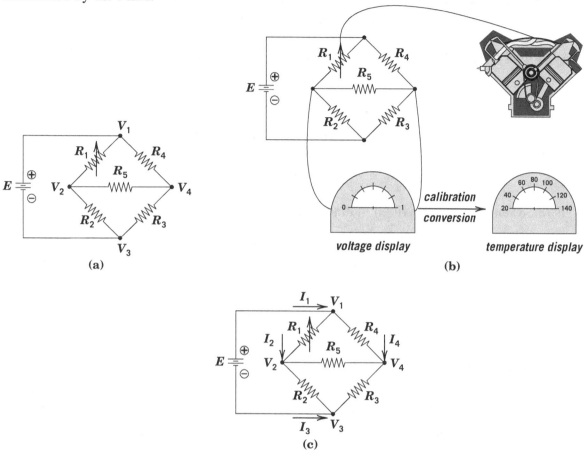

Figure 8.24 Wheatstone Bridge: (a) basic circuit; (b) circuit employed as a temperature measuring device; and (c) FEM discretization.

In the finite element formulation, the DOFs are the voltages at the circuit nodes. In the case of the Wheatstone Bridge, there are four DOFs, namely, V_1, V_2, V_3, and V_4, as shown in **Figure 8.24c**. At each DOF, global currents are introduced pointing into each node, as indicated by I_1, I_2, I_3, and I_4. By convention in the FEM, currents pointing into nodes are positive. Furthermore, currents emanating from a source, such as a battery, are nonzero, in this case, I_1 and I_3. Currents not emanating from a source are set to zero, in this case I_2 and I_4.

Local element conductance matrix

The resistor local element conductance matrix is developed for nodes, V_n and V_m, as shown in **Figure 8.25**, with two local currents, i_n and i_m pointing toward the nodes. The current direction is consistent with the sign convention of global currents. The element resistance is R.

Figure 8.25 Resistor element between two local nodes V_n and V_m, with currents into the nodes consistent with sign convention.

A positive current flows from high to low voltage; thus, from Ohm's law, $V_n - V_m = i_n R$, or

$$i_n = (V_n - V_m)G \tag{8.58}$$

where $G = \dfrac{1}{R}$ is defined as the conductance. The unit of G is *mho*. Similarly,

$$i_m = (V_m - V_n)G \tag{8.59}$$

Equations (8.58) and (8.59) are written in matrix form as

$$\begin{bmatrix} i_n \\ i_m \end{bmatrix} = \begin{bmatrix} G & -G \\ -G & G \end{bmatrix} \begin{bmatrix} V_n \\ V_m \end{bmatrix} \tag{8.60}$$

where $\begin{bmatrix} G & -G \\ -G & G \end{bmatrix}$ is defined as the local element conductance matrix.

Global matrix assembly

As in the force-displacement FEM, the relationship between global currents I_1, I_2, ..., I_N and DOF V_1, V_2, ..., V_N is a global matrix formulated as $I = KV$, where K is the conductance matrix. For the Wheatstone Bridge,

$$\begin{bmatrix} I_1 \\ I_2 \\ I_3 \\ I_4 \end{bmatrix} = \begin{bmatrix} - & - & - & - \\ - & - & - & - \\ - & - & - & - \\ - & - & - & - \end{bmatrix} \begin{bmatrix} V_1 \\ V_2 \\ V_3 \\ V_4 \end{bmatrix} \tag{8.61}$$

where the 4 x 4 K matrix is assembled from local element conductance matrices, as we did in the force-displacement matrix. However, unlike the force-displacement DOFs that are ordered sequentially from left to right, the ordering of local conductance matrices in K should be carried out hand in hand with the schematic under consideration, in this case, **Figure 8.24c**. For five elements, we expect five local conductance matrices. For element R_1 between nodes V_1 and V_2, the local conductance matrix is $\begin{bmatrix} G_1 & -G_1 \\ -G_1 & G_1 \end{bmatrix}$ that is ordered in K on a lattice consisting of columns 1 and 2, and rows 1 and 2, as shown in **Figure 8.26a**.

$$
\begin{bmatrix} G_1 & -G_1 & | & | \\ -G_1 & G_1 & | & | \\ | & | & & \\ | & | & & \end{bmatrix}
\quad
\begin{bmatrix} & | & | & \\ - & G_2 & -G_2 & - \\ - & -G_2 & G_2 & - \\ & | & | & \end{bmatrix}
\quad
\begin{bmatrix} & & | & | \\ & & | & | \\ - & - & G_3 & -G_3 \\ - & - & -G_3 & G_3 \end{bmatrix}
\quad
\begin{bmatrix} G_4 & - & - & -G_4 \\ | & & & | \\ | & & & | \\ -G_4 & - & - & G_4 \end{bmatrix}
$$

$$\textbf{(a)} \qquad\qquad \textbf{(b)} \qquad\qquad \textbf{(c)} \qquad\qquad \textbf{(d)}$$

$$
\begin{bmatrix} & | & & | \\ - & G_5 & - & -G_5 \\ & | & & | \\ - & -G_5 & - & G_5 \end{bmatrix}
$$

$$\textbf{(e)}$$

Figure 8.26 Local conductance matrices for the Wheatstone Bridge with contributions from resistors: (a) R_1 between nodes 1 and 2; (b) R_2 between nodes 2 and 3; (c) R_3 between nodes 3 and 4; (d) R_4 between nodes 1 and 4; and (e) R_5 between nodes 2 and 4.

Element R_2 is between V_2 and V_3. Thus, its local matrix is ordered on a lattice consisting of columns 2 and 3, and rows 2 and 3, as shown in **Figure 8.26b**. The same applies to element R_3 between V_3 and V_4, as shown in **Figure 8.26c**. Element R_4 is between V_1 and V_4. Thus, its local matrix is ordered on a lattice consisting of columns 1 and 4, and rows 1 and 4, as shown in **Figure 8.26d**. Finally, Element R_5 is between V_2 and V_4. Thus, its local matrix is ordered on a lattice consisting of columns 2 and 4 and rows 2 and 4, as shown in **Figure 8.26e**. Adding the five matrices in **Figure 8.26**, element by element, yields

$$
\begin{bmatrix} I_1 \\ I_2 \\ I_3 \\ I_4 \end{bmatrix} =
\begin{bmatrix}
G_1+G_4 & -G_1 & 0 & -G_4 \\
-G_1 & G_1+G_2+G_5 & -G_2 & -G_5 \\
0 & -G_2 & G_2+G_3 & -G_3 \\
-G_4 & -G_5 & -G_3 & G_3+G_4+G_5
\end{bmatrix}
\begin{bmatrix} V_1 \\ V_2 \\ V_3 \\ V_4 \end{bmatrix}
\tag{8.62}
$$

Example 8.10 Wheatstone Bridge for an Automobile Temperature Circuit

Consider the Wheatstone Bridge circuit for the automobile engine temperature, as shown in **Figure 8.24**, where $R_2 = R_3 = R_4 = 1\,ohm$ and $R_5 = 10\,ohms$. The battery voltage is $E = 12V$.

1. Apply Eq. (8.62) and find the range of R_1 for the resistance temperature detector, so that $0 \le V_b \le 1$, where $V_b = V_4 - V_2$.

Solution:

For $R_2 = R_3 = R_4 = 1\,ohm$, $G_2 = G_3 = G_4 = 1\,mho$. For $R_5 = 10\,ohms$, $G_5 = 0.1\,mho$. Inserting the G values in Eq. (8.62) yields

$$\begin{bmatrix} I_1 \\ I_2 \\ I_3 \\ I_4 \end{bmatrix} = \begin{bmatrix} G_1+1 & -G_1 & 0 & -1 \\ -G_1 & G_1+1.1 & -1 & -0.1 \\ 0 & -1 & 2 & -1 \\ -1 & -0.1 & -1 & 2.1 \end{bmatrix} \begin{bmatrix} V_1 \\ V_2 \\ V_3 \\ V_4 \end{bmatrix} \tag{8.63}$$

In this circuit, node V_1 is the same as the positive terminal of the battery; thus, $V_1 = E = 12$. On the other hand, node V_3 is the same as the negative terminal of the battery or ground; thus, $V_3 = 0$. Furthermore, $I_2 = I_4 = 0$ because their nodes are not connected directly to a source. Substituting the foregoing values in Eq. (8.63) yields

$$\begin{bmatrix} I_1 \\ 0 \\ I_3 \\ 0 \end{bmatrix} = \begin{bmatrix} G_1+1 & -G_1 & 0 & -1 \\ -G_1 & G_1+1.1 & -1 & -0.1 \\ 0 & -1 & 2 & -1 \\ -1 & -0.1 & -1 & 2.1 \end{bmatrix} \begin{bmatrix} 12 \\ V_2 \\ 0 \\ V_4 \end{bmatrix} \tag{8.64}$$

Reading the second and fourth rows in Eq. (8.64) yields $\begin{bmatrix} 12G_1 \\ 12 \end{bmatrix} = \begin{bmatrix} G_1+1.1 & -0.1 \\ -0.1 & 2.1 \end{bmatrix} \begin{bmatrix} V_2 \\ V_4 \end{bmatrix}$ so that

$$\begin{bmatrix} G_1+1.1 & -0.1 \\ -0.1 & 2.1 \end{bmatrix}^{-1} \begin{bmatrix} 12G_1 \\ 12 \end{bmatrix} = \begin{bmatrix} V_2 \\ V_4 \end{bmatrix} \tag{8.65}$$

Using CAS,

$$V_2 = 12 \frac{21G_1+1}{21G_1+23} \tag{8.66}$$

and

$$V_4 = 132 \frac{G_1+1}{21G_1+23} \tag{8.67}$$

From these expressions, derive the bridge voltage

$$V_b = V_4 - V_2 = 120 \frac{1-G_1}{21G_1+23} \tag{8.68}$$

When $V_b = 0$, Eq. (8.68) yields $G_1 = 1 mho$ so that $R_1 = \dfrac{1}{G_1} = 1 ohm$. When $V_b = 1 volt$, Eq. (8.68) yields $G_1 = 0.688 mho$ so that $R_1 = 1.45 ohms$. Thus, for $0 \le V_b \le 1$, the resistance of the temperature detector should be in the range $1 \le R_1 \le 1.45 ohms$.

Notice that when $G_1 = 1mho$, Eqs. (8.66) and (8.67) yield $V_2 = V_4 = 6volts$ so that the first and third rows in Eq. (8.64) yield $I_1 = 24 - V_2 - V_4 = 12$, and $I_3 = -V_2 - V_4 = -12$. $I_1 > 0$ implies that the sign of I_1 is correct; that is, current I_1 enters node 1 from the battery. $I_3 < 0$ implies that the sign of I_3 is incorrect; that is, current I_3 leaves node 3 into the negative terminal of battery, which is consistent with the physics of the battery, and current flows through the battery in one direction with the same magnitude, $I_1 = |I_3| = 12amps$. You also can show that the same conclusion applies to the case when $G_1 = 0.688mho$, which yields $V_2 = 4.95volts$, $V_4 = 5.95volts$, $I_1 = 10.9amps$, and $I_3 = -10.9amps$.

In this engineering example, you learned how to apply the FEM to a fairly complex resistive circuit.

■

HOMEWORK PROBLEMS FOR CHAPTER EIGHT

Problems for Section 8.2

1. Revisit Example 8.2: Tank Flow Governed by a Linear ODE.

   a. Take $\Delta t = 0.1\tau$ and follow up on the example to calculate manually h_3 at t_3 and h_4 at t_4 by the forward Euler method, backward Euler method, and improved Euler method.
   b. Compare all methods with the exact solution using absolute percent error in a table similar to **Table 8.1**.

2. Revisit Example 8.3: Tank Flow Governed by a Nonlinear ODE.

   a. Take $\Delta t = 6s$ and follow up on the example to find h_2 at t_2 by the backward Euler method in two ways:
      1. Develop an expression $r(h_2)$ similar to Eq. (8.10), plot $r(h_2)$ versus h_2, and estimate h_2 to one significant digit.
      2. Use a root-finding function in CAS to find h_2 to double precision.
   b. Find the exact value of h_2 by rearranging your discrete equation to a quadratic equation. See the paragraph following **Figure 8.4**.
   c. Compare the exact value of h_2 with that found by a root-finding function in CAS.

3. Revisit Example 8.3: Tank Flow Governed by a Nonlinear ODE. Take $\Delta t = 6s$ and calculate manually h_1 at t_1 and h_2 at t_2 by the improved Euler method.

4. Revisit Example 8.2: Tank Flow Governed by a Linear ODE.

   a. Take $\Delta t = 0.1\tau$ and calculate manually h_1 at t_1 and h_2 at t_2 by the RK4 method.
   b. Compare your results with the exact solution using absolute percent error.

5. Revisit Example 8.3: Tank Flow Governed by a Nonlinear ODE. Take $\Delta t = 6s$ and calculate manually h_1 at t_1, and h_2 at t_2 by the RK4 method.

6. Revisit Example 8.2: Tank Flow Governed by a Linear ODE. Take $\Delta t = 0.1\tau$. Find $h(t)$ numerically by the forward Euler method over $0 \le t \le 400s$. You should be able to develop the algorithm in a spreadsheet.

7. Revisit Example 8.2: Tank Flow Governed by a Linear ODE. Take $\Delta t = 0.1\tau$. Find $h(t)$ numerically by the improved Euler method over $0 \le t \le 400s$. You should be able to develop the algorithm in a spreadsheet.

8. Revisit Example 8.2: Tank Flow Governed by a Linear ODE. Take $\Delta t = 0.1\tau$. Find $h(t)$ numerically by the RK4 method over $0 \le t \le 400s$. You should be able to develop the algorithm in a spreadsheet.

9. Revisit Example 8.3: Tank Flow Governed by a Nonlinear ODE. Take $\Delta t = 1s$.

a. Experiment with different time ranges to find the steady state of the fluid level by solving $h(t)$ numerically by the improved Euler method. You should be able to develop the algorithm in a spreadsheet.

b. Report the time in s and the value of h in m at steady state.

For Problems 10–11, consider the RC low-pass filter, as shown in **Figure 8.27**, which is governed by $\tau \dfrac{dV_o}{dt} + V_o = V(t)$, where $\tau = RC = 0.05s$, $V_o(t)$ is the output voltage across the resistor, or response, and $V(t) = 2\cos(0.5\pi t)$ is the input voltage to the circuit, or forcing function. The low-pass filter should output a response almost equal to the forcing function, except for a phase shift; see Example 1.10: The Low-Pass Filter, Chapter 1. The IC is $V_o(0) = 0$.

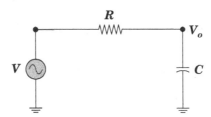

Figure 8.27 RL circuit, a low-pass filter.

10.
a. Take $\Delta t = 0.1\tau$ and find $V_o(t)$ numerically by the improved Euler method over $0 \le t \le 4s$.
b. Compare $V_o(t)$ with $V(t)$ graphically over $0 \le t \le 4s$ to validate that the circuit is a low-pass filter.

11.
a. Take $\Delta t = 0.1\tau$ and find $V_o(t)$ numerically by the RK4 method over $0 \le t \le 4s$.
b. Compare $V_o(t)$ with $V(t)$ graphically over $0 \le t \le 4s$ to validate that the circuit is a low-pass filter.

12. Revisit Example 8.4: Positioning Table Governed by Second-Order ODE: RK4 Method. Take $\Delta t = 0.02s$.

a. Find ω_2 and α_2 manually from the values of ω_1 and α_1 found in the first iteration.
b. Compare ω_2 with the exact solution using absolute percent error.

13. Revisit Example 8.4: Positioning Table Governed by Second-Order ODE: RK4 Method. Take $\Delta t = 0.015s$.

   a. Find $\omega(t)$ by the RK4 method over $0 \le t \le 1s$. You should be able to develop the algorithm in a spreadsheet.

   b. Compare $\omega(t)$ with the exact solution in a table similar to that in **Table 8.5**.

   c. Compare $\omega(t)$ with the exact solution graphically similar to that in **Figure 8.6**.

14. Consider the RLC circuit as shown in **Figure 8.28**.

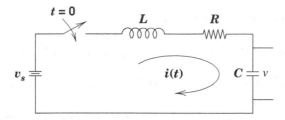

Figure 8.28 RLC circuit activated by a constant-voltage source.

The instantaneous current $i(t)$ is described by Kirchhoff's voltage loop law as

$L\dfrac{di}{dt} + Ri + v = v_S$, where v is the instantaneous voltage across the capacitor. By Ohm's law,

the current across the capacitor is given by $i = C\dfrac{dv}{dt}$. The two foregoing ODEs can be cast in

the form of Eq. (8.2) as two-coupled ODEs

$$v' = \frac{i}{C} = f_1(i)$$
$$i' = \frac{1}{L}(-Ri - v + v_S) = f_2(i,v)$$

(8.69)

where $v' = \dfrac{dv}{dt}$ and $i' = \dfrac{di}{dt}$.

   a. Take $\Delta t = 0.004s$ and find $v(t)$ and $i(t)$ numerically from Eq. (8.69) by the improved Euler method over $0 \le t \le 0.4s$. The ICs are $i(0) = 0$ and $v(0) = 0$.

   b. Plot $v(t)$ and $i(t)$ over $0 \le t \le 0.4s$.

Data: $L = 10\,henries$, $R = 1000\,ohms$, $v_s = 100\,volts$, and $C = 4(10^{-6})\,farad$.

For Problems 15–18, consider $\dfrac{dy}{dt} = -10y$ with IC $y(0) = 1$. The exact solution of the ODE and its IC is $y(t) = e^{-10t}$, which is a decaying function. As stated in **Table 8.6**, the forward Euler method has limited stability.

a. Find $y(t)$ numerically over $0 \le t \le 10$.

b. Plot the numerical solution and the exact solution over $0 \le t \le 10$.

c. Report if the solution is stable, neutrally stable, or unstable.

15. $\Delta t = 0.201$.
16. $\Delta t = 0.2$.
17. $\Delta t = 0.198$.
18. $\Delta t = 0.1$

For Problems 19–21, consider $\dfrac{dy}{dt} = -10y$ with IC $y(0) = 1$. The exact solution of the ODE and its IC is $y(t) = e^{-10t}$, which is a decaying function. As stated in **Table 8.6**, the backward Euler method is unconditionally stable.

a. Find $y(t)$ numerically over $0 \le t \le 10$.

b. Plot the numerical solution and the exact solution over $0 \le t \le 10$.

c. Discuss the behavior of the numerical solution.

19. $\Delta t = 1$.
20. $\Delta t = 0.5$.
21. $\Delta t = 0.2$.

Problems for Section 8.3

22. Revisit Example 8.5: Cooling Rod-Mixed Boundary Condition.

a. Discretize the rod with 10 nodes, including the boundaries, and solve Eq. (8.29) with the BCs, Eqs. (8.30) and (8.31) by the direct method.

b. Compare the numerical results with the exact solution in a table similar to **Table 8.9**.

For Problems 23–24, revisit Example 8.5: Cooling Rod-Mixed Boundary Condition. Discretize the rod with 10 nodes, including the boundaries, and solve Eq. (8.29) with the given BCs by the direct method.

23. $T(0) = 80$ and $T(0.0254) = 20$.

24. $T(0) = 80$ and $T(0.0254) + 10\dfrac{dT(0.0254)}{dx} = 25$.

25. A 3-m-long wooden beam, pinned at both ends, supports a load distributed uniformly, $w = 1500\,\dfrac{N}{m}$, as shown in **Figure 8.29a**. The distributed load causes beam deflection, $y(x)$, as shown in **Figure 8.29b**, which is governed by the second-order ODE,

$$EI\frac{d^2y}{dx^2} = \frac{w\,x}{2}(L-x),$$ over $0 \le x \le L$, where $EI = 4\left(10^5\right) N.m^2$. The BCs are $y(0)=0$ and $y(L)=0$, where $L=3m$.

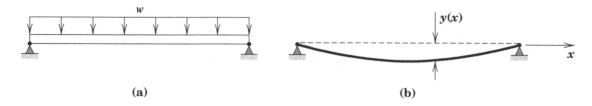

(a)   (b)

Figure 8.29 A 3-*m*-long wooden beam: (a) uniformly loaded pinned beam; and (b) beam deflection.

a. Discretize the beam with 10 nodes, including the boundaries, and solve the ODE and its BCs by the direct method.
b. Tabulate the numerical results with the exact solution, and list the percent error between the two solutions. The exact solution can be found by direct integration using Eq. (3.9) in Section 3.3, Chapter 3.
c. Compare the solutions graphically using a solid line for the exact solution and discrete points for the numerical solution.

Problems for Section 8.4

Solve Problems 26–27 as follows:

a. Model the springs, as shown in **Figure 8.9**.
b. Assemble the global stiffness matrix using Eq. (8.46).
c. Report the displacement and force at each node.

26. The spring system consists of three nodes and two elements, as shown in **Figure 8.30**. At node 3, a force $F_3 = 10N$ is applied in tension. For elements 1 and 2, $k_1 = 1000\frac{N}{m}$ and $k_2 = 500\frac{N}{m}$, respectively.

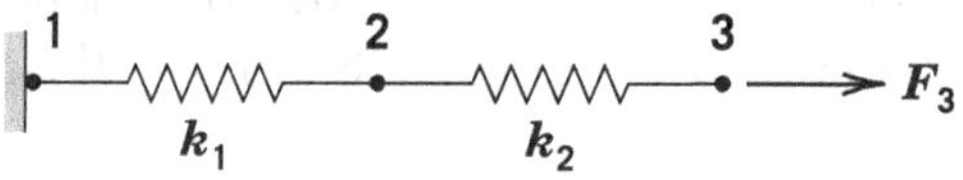

Figure 8.30 Spring system in tension with three nodes and two elements.

27. The spring system consists of three nodes and two elements, as shown in **Figure 8.31**. At node 2, a force $F_2 = 10N$ is applied as shown. For elements 1 and 2, $k_1 = 1000\frac{N}{m}$ and $k_2 = 2000\frac{N}{m}$, respectively.

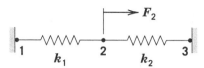

Figure 8.31 Spring system in tension with three nodes and two elements.

For Problems 28–29,

a. Model the springs, as shown in **Figure 8.9**.
b. Assemble the global stiffness matrix using Eq. (8.46). Neglect the weight of the beam.
c. Report the displacement and force at each node.
d. Does the beam meet the design criterion?

28. Two I-beams that rest on a rigid foundation, as shown in **Figure 8.32a**, are used to support a laser system. The laser causes a pressure load $P = 7\left(10^4\right)\dfrac{N}{m^2}$ on the top flange of the beams. The height of the laser is critical for its operation; thus, a design criterion is established that each beam should not deflect more than $2.54\left(10^{-6}\right)m$.

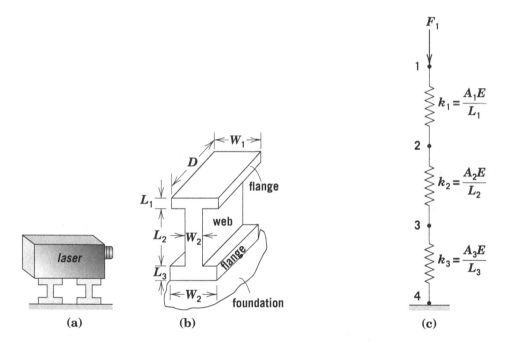

Figure 8.32 Laser foundation: (a) Laser and I-beam support system; (b) I-beam geometry; and (c) three-spring model.

A typical beam is shown in **Figure 8.32b**, which is modeled as a three-spring system, as shown in **Figure 8.32c**, where the top and bottom springs represent the beam flanges and the middle spring represents the beam web. Each spring stiffness is $k = \dfrac{AE}{L}$, where L is the height of a beam segment, A is the cross-sectional area of a beam segment perpendicular to the load, for example,

$A_1 = DW_1$, and E is Young's modulus, a material property. The force load per beam at node 1 is $F_1 = PA_1$, as shown in **Figure 8.32c**. Node 4 is fixed.

Data: $L_1 = L_3 = 0.0125m$; $L_2 = 0.175m$; $W_1 = W_3 = 0.1m$; $W_2 = 0.125m$; $D = 0.635m$;

$E = 7(10^{10})\dfrac{N}{m^2}$.

29. A TV camera system for sporting events consists of a camera, beam, hoist, and cable, as shown in **Figure 8.33a**.

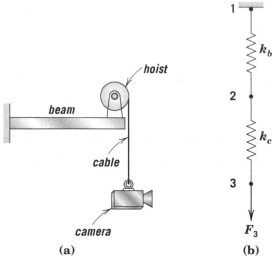

Figure 8.33 Camera system: (a) camera beam support system; and (b) beam and cable spring model.

The camera travels along the beam to capture the sporting event from different angles. If the beam deflection is excessive, then the camera vibration causes image distortion. Thus, a design criterion is established that the cable and beam should not deflect more than $1mm$ when the hoist is at the end of the beam. The beam and cable system are modeled as a two-spring system, as shown in **Figure 8.33b**, where the top spring represents the beam and the bottom spring

represents the cable. The beam stiffness is $k_b = \dfrac{Es^4}{4L_b{}^3}$, where s is the side of the beam square

cross section, L_b is the beam length, and E is Young's modulus, a material property. The cable

stiffness is $k_c = \dfrac{AE}{L_c}$, where L_c is the cable length and A is the cable cross-sectional area. The

camera weight is represented by force $F_3 = mg$. Node 1 is fixed.

Data: $L_b = L_c = 2m$; $s = 0.1m$; $A = (1)10^{-4}m^2$; $E = 7(10^{10})\dfrac{N}{m^2}$; $m = 4kg$; $g = 9.81\dfrac{m}{s^2}$.

For Problems 30–31,

a. Discretize the domain as a FEM conduction model.
b. Formulate the global conduction matrix.
c. Report the temperatures at all nodes.
d. Report the heat fluxers at all nodes and their directions.
e. Plot the temperature distribution between nodes.

Data: Material A, $k = 0.5 \dfrac{W}{m^oC}$; Material B, $k = 1 \dfrac{W}{m^oC}$; Material C, $k = 2 \dfrac{W}{m^oC}$.

30. The composite material consists of three nodes and two elements with BCs, as shown in **Figure 8.34**.

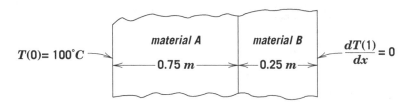

Figure 8.34 Heat transfer in a composite material with Dirichlet and Neumann boundaries.

31. The composite material consists of four nodes and three elements with given BCs, as shown in **Figure 8.35**. In addition, there is a heat source between materials A and B.

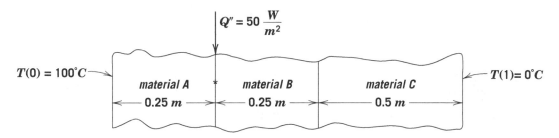

Figure 8.35 Heat transfer in a composite material with Dirichlet boundaries and a heat source.

32. Revisit Example 8.9: Double-Pane Window by FEM. Resolve the system as a triple-pane window, as shown in **Figure 8.36.** Here, there are six nodes and five elements. Use the same material properties for glass and air space.

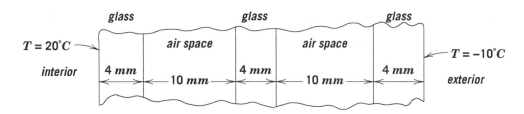

Figure 8.36 Triple-pane window.

a. Discretize the triple-pane window as a FEM conduction model.
b. Formulate the global conduction matrix.
c. Report the temperatures at all six nodes.
d. Report the heat fluxers at the outer nodes and their directions.
e. Plot the temperature distribution between nodes.

33. The wall system of a trailer consists of vinyl siding, plywood sheathing, insulation, and wallboard, as shown in **Figure 8.37a**. You are assigned by a manufacturer the task of determining the heat transfer from the trailer during winter. For that purpose, model the wall as four conduction elements with five nodes, as shown in **Figure 8.37b**.

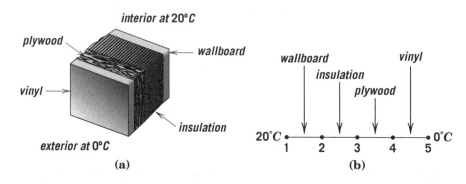

Figure 8.37 Wall system of a trailer: (a) materials composition; and (b) FEM conduction model.

Material	Thermal conductivity, $k\left(\dfrac{W}{m^o C}\right)$	Thickness (m)
vinyl siding	0.25	0.004
plywood sheathing	0.11	0.02
insulation	0.035	0.1
wall board	0.15	0.02

a. Discretize the wall system as a FEM conduction model.
b. Formulate the global conduction matrix.
c. Report temperatures at all five nodes.
d. Report the heat fluxers at the outer nodes and their directions.
e. Plot the temperature distribution between nodes.
f. What is the heat flux from the trailer?

34. Consider a circuit that consists of four nodes and three elements, as shown in **Figure 8.38**. Note that $V_1 = 12$, $V_3 = 9$, and $V_4 = 0$ (ground), and global current I_2 does not come from a direct source.

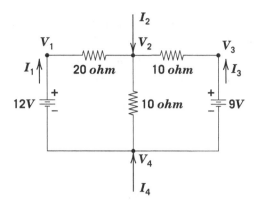

Figure 8.38 A circuit modeled by FEM using four nodes and three elements.

a. Approach the problem by showing how to order the local matrices on appropriate lattices in the 4 x 4 global conductance matrix, similar to **Figure 8.26**. Between nodes 1 and 2, designate the conductance, G_1, and between nodes 2 and 3 and nodes 2 and 4, G_2.

b. Assemble the matrices from part a into a single global matrix as in Eq. (8.62).

c. Find V_2, I_1, I_3, and I_4.

35. Consider a circuit that consists of three nodes and three elements, as shown in **Figure 8.39**. Note that $V_1 = 0$ (ground), $V_3 = 12$, and global current I_2 does not come from a direct source.

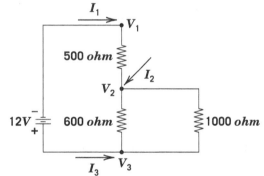

Figure 8.39 A circuit modeled by FEM using three nodes and three elements.

a. Approach the problem by showing how to order the local matrices on appropriate lattices in the 3 x 3 global conductance matrix, similar to **Figure 8.26**. Between nodes 1 and 2, designate the conductance, G_1, and between nodes 2 and 3, G_2 for the left resistance, and G_3 for the right resistance.

b. Assemble the matrices from part a into a single global matrix as in Eq. (8.62).

c. Find V_2, I_1 and I_3.

36. Consider the music system that consists of a CD player and speakers, as shown in **Figure 8.40a**. The CD player sends an electrical signal to the speakers in digital form. However, the speakers receive an analog signal via a digital signal to analog (DAC) circuit, as shown in **Figure 8.40b**, where $R = 800 ohms$. The circuit requires certain power given by $P = VI_T$ in *Watts*, where $V = 8 volts$. However, the current I_T is unknown, which you shall find by the FEM.

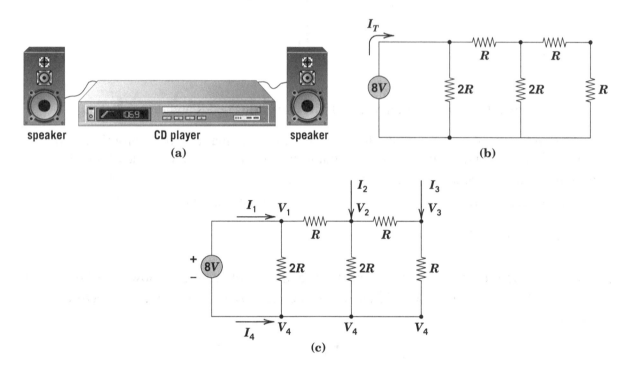

Figure 8.40 Music system: (a) CD player and speaker; (b) DAC circuit; and (c) FEM circuit model using four nodes and five elements.

Modeling the circuit by the FEM is shown in **Figure 8.40c**, which consists of four DOFs. Note that $V_1 = 8$ and $V_4 = 0$ (ground). The global currents I_2 and I_3 do not come from a direct source.

a. Approach the problem by showing how to order the local matrices on appropriate lattices in the 4 x 4 global conductance matrix, similar to **Figure 8.26**.
b. Assemble the matrices from part a into a single global matrix as in Eq. (8.62).
c. Find I_T and the power of the power supply.